Vernacular Architecture in the Twenty-First Century

At the dawn of a new millennium, in a time of rapid technological developments and globalization, vernacular architecture still occupies a marginal position. Largely ignored in architectural education, research and practice, recognition of the achievements, experience and skills of the world's vernacular builders remains limited. Faced with the persistent denial of the importance of vernacular architecture, questions about its function and meaning in the twenty-first century present themselves.

This book, written by authors from a variety of disciplinary backgrounds, aims to give the initial impetus to discussions about the way in which the vernacular can play a part in the provision of future built environments. Analysing the value of vernacular traditions to such diverse fields as housing, conservation, sustainable development, disaster management and architectural design, the contributors argue that there are valuable lessons to be learnt from the traditional knowledge, skills and expertise of the vernacular builders of the world.

The contributors argue for a more processual, critical and forward-looking approach to vernacular research, education and practice. Drawing on case studies from around the world, they aim to show that such an approach will enable the active implementation of vernacular know-how in a contemporary context, and will show that there still is a place for vernacular architecture in the twenty-first century.

Lindsay Asquith worked with Paul Oliver on the compilation of the *Encyclopedia of Vernacular Architecture of the World* (1997) and subsequently as a Research Associate in the Department of Architecture at Oxford Brookes University. She completed her PhD in 2003 and now works as an Architectural Design Consultant in the housing sector.

Marcel Vellinga is Research Director of the International Vernacular Architecture Unit at Oxford Brookes University. He is the author of *Constituting Unity and Difference: Vernacular Architecture in a Minangkabau Village* (2004) and of various articles dealing with the anthropology of architecture in Indonesia. He is currently co-editing, with Paul Oliver, the *Atlas of Vernacular Architecture of the World*.

Vernacular Architecture in the Twenty-First Century

Theory, education and practice

Edited by Lindsay Asquith
and Marcel Vellinga

Taylor & Francis
Taylor & Francis Group
LONDON AND NEW YORK

First published 2006
by Taylor & Francis
2 Park Square, Milton Park, Abingdon, Oxon OX14 4RN

Simultaneously published in the USA and Canada
by Taylor & Francis
270 Madison Ave, New York, NY 10016

Taylor & Francis is an imprint of the Taylor & Francis Group

© 2006 Lindsay Asquith and Marcel Vellinga, selection and editorial material;
individual chapters, the contributors

Typeset in Univers by
Florence Production Ltd, Stoodleigh, Devon
Printed and bound in Great Britain by
TJ International Ltd, Padstow, Cornwall

All rights reserved. No part of this book may be reprinted or reproduced or
utilized in any form or by any electronic, mechanical, or other means, now known
or hereafter invented, including photocopying and recording, or in any information
storage or retrieval system, without permission in writing from the publishers.

British Library Cataloguing in Publication Data
A catalogue record for this book is available from the British Library

Library of Congress Cataloging in Publication Data
A catalog record for this book has been requested

ISBN10: 0–415–35781–0 (hbk)
ISBN10: 0–415–35795–0 (pbk)

ISBN13: 9–78–0–415–35781–4 (hbk)
ISBN13: 9–78–0–415–35795–1 (pbk)

Dedicated to Paul and Valerie Oliver

Contents

List of illustration credits	ix
List of contributors	xi
Preface	xv
Foreword Nezar AlSayyad	xvii
Introduction Lindsay Asquith and Marcel Vellinga	1

Part I   The vernacular as process — 21

1   Building tradition: control and authority in vernacular architecture — 23
Simon J. Bronner

2   Endorsing indigenous knowledge: the role of masons and apprenticeship in sustaining vernacular architecture – the case of Djenne — 46
Trevor H.J. Marchand

3   Forms and meanings of mobility: the dwellings and settlements of sedentarized Irish Travellers — 63
Anna Hoare

4   Engaging the future: vernacular architecture studies in the twenty-first century — 81
Marcel Vellinga

Part II   Learning from the vernacular — 95

5   Traditionalism and vernacular architecture in the twenty-first century — 97
Suha Özkan

6   Learning from the vernacular: basic principles for sustaining human habitats — 110
Roderick J. Lawrence

Contents

7	Lessons from the vernacular: integrated approaches and new methods for housing research *Lindsay Asquith*	128
8	Sheltering from extreme hazards *Ian Davis*	145
9	A journey through space: cultural diversity in urban planning *Geoffrey Payne*	155

Part III Understanding the vernacular — 177

10	Vernacular design as a model system *Amos Rapoport*	179
11	'Generative concepts' in vernacular architecture *Ronald Lewcock*	199
12	The future of the vernacular: towards new methodologies for the understanding and optimization of the performance of vernacular buildings *Isaac A. Meir and Susan C. Roaf*	215
13	Architectural education and vernacular building *Howard Davis*	231
14	Educating architects to become culturally aware *Rosemary Latter*	245

Afterword: raising the roof — 262
Paul Oliver

Bibliography — 269
Index — 288

Illustration credits

The authors and the publishers would like to thank the following individuals and institutions for giving permission to reproduce material in this book. We have made every effort to contact copyright holders, but if any errors have been made we would be happy to correct them a a later printing

Chapter 1
© Simon J. Bronner: 1.1–1.3, 1.6–1.9
© Mel Horst: 1.4, 1.5

Chapter 2
© Trevor H.J. Marchand: 2.1–2.4

Chapter 6
© Roderick J. Lawrence: 6.1–6.5

Chapter 7
© Lindsay Asquith: 7.1–7.8; Tables 7.1–7.2

Chapter 9
Courtesy of Culpin Planning: 9.8–9.9
Courtesy of Halcrow Fox and Associates: 9.10a, 9.10b
Gunter Nitschke: 9.1–9.4
© Geoffrey Payne: 9.5–9.7, 9.11

Chapter 10
Based on Rapoport (1990c: Fig. 4.18, p. 100; 1999a: Fig. 1. p. 57): 10.1; (1990a: Fig. 3.12, p. 111; 1998: Fig. 7. p. 14; 2001: Fig. 2.1; 2004: Figs 37–39): 10.2
Rapoport (1990e: Fig. 2.2, p. 12; 2004: Fig. 47): 10.3; (1998: Fig. 6. p. 11; 2000a: Fig. 4. p. 149; 2000b: Fig. 2, p. 129; 2004: Fig. 45; in press, b: Fig. 4; in press, c: Fig. 5): 10.4

Chapter 11
Clark, G. and Piggott, Stuart (1970: *Prehistoric Societies*, Harmondsworth: Penguin [source of original not identified]): 11.1b
Crème, J. (1959: 'L'Architectura Romana' in *Encyclopedia Classica*, III, Vol. 2.1. p. 1. Turin): 11.5d
Encyclopedia of World Art, IX (1958: New York: McGraw Hill, pl. 427, bottom left): 11.4d

Illustration credits

Griaule, M. (1948: *Dieu d'Eau: entretiens avec Ogotemmêli*, Paris: Éditions du Chêne. Translated as *Conversations with Ogotemmeli*. Oxford: Oxford University Press, 1965, p. 95, Fig. 8. Copyright © International African Institute): 11.8c, 11.8d

Griaule, M. (1949: 'L'Image du Monde au Sudan' in *Journal de la Société des Africanistes*, XIX, ii, Société des Africanistes): 11.8a

Haberland, Wolfgang (1964: *Art of the World, the Art of North America*, New York: Crown Publishers, p. 117): 11.1e

Hitchcock, H.R. (1942: *In the Nature of Materials, 1887–1941: The Buildings of Frank Lloyd Wright*, New York: Duell, Sloan & Pearce, p. 107): 11.9d

© Honda, Tomotsune: 11.2 a–c

Kroeber, A.L. (1972: *Handbook of the Indians of California*, Dover, New York: Bulletin Smithsonian Institution, Bureau of American Ethnology 78): 11.1d

Leacroft, H. and R. (1966: *The Building of Ancient Greece*, New York: W.R. Scott, p. 3): 11.3a; (p. 8) 11.3c; (p. 6) 11.3d

© Ronald Lewcock: 11.4b, 11.4c, 11.5a, 11.6b, 11.6c, 11.7c, 11.9a, 11.9b

MacKendrick, P. (1962: *The Greek Stones Speak: The Story of Archaeology in Greek Lands*, New York: St. Martin's Press, p. 20, Fig. 1.8): 11.3b

Mellaart, J. (1967: *Catal Huyuk: A Neolithic Town in Anatolia*, London: Thames & Hudson): 11.1c

Oliver, Paul (1975: *Shelter Sign and Symbol,* London: Barrie & Jenkins, p. 208, upper left): 11.8b

Piggott, Stuart (1966: *Ancient Europe from the Beginning of Agriculture to Classical Antiquity*, Chicago: Aldine Publishing Company, p. 29): 11.1a ; (p. 98, Fig. 51 right) 11.4a [sources of originals not identified]

Ragette, F. (1974: *Architecture in Lebanon*, Beirut: American University of Beirut, p. 87, Fig 14): 11.5c; (p. 70, lower left) 11.6a

Renfrew, Colin (ed.) (1988: *An Atlas of Archaelogy*, London: Times Books; New York: HarperCollins, p. 98): 11.5b

Semper, G. (1963: *Der Stil II*. Reproduced in: Mallgrave, H.F., 1996, *Gottfried Semper*, New Haven and London: Yale University Press, p. 199): 11.9c

Volwahsen, A., (1969) *Living Architecture: Indian*, New York: Grosset & Dunlap, p. 44): 11.7a; (p. 87) 11.7b [sources of originals not identified]

Wiegand, Priene (1904: *Ergebnisse der Ausgrabungen und Untersuchungen in den Jahren 1895–1898*, Berlin: G. Reimer): 11.3e

Chapter 12

© Rajat Gupta: 12.6
© Isaac A. Meir: 12.1, 12.3, 12.5
Meir, Gilead, Runsheng, Mackenzie Bennett and Roaf, 2003: Table 12.1
Meir, Mackenzie Bennett and Roaf, 2001: Table 12.2
© Susan C. Roaf: 12.2

Chapter 14

© Foundation Jaume II el Just: 14.4
© ISVA 1999: 14.2
© Rosemary Latter: 14.1, 14.3

Contributors

Lindsay Asquith worked with Paul Oliver on the compilation of the *Encyclopedia of Vernacular Architecture of the World* (1997) and subsequently as a Research Associate in the Department of Architecture at Oxford Brookes University. She completed her PhD in 2003 and now works as an Architectural Design Consultant in the housing sector.

Simon J. Bronner is Distinguished University Professor of American Studies and Folklore at the Pennsylvania State University, Harrisburg. He has published many books on material culture, including *American Material Culture and Folklife* (1985), *Grasping Things: Folk Material Culture and Mass Society in America* (1986) and *Folk Art and Art Worlds* (1986). He is editor of the *Material Worlds Series* for the University Press of Kentucky and *Pennsylvania–German History and Culture Series* for Penn State Press.

Howard Davis is Professor of Architecture at the University of Oregon. He is the author of *The Culture of Building* (1999). He also contributed eighteen entries in the *Encyclopedia of Vernacular Architecture of the World*. His research interests include contemporary European architecture and its production, mixed-use buildings, community-based initiatives in south and southeast Asia, and American architectural education.

Ian Davis is a Visiting Professor in the Resilience Centre of Cranfield University and has worked in Disaster Management since 1972. He has worked as a senior advisor to various NGOs, governments and UN organizations and has experience in research, advocacy, higher education and consultancy. He has published widely on disaster related themes, including, most recently, as co-author, *At Risk: Natural Hazards, People's Vulnerability and Disasters* (2003).

Anna Hoare is studying for the MRes in Anthropology at University College, London, specializing in the anthropology of dwelling and settlement. She is studying the sedentary settlement forms of Irish Travellers, following earlier research into Irish Travellers' nomadic movement patterns. She is an MA graduate in International Vernacular Architecture Studies led by Paul Oliver at Oxford Brookes University.

Rosemary Latter is the co-chair of the Master's course in International Studies in Vernacular Architecture at the Department of Architecture, Oxford Brookes

Contributors

University. Qualified as an architect since 1988, she has practised in London, Paris and Seville. Her research interests stem from work on the *Comparative Lexicon* in the *Encyclopedia of Vernacular Architecture of the World*.

Roderick J. Lawrence is Professor in the Faculty of Social and Economic Sciences at the University of Geneva and works in the Centre for Human Ecology and Environmental Sciences. In January 1997 he was nominated for the New York Academy of Science. He is the director of a continuing education course on sustainable development and Agenda 21 at the University of Geneva.

Ronald Lewcock was Aga Khan Professor in Architecture at MIT from 1984 to 1992. He has previously taught at Natal, Columbia, Cambridge and the Architectural Association. He is now Professor in the Doctoral Program in Architecture, Georgia Institute of Technology. In addition to practice, he has been technical coordinator of two UNESCO International Campaigns, and consultant on conservation in many Asian and African countries.

Trevor H.J. Marchand studied architecture at McGill University and anthropology at the School of Oriental and African Studies, London. He is currently a lecturer in anthropology at SOAS, specializing in building crafts and skill-based knowledge. He is the author of *Minaret Building and Apprenticeship in Yemen* (2001) and *The Masons of Djenne* (forthcoming).

Isaac A. Meir studied architecture and town planning, and archaeology, and is a senior researcher (since 1986) at the Desert Architecture and Urban Planning Unit, Institute for Desert Research, Israel. He is involved in research and design projects concerned with appropriate design and planning, energy conservation and adaptation to desert conditions, and has authored and co-authored over eighty papers, chapters and technical reports.

Suha Özkan studied architecture at the Middle East Technical University (METU) in Ankara, and theory of design at the Architectural Association in London. Dr Özkan has served as secretary general of the Aga Khan Award for Architecture since 1991 and as a jury member on various international awards, and has organized international architecture competitions.

Geoffrey Payne is a housing and urban development consultant with more than thirty years experience. He has undertaken consultancy, research and training assignments in most parts of the world, taught in several universities, written, edited or contributed to many publications and participated in numerous international conferences and workshops. His latest book (with Michael Majale) is *The Urban Housing Manual: Making Regulatory Frameworks Work for the Poor* (2004).

Amos Rapoport is Distinguished Professor Emeritus of Architecture at the University of Wisconsin-Milwaukee. He is one of the founders of Environment

Behaviour Studies. His work has focused mainly on the role of cultural variables, cross-cultural studies and theory development and synthesis. His six books and many other publications have been translated into a number of languages.

Susan C. Roaf is Professor in the Department of Architecture at Oxford Brookes University. During ten years in Iran and Iraq she studied aspects of traditional technologies, including nomadic tents, windcatchers, ice-houses, landscape and water wheels. In 1995 she built her own Ecohouse in Oxford with the first UK photovoltaic roof. She is widely published on a range of subjects from traditional technologies, passive building design and renewable energy systems.

Marcel Vellinga is Research Director of the International Vernacular Architecture Unit at Oxford Brookes University. He is the author of *Constituting Unity and Difference: Vernacular Architecture in a Minangkabau Village* (2004) and of various articles dealing with the anthropology of architecture in Indonesia. He is currently co-editing, with Paul Oliver, the *Atlas of Vernacular Architecture of the World*.

Preface

The compilation of this book began as a tribute to Paul Oliver. We have both had the privilege of working with Paul for a number of years and his knowledge and experience have been not only an inspiration, but an essential part of our education in the field of vernacular architecture studies. His dedication, commitment and enthusiasm for a subject which is often marginalized, if not ignored, is unparalleled.

Paul's late wife Valerie was instrumental in the planning stages of this volume, and we are indebted to her for her knowledge of the peoples, places and events that all contributed to the journey she and Paul undertook to heighten understanding and knowledge of the building traditions that exist among the many cultures and habitats throughout the world. She is greatly missed by us as a colleague, mentor and friend.

The authors represented in this book by no means make up an exhaustive list of specialists in the field of vernacular architecture. Many have written on the subject for a number of years and some are just beginning their work in the field, but all are committed to the education and transfer of knowledge that is imperative if the field of vernacular architecture studies is to grow and achieve the importance it deserves. Working in such diverse fields as architecture, planning, housing, urban studies, anthropology and folklore studies, they also represent the multi-disciplinary nature of the discourse.

We would like to thank all of the authors for their valuable contributions to this book and especially want to acknowledge the intended contribution by Jeffrey Cook, who sadly passed away before the completion of his chapter on climate change and the vernacular. We also thank Nezar AlSayyad for agreeing to write the foreword to this volume, and of course Paul Oliver for the afterword. We would also like to acknowledge the support of our families in this endeavour and especially Zita Vellinga for the compilation of the bibliography and index. We also express our gratitude to Caroline Mallinder and her team at Taylor & Francis, for supporting the project from the beginning and the valuable assistance they have given us.

Our aim in all of this is to further the debate on the importance of vernacular architecture studies now and throughout the twenty-first century, not as a study of past traditions, but as a contribution to new methods, solutions and achievements for the future built environment. If this is achieved it will be to the

Preface

credit of not only those that contributed to this work, but to all those involved in the field, academics and practitioners alike, whose integrated and dynamic approaches will ensure that the vernacular does not stand still, but continues to influence and enhance the world we live in.

Lindsay Asquith and Marcel Vellinga, February 2005

Foreword

Nezar AlSayyad

Vernacular architecture is a nineteenth-century invention. As a category of scholarship, its presence has been consolidated in courses and research programmes in the academy in the last two decades of the twentieth century. However, it has remained a considerably unknown subject in the arenas of public and policy discourses.

What would a vernacular architecture for the twenty-first century look like? Will the discussion of it be different or more influential? Will vernacular architecture simply disappear? Or will everything simply be classified as vernacular? These are the challenges of vernacular architecture in the twenty-first century and this important book attempts to address some of them.

As someone who has been involved in the study of the vernacular for a quarter of a century, I am often asked to define the vernacular. But while there are many operational definitions of the vernacular, the first challenge we must confront is the etymological and epistemological limitations of the concept. The many members of the International Association for the Study of Traditional Environments (IASTE), which I co-founded eighteen years ago, have taken on this particular challenge.

Etymologically, for anything to be considered vernacular, it has always been assumed that it must be native or unique to a specific place, produced without the need for imported components and processes, and possibly built by the individuals who occupy it. In the twenty-first century, as culture and tradition are becoming less place-rooted and more information-based, these particular attributes of the vernacular have to be recalibrated to reflect these changes.

Epistemologically, or with regard to our ways of knowing and classifying, the *meaning* of the vernacular also has to change. For example, the idea of modern knowledge as different from, and possibly opposite to, vernacular knowledge should be abandoned as we recognize that the vernacular in some instances may in fact be the most modern of the modern. Many years ago at the first IASTE conference, Y.-F. Tuan argued that tradition is often a product of the absence of choice. As such, we must come to terms with the nature of constraint in the practices of the vernacular. We must accept that the gradual change that occurs in vernacular architecture over long periods of time is not a result of conservative practices and aesthetics but simply of geographic or economic limitations that cannot be overcome by a segment of the local population of a region.

Another major challenge we have to face in the twenty-first century concerns our methods. Here I am not only referring to what we do as scholars in

Foreword

various disciplines when we go out in the field, but also what we choose to focus on in our study. Again, at the first IASTE conference, Paul Oliver argued that there is no such thing as a traditional building but rather buildings that embody certain vernacular traditions. He urged us to focus our attention on the practice of transmission as a way of understanding the vernacular and maintaining it. Oliver's advice is still valid; the only difference, perhaps, is that the practices of transmission have changed considerably in an era of technological advancement and increased communication. We should no longer assume that vernacular builders are unskilled, illiterate, technologically ignorant or isolated from the world of global communication.

The last but most important challenge that we as scholars of the vernacular have to face concerns the utility of our labour. Of course, research about the vernacular is valuable in and of itself as a field of humanistic discourse. But researchers of the vernacular, as attested by the contributors to this book, are not simply satisfied with the status quo. If earlier work on the vernacular consisted primarily of object-oriented, socially-oriented or culturally-oriented studies, the direction outlined by this book is a new one that I could call *activist-oriented* studies. With such studies, there is a recognition of the limitations of the vernacular, and of the reasons why the professional community continues to shun it. More research on these issues will be needed.

More research also needs to be done on the assumed utility of vernacular knowledge in the field of housing, particularly in relation to solving the problems of urban squatters. The connection between the two areas of knowledge is not yet well established. In addition, further research should be done on the sustainability of vernacular settings. As some of the contributors to this book suggest, some vernacular forms are neither sustainable nor efficient and I would indeed add that they are often unaffordable. We also need to know the significance of our own classification of emerging forms of squatting as a new vernacular.

We should not be left with the belief that everything is vernacular yet nothing is vernacular any more. As I have argued in the last few years, the vernacular is not dead, and it has not ended. What has ended, or should end, is our conception of it as the only harbinger of authenticity, as the container of specific determined cultural meaning, as a static legacy of a past. What will emerge, I hope, is a vernacular as a political project, a project whose principal mission is the dynamic interpretation and re-interpretation of this past in light of an ever-changing present. This, I believe, would be a vernacular architecture worthy of the twenty-first century.

Berkeley, September 2005

Introduction

Lindsay Asquith and Marcel Vellinga

In his Hepworth Lecture entitled 'Vernacular architecture in the twenty-first century', read for the Prince of Wales Institute in 1999, Paul Oliver draws attention to the fact that, at the dawn of the new century, vernacular architecture still occupies a marginal position (Oliver 1999). Recognition and support from professionals and policymakers involved in the fields of architecture and housing is still not forthcoming. Vernacular architecture continues to be associated with the past, underdevelopment and poverty, and there seems to be little interest among planners, architects and politicians in the achievements, experience and skills of the world's vernacular builders or the environmentally and culturally appropriate qualities of the buildings they produce. Native American pueblos, Indonesian longhouses or West African family compounds may be admired for their conspicuous design, functionality or aesthetic qualities, but they are hardly ever regarded as relevant to current housing projects. More often than not, vernacular houses are regarded as obstacles on the road to progress, which should be replaced by house types and living patterns that fit western notions of basic housing needs but which are adverse to the norms, wishes and values of the cultures concerned.

As Oliver points out, this attitude towards vernacular architecture is short-sighted. At the beginning of a new century, a major challenge facing the global community is to house the billions of people that inhabit the world, now and in the future, in culturally and environmentally sustainable ways. Current estimates predict an increase of the world's population to approximately 9 billion people in 2050, all of whom will need to be housed. Though actual numbers do not exist and estimates vary, vernacular dwellings, built by their owners and inhabitants using locally available resources and technologies, according to regulations and forms that have been handed down and adapted to circumstances through local traditions, are presently believed to constitute about 90 per cent of the world's total housing stock (Oliver 2003: 15). The problem of housing the world has not yet attracted the amount of attention paid to issues of health, food, climate change or the depletion of biodiversity, Oliver notes, yet it is one that will have to be recognized and faced by governments and other policymakers if the future well-being of the global population is to be ensured. In order to meet the unprecedented demand for houses, he writes, it is essential that vernacular building traditions are supported; to assist local builders in matters of sanitation and disaster preparedness, while at the same time learning and benefiting from their experience, knowledge and skills (1999: 11; see also Oliver 2003: 258–63).

Lindsay Asquith and Marcel Vellinga

The serious issue of global housing and, more particularly, the positive contribution that vernacular architecture could make to it cannot be ignored by those professionally involved in the fields of planning and housing. Furthermore, it should not be ignored by the media and academia. As in politics and policy-making, much media attention is paid to environmental, food and health issues but, as the agenda and news coverage of the 2002 World Summit on Sustainable Development in Johannesburg has shown (e.g. United Nations 2001), comparatively little thought is given to matters of housing. At an academic level, the inter-disciplinary field of vernacular architecture studies has admittedly grown significantly in the last decades of the twentieth century, with important and sometimes pioneering work done by architects, anthropologists and geographers. Yet, as Oliver notes, there are few academic courses or educational resources available to students, and formal recognition among scholars involved in these and other fields is still lacking.

Oliver's paper addresses an important issue and asks for a further elaboration of the future role and importance of vernacular traditions. The problem of housing a rapidly increasing world population constitutes a major concern for mankind, yet it is not the only challenge to be faced in the twenty-first century. Environmental crises and climate change; processes of economic and political globalization; cultural interaction and conflicts caused by migration, tourism and war; and rapid technological developments constitute some of the other major issues that profoundly effect the way in which the world is perceived, organized and lived in at the beginning of the new millennium. Each of these issues has major social and cultural implications, and all of them relate in one way or another to vernacular traditions. Important questions regarding the function and meaning of vernacular traditions therefore arise for those involved in the field of vernacular architecture studies: how, for instance, will vernacular traditions be affected by the ecological, cultural and technological changes? What part can they play in them? Will they be able to respond or adapt in order to come to terms with the new ecological and cultural circumstances, or will they be forced to disappear, as so many traditions already have done in the course of the last century? Are particular elements of traditions more susceptible to change or preservation than others? Can certain changes more easily be incorporated than others, and will there be regional or cultural differences? In short: is there still a place for vernacular architecture in the twenty-first century?

It will be noted that such questions are not new or specific to the beginning of the twenty-first century, because cultures and vernacular building traditions have always been dynamic and changing. Yet, it is important that they are addressed, not only because of the global scope and unparalleled pace of the changes, but also because doing so will help to increase the academic, professional, political and public awareness of the importance and relevance of vernacular architecture, and as such may lead to the disposal of its stigma of a backward past, poverty and underdevelopment. Despite popular conceptions to the contrary, vernacular building traditions are not remnants of an underdeveloped or romantic past, but are of importance and relevance to many cultures and peoples in the world, past, present and future. From a purely academic point of view, an understanding of the way in which vernacular traditions respond and react to ecological, technological and cultural changes will offer better insights in the nature of traditions and the processes of change that at different times and in various parts of the world have led to the disappearance, adaptation, revival or endurance of such traditions. From a more practical and professional perspective, such insights may help us to identify how vernacular architecture may best play a part in current and future attempts to create an appropriate and sustainable built environment for all.

As the contributions to follow show, the issues surrounding the potential, function and meaning of vernacular architecture in the twenty-first century are complex and extensive. This book is not intended to address them all, but aims to give the initial impetus to discussions that will hopefully result in a greater understanding and acknowledgement of the future importance of vernacular traditions. The subject is explored through chapters that each address an important aspect of vernacular architecture or reflect on the academic and professional discourse on it, dealing with issues such as theory, education, sustainability, housing, disaster management, conservation and design. Written by authors from a variety of disciplinary backgrounds, the chapters aim to provide an overview of the current state of affairs regarding the study of vernacular architecture and attempt to indicate how vernacular traditions relate to, and perhaps may contribute to the way one may deal with, some of the major issues facing the global community in the coming age. As such, they are hoped to become a starting-point for future research and education. Although vernacular architecture studies is emerging as a promising and fascinating area of study in recent years, much still needs to be done, both theoretically, methodologically and through recording and documentation, before the relevance of vernacular architecture in the twenty-first century can be acknowledged and understood.

Vernacular architecture studies and the future
As Oliver (1997b: xxiii) has noted, 'research in vernacular architecture may have to wait some time before it has an historian'. Still, generally speaking, it can be said that the interest in vernacular architecture in the sense of non-classical and non-western buildings can be traced back to the eighteenth century, while the first scholarly analyses of vernacular architecture as rural, non-monumental and pre-industrial traditions started to appear in the late nineteenth century (Upton 1990 and 1993; Oliver 1997b). Many of these early studies were made in Europe and the US, often by antiquarians and architects who were influenced by the Arts and Crafts movement. Writings of this period on the vernacular traditions of the non-western world were often embedded in the accounts of the many travellers, missionaries and colonial officials who, in a time of rapid colonization and scientific exploration, were scattered around the world and encountered buildings that were often fundamentally different from the ones they were familiar with back home. In both cases, many of the studies were 'tinged by nostalgia' for traditions that, though often in decline, were regarded as examples of functionalist aesthetics and that, consequently, were often seen to serve as sources of inspiration for contemporary

design (Oliver 1997b: xxiii). Not infrequently in such accounts, vernacular buildings were seen as 'more innocent, natural or spontaneous, and therefore truer' than, if not superior to, their later counterparts (Upton 1990: 200).

Although many of these early studies, particularly those dealing with western traditions, were used as a means to evade and criticise contemporary architectural practice, they usually did not pay much explicit attention to the way in which the traditions concerned might contribute to the creation of future built environments. Nor (leaving aside notable exceptions such as Morgan (1965)) did they focus much on the ways in which the vernacular traditions related to the cultures of which they formed part. On the whole their interest was in the documentation, classification and naming of historic or traditional forms, plans, materials and styles, most of which (especially in the case of non-western traditions) were regarded as destined to disappear. This tendency to focus on the documentation and preservation of traditions that were regarded as more spontaneous, instinctive and true, without paying much attention to cultural context or, indeed, the future potential of the traditions concerned, persisted well into the twentieth century (e.g. Rudofsky 1964). The late 1960s, however, saw the publication of a number of seminal works that, by stressing the importance of studying the vernacular within its historic and cultural context, sought to free the vernacular from its associations with anonymity, nostalgia and the past, and explicitly stressed the part that vernacular traditions may play in the provision of more sustainable settlements and buildings for the future (Oliver 1969; Rapoport 1969). Written in a time of rapid modernization, these studies provided a major research impetus and contributed greatly to the increased recognition of vernacular architecture studies as a multi-disciplinary field of academic and professional interest.

The ever-growing number of studies that has continued to appear since this time has greatly increased our knowledge and understanding of historic and contemporary vernacular traditions, both in the western and non-western world (e.g. Glassie 1975; Upton and Vlach 1986; Bourdier and AlSayyad 1989; Turan 1990). Yet despite this undeniable increase, it can be argued that at the beginning of the twenty-first century interest in vernacular traditions is still rather marginal and more obvious in some parts of the world than others, while the numerical representation of scholars from different disciplines remains uneven. Besides, ideas differ about what kind of traditions the category of the vernacular is supposed to consist of, making it difficult to actually speak of *a* field of vernacular architecture studies. Because of the enormous diversity of building traditions classified under the umbrella of 'the vernacular' and the varied disciplinary and national backgrounds of those studying them, it may be said that two very different scholarly discourses (one, generally speaking, dealing with historical western traditions, and the other with contemporary non-western ones) exist, each with their own concepts, perspectives and interests. As Dell Upton (1993: 10) has noted, scholars taking part in these discourses 'work to all intents and purposes ignorant of one another'. What is more, although both are interested in processes of change, their attitude towards the status of the vernacular in the future is different.

Without wanting to oversimplify too much, the discourse on western vernacular architecture can be said to be largely concerned with the documentation and understanding of the historical, rural and pre-industrial building heritage. Dominated by architectural historians, preservationists, folklorists and geographers, many of the publications on European or North American vernacular architecture focus on the classification and dating of individual buildings, or of specific forms, materials or plans, tracing distribution and diffusion patterns as well as changes in type within the context of social history. Though it is never expressed in so many words, an implicit assumption of this discourse is that there can be no real future for the vernacular, as the ongoing processes of modernization and globalization leave the pre-industrial building heritage of farmhouses, barns and mills that form the core of research ever more out of touch with the present. The reaction of many scholars who form part of this discourse has been to withdraw into the past, to look at the historic meaning, use and construction of buildings while ignoring the active re-use, re-interpretation or adaptation of the same or similar traditions in the present. As a result, studies of particular buildings, building types or time periods keep increasing in number and become ever more detailed, though arguably this goes at the expense of any real understanding of the processual nature of the vernacular traditions concerned.

The second discourse on vernacular architecture, which exists in parallel to the first one but with little academic overlap or communication, is more concerned with non-western traditions, although studies of western traditions are included as well. Studies that document particular vernacular building traditions within their cultural and historic context are also common in this discourse, but these are increasingly complemented by analyses of the ways in which the design, use and meaning of these traditions change within the context of contemporary processes of modernization and globalization. More theoretically oriented, paying attention to rural as well as urban traditions, these analyses look at the impacts of current trends like consumerism, the manufacturing of heritage, deterritorialization and ethnic revitalization on vernacular traditions, and discuss what the implications of such impacts are in terms of the negotiation of identity and the definition and value of key concepts like tradition, modernity and place. Instead of withdrawing into the past, many of these studies thus actively engage the present, increasingly arguing that vernacular traditions should be seen as processes that dynamically, and inter-dependently relate to identities, evolving and transforming over time. Nonetheless, as ongoing discussions on the 'end of tradition' and 'post-traditional environments' exemplify (e.g. AlSayyad 2004), it can be argued that, among many of those involved in this discourse, there is still an underlying concern for the future survival of the 'true' vernacular in an increasingly global world.

Largely focusing on the active part that vernacular traditions can play in the provision of appropriate and sustainable architecture for the future, the chapters in this book build on the work that has been done so far, particularly that in the discourse on the non-western vernacular. But they also intend to expand its scope by looking at issues that, until now, have not received much attention.

Lindsay Asquith and Marcel Vellinga

Though stimulating and sometimes challenging, many of the studies that have been carried out in recent years have focused on the impacts of the process of globalization on the constitution of local and regional identities, and the way in which these relate to design and the concepts of tradition and modernity. But, as noted above, apart from the effects of globalization on issues of cultural identity, the global community faces many other challenges at the beginning of the new millennium, including climate change, the depletion of resources, mass migrations, the impacts of natural disasters and ever-growing demands for housing. Though equally important issues in terms of the future role and sustainability of the vernacular, the way in which vernacular knowledge and experience may be used to respond to these challenges has so far, despite some notable exceptions (see Afshar and Norton 1997), not been the subject of much discussion (Oliver 2003: 14). This book, though by no means covering all the issues concerned, aims to stimulate some of this discussion.

The message of the book is a forward-looking and positive one. In contrast to the current stereotypes of a backward and old fashioned past, 'disappearing worlds', underdevelopment and poverty, all authors argue that in this time of rapid technological development, urbanization, mass consumption and the internationalization of power and wealth, there is still a lot that can be learned from the traditional knowledge, skills and expertise of the vernacular builders of the world. What is needed to make the active implementation of such vernacular know-how come true in a modern or development context, is an investment in research and education that explicitly stresses the dynamic nature of vernacular traditions. By critically investigating the achievements *and* shortcomings of vernacular traditions, and examining the ways in which that which is valuable in the vernacular may be integrated with that which is valuable in modern architectural practice, it will be possible to develop, through upgrading and adaptation, those aspects of contemporary built environments that are currently unsustainable or culturally inappropriate. Only when such a processual, critical and forward-looking perspective is adopted, will it be possible to dismantle the stereotypes that continue to cling to the vernacular and to say that there still is a place for vernacular architecture in the twenty-first century.

In the next section, we will shortly introduce the chapters in the book, which has been divided into three parts. The chapters in Part I deal with the importance of viewing the vernacular as a process. Those in Part II discuss the ways in which the vernacular may provide lessons for contemporary design. Finally, the chapters in Part III identify important areas for future research and education.

Part I: The vernacular as process

As has recently been summarized by AlSayyad (2004: 6–12), the concept of tradition has been a major theme in writings on vernacular architecture, most especially in the discourse on non-western traditions. A number of conferences have been devoted to its theme, and many important studies published (e.g. Bourdier and AlSayyad 1989; Abu-Lughod 1992; Upton 1993; AlSayyad 2004). In line with ideas prevalent in the contemporary fields of anthropology, cultural geography, history and archaeology, these studies have increasingly stressed the dynamic and processual nature of tradition. Traditions can be seen as creative processes through which people, as active agents, interpret past knowledge and experiences to face the challenges and demands of the present. Arguably then, as Oliver (1989) has observed, it is this active process of the transmission, interpretation, negotiation and adaptation of vernacular knowledge, skills and experience that should form the focus of research and teaching, as much as the actual buildings that form their objectification.

The chapters in Part I reiterate and emphasize the importance of regarding tradition as a creative process, and indeed many contributors to the book, including Marchand, Lawrence, Vellinga and Howard Davis, identify the dynamics and transmission of vernacular traditions as a crucial focus for future research, practice and education. That tradition is changing and varied and should form a central point of attention in vernacular research is perhaps most explicitly discussed by Simon Bronner. Defining tradition as 'a reference to the learning that generates cultural expressions and the authority that precedent holds', Bronner provides a concise yet insightful treatise on the processual nature of tradition, focusing especially on the issues of creativity, innovation and authority that its study raises. Tradition, in Bronner's view, is about expectation and social acceptance rather than, as is often noted, constraint. It is 'as the local saying that gains credit by long and frequent use'. As a reference to precedent and a social construction, tradition invites commentary and interpretation and is often continuously re-negotiated, from generation to generation. As such it allows for creativity, and for adaptations and innovations that may ultimately, when they have been socially accepted, be integrated and become part of the tradition. 'Creativity and tradition', writes Bronner, 'are intertwined, and represent the complex processes of humans expressing themselves to others in ways that carry value and meaning'.

Taking into account this dynamic balance of social custom and individual innovation leads to important questions of authority and control. When studying vernacular traditions, Bronner notes, one should not just ask why buildings look the way they do, why they came into being and how they changed along the way, but also by whose standards, by what precedents and with whose skills creation, transmission and change occur. Focusing on three case studies (Jewish *sukkah*, Amish barn-raisings and 'recycled' houses in Houston), he briefly discusses the dynamics of vernacular building traditions in a complex society such as twenty-first-century America and notes significant differences in how and why such traditions are transmitted and continued. In all three cases though, negotiation is inherent to the tradition, and change can be identified as a constant. Because they allow for creativity, innovation and change, Bronner notes, building traditions like the Jewish *sukkah* and Amish barn-raisings will continue to evolve and change, while new ones like the recycled houses will arguably keep emerging. Though such new 'grassroots traditions' may not be as established as the *sukkah* or Amish barn-raisings, Bronner notes that they may well represent the future of the vernacular in industrialized societies, and shows how studying them may teach us much about what a tradition is, as well as, importantly, about what a tradition does.

A good ethnographic example of the way in which traditions are continuously re-negotiated through the process of transmission, as well as of how issues of authority and individual creativity play an important part in this dynamic process, is provided by Trevor Marchand. In his discussion of the apprenticeship system operated by vernacular masons in Djenne (Mali), Marchand shows how this age-old form of education simultaneously allows for the gestation and transmission of technical knowledge, as well as for the operation of a process of socialization that helps to forge the professional identity of aspiring masons. Working within an established system of authority, under the guidance of an established master mason, young apprentices are able to obtain the technical, ritual, economic and political skills needed to become a publicly recognized mason. At the same time, in taking part in the building process and observing the activities of the master mason they work with, they develop a sense of professional and social identity and, significantly, learn to negotiate the boundaries of the tradition. Again, as was also noted by Bronner, these boundaries are not rigid or static. Indeed, Marchand observes, creativity and innovative intervention are awarded and esteemed in Djenne, as it is those elements that ensure that the local built environment, combining tradition *and* innovation, remains meaningful.

The transmission of knowledge and the negotiation of identities and boundaries that takes place through the system of apprenticeship allow the masons of Djenne to sustain standards and enables them to continuously create a meaningful built environment. Such a built environment, Marchand notes, is inherently dynamic, 'while remaining rooted in a dialogue with history and place'. Crucially, he argues, this dynamism needs to be taken into account when considering the sustainability of the building tradition. As a UNESCO World Heritage Site, the conservation of the distinctive *style-Soudanaise* architecture is a major concern in Djenne, involving many actors including architects, conservationists, government officials, funding bodies and anthropologists. Recalling the issue of authority discussed by Bronner, all these parties (and more) compete for control over the meaning of Djenne's architectural heritage. Marchand argues nonetheless that it is crucial that the expert status of the masons is acknowledged and given centre stage in all this, as it is their knowledge and system of education that ultimately defines the tradition. Djenne's building tradition, he writes, should be understood 'as a set of meaning-making practices rather than a landscape of physical objects to be conserved for their unique forms or some inherent historic value'.

Another example of the way in which a vernacular tradition may change and adapt to new socio-political or environmental circumstances, but in the process will maintain certain features that are distinctive and specific, having arisen from a unique cultural, social and economic history, is provided by Anna Hoare. Discussing the settlements and dwellings of sedentarized Irish Travellers, Hoare criticises the academic discourse on nomadism, which in general has tended to regard mobility as an ecological adaptation. This view, she notes, does not do justice to the cultural dynamics, flexibility and variety of nomadic groups throughout the world and ignores the fact that mobility may be motivated by a variety of cultural, social, economic and political factors, serving different purposes in different contexts. It undeservedly 'suggests passivity on the part of people in their relations towards external environments and changing circumstances, to which they apparently merely react'. Besides, she writes, it leads to the prevalent notion that, at the beginning of a new millennium, nomadism has outlived its usefulness, and that the way of life and behaviour associated with it have become redundant and inappropriate.

Questioning this verdict, Hoare sets out to show that nomadic 'social life and mobility are creative, affective factors in themselves rather than dependent corollaries of economic and ecological adaptation', and that the vernacular skills, understandings and values of nomadic cultures are therefore likely to find new forms of social expression, and distinctive ways of living. Focusing on a number of Irish Traveller groups in the UK and Ireland, all of whom in the second half of the twentieth century have been forced to sedentarize, she shows how mobility is still a distinctive element of Traveller culture, despite the fact that the groups are now no longer 'on the road'. Its social and political importance is evident, for instance, from the way in which it continues to inform the social composition of the settlements, to shape the constitution of relationships and identities, to influence the production and built forms of dwellings, and to frame the experience of a larger world. Despite the forced changes to the vernacular way of living and building, then, life among Irish Travellers continues to represent a distinctive cultural trajectory.

The argument that the vernacular should not be regarded as an architectural category consisting of static buildings that need to be carefully safeguarded, but as a concept which identifies dynamic building traditions that continuously evolve while remaining distinctive to a specific place, is central to Marcel Vellinga's chapter. Noting that discussions of change in current discourses on vernacular architecture tend to emphasize processes of loss and decline, Vellinga calls for a critical reconsideration of the concept of the vernacular. Following Upton (1993) he argues that, in the pursuit of recognition for non-western and non-monumental buildings, the concept has unintendedly become reified. 'A name', he writes, 'has become a thing'. Essentially, the vernacular has been defined in opposition to the modern. In so doing, Vellinga observes, the vernacular has effectively been relegated to the past, to a distinct 'traditional' period of time that somehow existed before the modern era, while at the same time it has been denied both a history and future. Because of its non-modern status, any changes introduced to the vernacular by the encounter with modernity are automatically seen to represent cultural decline and a loss of authenticity. 'The vernacular and the modern, it seems, cannot go together.'

Still, Vellinga notes, many vernacular and modern traditions nowadays do merge, throughout the world, often in creative and unexpected ways. The building traditions that are the result of such amalgamations (such as, for instance, the 'counter culture' traditions discussed by Bronner, or modernized 'replicas' of traditional buildings) are nonetheless largely ignored in the field of vernacular architecture studies because they are regarded as being not, or no longer, truly or 'authentically' vernacular. Yet to do so, Vellinga argues, is to deny the dynamic nature of building traditions and the application of meaning, and effectively restricts the development of the field of vernacular architecture studies. Calling instead for research that explicitly focuses on the way in which vernacular and modern traditions merge, he proposes widening the vernacular concept so that it includes all those buildings that are 'distinctive cultural expressions of people who live in or feel attached to a particular place or locality'. Aware that such a conceptualization ultimately makes the category of the vernacular redundant, he argues that it would help those building traditions that are now called vernacular to get rid of the stigma of underdevelopment and a backward past, as such enabling them, as sources of architectural know-how, to assume an active part in the provision of sustainable architecture for the future.

Part II: Learning from the vernacular

Marchand and Vellinga's assertion that vernacular traditions may have an important contribution to make to the development of sustainable future built environments is shared by most contributors to the book, and is elaborated upon in the chapters in Part II. As a form of what in more general terms has usually been referred to as 'indigenous knowledge' (Ellen, Parkes and Bicker 2000; Sillitoe, Bicker and Pottier 2002), much may still be learned from vernacular know-how, skills and experience. Of course this notion is not a fundamentally new one. As Afshar and Norton (1997) have summarized, there already exists a long-established, though still rather marginalized discourse that focuses on the ways in which vernacular traditions may be integrated into contemporary building practices in order to create more appropriate settlements and buildings. Now, at a time when concerns over sustainability and cultural identity continue to cast doubts over the processes of modernization and globalization, and alternative approaches to development are increasingly being looked for, it seems more timely and urgent than ever to build upon the achievements of this research.

A concise history of the discourse dealing with the incorporation of vernacular traditions in contemporary architectural practice is provided by Suha Özkan in his chapter on, what he calls, the 'traditionalist' approach to architecture. Reiterating briefly how the history of architecture may be written in terms of a succession of periods in which theoretical, professional and aesthetic principles are at first agreed upon and then challenged, he focuses in particular on the various reactions that sprung up in response to the dominance of Modernism at the beginning of the twentieth century. Apart from, among others, post-modernism and the 'architecture of freedom' movement, one of the most influential reactions, Özkan notes, has been that of 'traditionalism'. This movement, which seeks to advance the integration of traditional skills and knowledge in contemporary building, has contributed to the emergence of the multi-disciplinary field of vernacular architecture studies which, he writes, has managed to 'fill the biggest vacuum within architectural theory' and has been increasingly, though slowly, recognized in academia. Discussing briefly how this movement relates to some of the other responses to Modernism, Özkan proceeds to discuss the work of some of its proponents such as, notably, Hassan Fathy, Paul Oliver, the Development Workshop and CRATerre, all of whom have demonstrated how vernacular technologies, materials and forms may be applied in contemporary design.

Although Özkan rightly notes that some of the work carried out by these pioneers (including, famously, Fathy's New Gourna project (Fathy 1973)) has been only partially successful, he also states that 'in time, the followers of Hassan Fathy and Paul Oliver are destined to be successful'. At the beginning of the twenty-first century, in a time of rapid ecological degradation, globalization and the destruction of much vernacular architectural heritage, concerns for the maintenance of local cultural identities and an awareness of the need to provide sustainable built environments are set to raise the interest in local vernacular traditions and their advantages in terms of cultural and environmental appropriateness. Importantly though, Özkan (echoing Marchand) notes that it is the cultural process of the transmission of traditions that needs to be looked at when we are considering the lessons that may be learned from the vernacular, rather than just the buildings or, what he calls, the 'physical shells' of those traditions.

This latter assertion, which entails that it is the appreciation and sustenance of vernacular knowledge, skills and experience that needs to be the focus of attention rather than the static preservation of actual buildings, also forms one of the central tenets of Roderick Lawrence's chapter. Discussing the way in which principles deduced from the vernacular may provide lessons to those involved in the contemporary provision of sustainable human settlements, Lawrence takes as an important starting point that 'it is unrealistic to consider an optimal sustainable state or condition of vernacular buildings, or any larger human settlement'. Because the vernacular, like all human constructs, results from the active and dialectic interrelation between ecological and cultural factors, and seeing that these factors are de facto dynamic because of the continuous mutual influencing that creates ever changing conditions, vernacular architecture and settlements by definition have to adapt in order to be sustainable. Therefore, Lawrence argues, sustaining human settlements involves an understanding of the mechanisms and principles involved in these adaptive processes. Since the relationship between the natural and human environments is mediated through knowledge, values, ideas and information, it is those aspects of a tradition, constituting 'a large warehouse of natural and cultural heritage', that can provide lessons for future generations.

After a concise but useful discussion of the main, though sometimes conflicting interpretations of what 'sustainability' is and a brief explanation of the premises of a human ecology perspective on vernacular architecture, Lawrence suggest a number of basic principles that may be applied in professional practice to increase the sustainability of future buildings and settlements. Using the vernacular architecture of the Valais (Wallis) valley in Switzerland as a case study to validate these principles, he stresses, among others, the need to consider ecological and cultural diversity, the importance of interrelations between different geographical scales (e.g. building, town, nation), the value of participatory approaches to development, and the critical need to raise public awareness of the issues concerned. In discussing these principles, Lawrence, like many before him (e.g. Oliver 1969; Amerlinck 2001), stresses the desirability of an integrated, multi-disciplinary approach to the study of vernacular architecture and settlements. More specifically, he notes the need for studies that pay attention to the reciprocal relationships between the material and non-material aspects of buildings, as well as for those that study actual social values and lifestyles to enable professionals and policy makers to predict and plan for social change.

A good example of a study that attempts to achieve this latter goal through the integration of a number of different disciplinary approaches and methodologies is provided by Lindsay Asquith. Concerned with bringing together the fields of vernacular architecture and housing (both of which currently assume a marginal position in architectural education and the architectural profession), Asquith notes how a vernacular approach to architecture that stresses the importance of intimate relationships between buildings and their inhabitants may contribute to the provision of the millions of homes that will be needed to house the rapidly growing world population. Like Marchand and Vellinga, she notes that an important condition for such a vernacular perspective to become a more urgent agenda point is that the vernacular is not seen as something static, but as 'constantly evolving, reacting to changes in the communities that shaped its form'. Furthermore, she argues that the use and application of vernacular knowledge needs to become a more urgent topic of discussion, and that new methodological approaches, combining tools from different disciplines, should be developed to assist housing research today and in the future. Research into housing needs, suitability and adaptability, Asquith argues, should not be the concern of developers, designers or architects only, but should be based on shared knowledge and should take account of insights gained by anthropologists, sociologists and behavioural scientists.

Asquith demonstrates the value of such an integrated approach with an example of research carried out in the UK (where, as she notes, an estimated 200,000 houses need to be built every year to reach current demands). Using a unique and innovative combination of qualitative and quantitative research methods, integrating interviews, time diaries, spatial mapping and spatial configuration diagrams, she shows how the influence of gender, age and time on the use of space may be measured and mapped, and how interesting and sometimes unexpected conclusions may be drawn from this regarding the way in which families in the UK nowadays claim and use space in their homes. As she concludes, these conclusions can and should be incorporated into future housing design to ensure that newly built homes will be more acceptable and appropriate to their intended inhabitants, as such avoiding the problems associated with mass housing in the past.

Another example of how vernacular know-how may be successfully applied in the contemporary provision of housing is provided by Ian Davis in his discussion of post-disaster housing practices. Briefly presenting some examples of the manifold post-disaster shelters proposed by 'legions of intrepid inventors, relief officials, architectural and industrial design undergraduates, and product manufacturers', Davis observes that, though the efforts of these designers have often (but not always) been motivated by humanitarian concerns, they have usually exclusively focused on the provision of protection from the elements. Failing to recognize local cultural constraints, local vernacular building traditions and, crucially, the many functions that housing performs apart from physical protection, such shelters, he notes, 'inevitably . . . appear to have more to do with the needs of those who generate the concepts and precious little with the harsh pressing shelter needs of survivors of disasters who need far more than physical protection'. As a result, many such shelters never see the light of day or, when they do, fail to meet the needs and wishes of the disaster survivors for whom they were intended.

Taking the *tsunami* that struck the coastal areas of Aceh, the Andaman and Nicobar Islands, Thailand, India and Sri Lanka in December 2004 as an example of the 'bewildering array of key issues that surround the issue of sheltering from extreme hazards', Davis shows how housing, anywhere in the world, has a variety of functions. These include physical protection as well as, for instance, the storage of belongings and property, the provision of emotional security and privacy, a staging point for future action, and an address for the receipt of services. What is needed, he notes, is a more holistic approach to post-disaster housing that recognizes all these functions of housing. Such an approach would let survivors participate in the process of decision-making and reconstruction, allowing them to use their vernacular skills and knowledge, while at the same time providing them with training to create safer buildings. Echoing, then, the arguments of Marchand, Davis stresses the needs for a process rather than a product approach to post-disaster housing, in which participants take centre stage in the efforts to reconstruct a built environment that will remain meaningful through being rooted in tradition and place.

Exploring the ways in which different societies have evolved rational and ingenious solutions to meet their housing needs and the lessons this may offer for professionals working in the field, Geoffrey Payne reiterates the point that vernacular knowledge may have a lot to contribute to the design of appropriate and sustainable housing and settlements. Although, as he notes, 'western models of planning and design based on commercial land markets are penetrating most parts of the world', societies throughout the world have developed a rich diversity of vernacular spatial languages and forms. Drawing upon practical and research experiences gained over a period of more than thirty years, Payne gives a number of examples of how such vernacular conceptions of space or systems of governance and land management can and have been successfully integrated in the planning or upgrading of informal settlements. He also notes, though, that by and large such integration is far from common, seeing that vernacular traditions have all too often been stigmatized and regarded as backward, and consequently been replaced by 'a half digested set of alien and largely inappropriate values'. Still, he shows, in those cases where local wishes and needs have been taken into account, the results have been encouraging and successful.

Payne's focus on urban and informal settlements as manifestations of vernacular traditions once more raises the issue, touched upon by Bronner and Vellinga, of how the vernacular should be defined. Like Lawrence, Payne also reminds us of the fact that, in studying vernacular traditions, it is important to focus on the way in which human settlements as a whole are organized and conceptualized, rather than on individual buildings only. The key challenge, he writes, for those involved in the field of (low-income) housing in the twenty-first century is to find ways of integrating the 'creative energy' inherent in vernacular spatial languages and forms into the development programs that are aimed at planning new settlements or at upgrading of existing ones. Like Lawrence and Asquith, he calls for a more 'holistic' and multi-disciplinary approach to housing that takes into account the perspectives and approaches of different disciplines like planning, architecture, anthropology and economics, and emphasizes the importance of local participation. Fundamentally, and anticipating the chapters of Howard Davis and Latter, he identifies an important role in this respect for education which, in contrast to current practices, should seek to loosen academic and professional boundaries rather than reinforce them.

Part III: Understanding the vernacular

As Payne notes, finding ways in which vernacular knowledge and expertise may be integrated into contemporary building design and practice constitutes one of the main challenges of those involved in the field of vernacular architecture studies in the twenty-first century. What is needed to enable such integration, it would seem, is the disposal of the stigmas of underdevelopment, poverty and the past that currently cling to the concept of the vernacular, as well as a greater understanding and awareness among members of the academy, the architectural professions, the media, policy makers and the general public of the characteristics and values of vernacular traditions. In order to achieve this, more research and, especially, education is needed. Such research and education, as the chapters in Part III emphasize, should focus on issues of process rather than product, identifying general principles and concepts rather than basic facts and figures. Most importantly, it should be critical and actively engage the realities of the present, rather than remaining focused on the past.

The assumption that those involved in contemporary design may learn from the vernacular forms the starting point of Amos Rapoport's chapter. Reiterating the need to look at vernacular environments rather than just buildings, Rapoport, echoing Lawrence's notion of a 'warehouse of heritage', regards the vernacular as 'an unequalled, and only possible, "laboratory" with a vast range of human responses to an equally vast range of problems' such as climate change, resources depletion, cultural change and technological development. Studying and analysing the dynamics, change, success and failure of these vernacular responses will enable the identification of general principles and mechanisms that may lead to important insights into the nature of design and, as such, may provide the lessons of use to those involved in the development of more sustainable future environments. What is required for such lessons to be learned, though, is a move into, what Rapoport calls the 'next stage' of vernacular research. So far, he notes, the field of vernacular studies has been in its 'natural history' stage, 'describing and documenting buildings, identifying their variety, classifying them and so on'. However, in order for the vernacular to teach lessons that are relevant to the future, a more problem-oriented, comparative and integrative stage that leads to explanatory theory needs to be entered.

One way in which this step may be taken, Rapoport suggests, is by looking at vernacular design as a model system. Most developed in biology and biomedical research, the use of model systems involves the use of one system (e.g. a fruit fly or a mouse) to study phenomena in another, apparently very different system (e.g. a human being). Briefly summarizing the use and role of model systems in biological research, he discusses how an approach that treats the vernacular as a model system for contemporary design may help to identify the general principles and mechanisms that characterize the relationship between humans and environments. Such mechanisms (including physiology, perception, meaning, cognition and culture) are constant, though differently expressed in different cultures, and provide the important lessons that may be learned from vernacular design. Rapoport's discussion of some of these mechanisms raises many questions and identifies many hypotheses and conceptual frameworks, illustrating that there is indeed still much research that can and needs to be done, while underlining that any useful lessons to be taught by the vernacular need to go beyond the simple copying of certain formal qualities of a romanticized tradition.

An example of a study that tries to identify some constant, though differently expressed principles and concepts is provided by Ronald Lewcock. Starting from the premise that works of architecture are fundamentally conceptual, Lewcock argues for the existence of certain 'mental models' or 'generative concepts' that act like archetypes and that explain some of the, sometimes unexpected, similarities that can be found between forms of vernacular architecture in many parts of the world, or between architectures of different time periods. Briefly discussing the recent work of cognitive scientists, many of whom argue that the human capacity to form mental models is genetically programmed, he argues that concepts that have been 'developed in the mind using the emotional connections of memory, particularly those that are simple and strong', have the potential to form the basis of an architectural or artistic expression. By establishing connections between ideas, emotions, forms and spaces, such 'generative' concepts (which are the result of genetically produced and learned processes) give rise to particular architectural ideas that have the strength of archetypes and that may either be common to all mankind, or belong more specifically to one society.

Drawing upon examples from places and periods as diverse as Palaeolithic Russia, Mycenaean Greece, eleventh-century India and contemporary New Guinea, Mali and Afghanistan, Lewcock attempts to demonstrate the existence, sometimes in combinations, of certain generative concepts in vernacular architecture, including the cave, the hearth, the covered courtyard and the anthropomorphic analogy. What is more, he argues that the strength and persistence of these generative concepts, exemplified by their widespread occurrence in many historical and contemporary vernacular traditions, is likely to persist into the twenty-first century, and can indeed also be traced in examples of modern architecture. Significantly, though, he notes that this persistence does not imply that the vernacular traditions concerned will not, or have never, changed. Like Bronner, Marchand and Vellinga, Lewcock stresses the point that tradition is a process in which innovation and precedent are dynamically combined and in which, as a result, change continuously takes place. What does remain the same though, he argues, is the underlying presence in the tradition of the generative concept. Studying such concepts and their impact on building design, he concludes, therefore is a promising area for future research.

A call for more research is also central to Isaac Meir and Sue Roaf's chapter. Discussing the performance of a number of generic house types common around the Middle East and the Mediterranean, they observe how many people will point to the vernacular when they are asked which buildings will be most resilient in the face of environmental change. Vernacular resources, technologies and forms such as adobe, windcatchers or courtyards are generally seen to be well adapted to local climatic conditions and are therefore often considered as appropriate bases for environmental design. Still, Meir and Roaf note, such assumptions are not always based on actual research, and the results in those cases where monitoring of their performance has taken place have sometimes been contradictory. Justifying the continued use of vernacular traditions on the basis of their perceived climatic advantages, or copying them for this reason as simple morphological emblems, therefore carries with it the danger of misunderstanding the underlying reasons for their evolution, and 'denies us the benefits of re-interpreting (understanding) rather than re-using (copying) the technology'. What is needed, they argue, are methods that enable us to systematically test the actual performance of vernacular traditions and generate an understanding of how they may be upgraded so as to provide truly sustainable buildings for the new millennium.

Briefly discussing the results of a project carried out by British and Israeli scholars and students, Meir and Roaf present examples of such methods, including *in situ* monitoring, infrared thermography and thermal and daylight simulations. They show how the results of the project have been 'counter-intuitive' in the sense that the buildings did not all perform as well as generally assumed, and suggest ways of improving their performance at realistic costs and with minimum auxiliary energy input. Meir and Roaf's chapter reminds us that a romantic image of the vernacular in which 'an inherent climatic suitability or superiority of materials and form' is assumed is not the way forward, nor is the straightforward documentation of existing traditions enough. What is needed is research that critically tests the performance of vernacular traditions in the face of the challenges of the twenty-first century. What is absolutely vital as well, they note, is a system of education that teaches architecture students about the way in which vernacular traditions relate to cultures and environments, and that in so doing will give them a better understanding of how buildings perform and may be improved so as to make them more resilient and regionally appropriate.

The need for a new kind of architectural education that pays attention to vernacular traditions and seeks to raise awareness and understanding of the ways in which vernacular knowledge may be integrated with new technologies forms the central argument of Howard Davis' chapter. Given the unprecedented demand for new buildings, he writes, especially in the non-western world, there is a need for 'the education of people who can help guide the complexities of building production in ways that lead to buildings that allow the life of people and their communities to flourish, and in ways that minimize negative environmental impacts'. In line with Meir and Roaf, he argues that such an education should engage vernacular traditions in a critical way, taking what is valuable from the traditions without shunning the application of advanced technologies to adapt that which is not working. The result of such an approach will be vernacular buildings that, combining traditional and new elements, do not look like those that existed before the process of industrialization, but which will nonetheless 'emerge out of the lives of people and groups they belong to'. Recalling the arguments of Bronner, Marchand, Vellinga and Lawrence that tradition is a process and that cultural exchange between traditions has always taken place, he thus stresses the need to dismantle the dichotomy of traditional and modern knowledge and, as such, again raises the issue of how the vernacular should be defined.

The contemporary system of architectural education, Davis notes, which developed during the nineteenth and twentieth centuries out of a variety of approaches, is not optimized to educate professionals in a way that may help such new vernacular architecture to emerge. Assuming that professional expertise is better than vernacular knowledge, valuing theory over practice, and emphasizing the importance of 'star designers', it puts too much emphasis on the role of the architect. Instead, what is needed is a system which teaches students that buildings are the result of building cultures which include many actors, none of whom have complete responsibility for the decisions that need to be taken. Incorporating practice-based ways of learning and paying attention to the cultural context of building, such a new system will educate architects who recognize the diversity of roles within a building culture and understand that professional expertise needs to be separated from professional dominance.

A good example of a course that attempts to raise students' awareness of the importance of vernacular architecture and in so doing hopes to educate architects who are more sensitive towards the cultural needs and wishes of their clients, is discussed by Rosemary Latter. Noting that contemporary models of large-scale planning and iconic high-rise buildings may be appropriate to the global industries of banking, insurance and tourism, but do not address the needs of local cultures, Latter writes that 'architects with knowledge of varying cultural contexts will be in a powerful position to influence and advocate sensitively on behalf of [such] communities'. Given the way in which vernacular traditions dynamically relate to cultural identities and the environmental context, a knowledge and understanding of the vernacular architecture of the world can only be helpful in achieving the aim of producing more sensitive practitioners. An inter-disciplinary course that connects such fields as architecture, development practice, geography and anthropology should therefore be an essential element of architectural education in the twenty-first century, she argues, and will be vital for architectural students in tackling the complexities of the world today.

Briefly summarizing the approach and curriculum of an existing postgraduate course in vernacular architecture, Latter offers a model of the kind of education that such an inter-disciplinary and international course could provide. Combining lectures and seminars that focus on the philosophy, theory, anthropology and cultural geography of vernacular architecture with more practical workshops, field trips and 'live projects' that encourage students to actually work with vernacular resources and technologies, and to make proposals for the conservation or regeneration of vernacular buildings or sites, an education in international vernacular architecture allows students to develop disciplinary and professional skills that will enable them to approach contemporary global architectural practice in a more sensitive way. '[W]ith an increasingly globalized and industrialized building industry' Latter (echoing Meir and Roaf, and Howard Davis) concludes, 'architects will be in a better position to assist with planning and housing issues, having had some education in vernacular architecture'.

Vernacular architecture in the twenty-first century

In contrast to the strong and persistent academic, professional and popular associations of the vernacular with the past and underdevelopment, the contributors to this book suggest that the knowledge, experience and skills of the vernacular builders of the world can still have an important contribution to make to the creation of the sustainable settlements and buildings needed in the future. As Oliver notes in his Hepworth Lecture: 'Much can be learned from traditional builders, who willingly pass on their know-how and skills as they have in the past' (Oliver 1999: 11). As a large 'warehouse' of know-how, the vernacular provides important opportunities to adapt modern, global building practices to local cultural and environmental circumstances, as such enabling the creation of forms of architecture that are more appropriate and sustainable, and hence better equipped to face the manifold challenges ahead. At the same time, though, those same vernacular builders may also have a lot to learn from the technologies, skills and knowledge common to contemporary, modern architectural practices. Confronted with the opportunities and problems generated by the process of globalization and with the increasing impacts of environmental change, vernacular builders all around the world will require new ideas and means to service their buildings in line with changing cultural needs and desires, and to protect them from the growing impacts of climate change and the increased risk of natural hazards.

It seems, then, that what is needed at the beginning of the new millennium is an architectural perspective in which valuable vernacular knowledge is integrated with equally valuable modern knowledge, so as to enable the development of settlements and buildings that are contemporary and modern, yet which build upon the characteristics of local vernacular traditions and as such fit within their cultural and ecological contexts. It is only in this way that the creation of a truly sustainable future built environment can be achieved. Whether one could, or indeed would want to call such a built environment 'vernacular' remains open to debate for now. What is clear, nonetheless, is that in order for such a dynamic cross-fertilization of modern and vernacular know-how to take place, an increased awareness, understanding and recognition of the achievements and significance of past and contemporary vernacular traditions is needed among academics, professionals, the media, politicians, NGOs and, indeed, the general public. Critically, it would seem that such an increase can only come about through further significant investments in research and education. Though the field of vernacular architecture studies has undeniably grown in recent decades, research projects that explicitly address the application and use of vernacular knowledge and skills in contemporary architectural practice are still rare. At the same time, there is a serious shortage in educational resources and programs dealing with vernacular traditions, especially at an international and inter-disciplinary level, and in terms of the application of vernacular know-how in a modern or development context. Clearly, much work still needs to be done.

The issues surrounding the value, meaning and recognition of contemporary vernacular traditions and their potential integration with modern architectural practices are extensive and complex. This book does not intend to deal with them all, nor does it aim to provide quick and straightforward answers, solutions or conclusions to the many questions and problems they raise, or the opportunities they suggest. Still, the chapters address many current and important topics that have a direct bearing on the future of the vernacular, and though they have been written from various disciplinary backgrounds, a number of synergies in terms of approaches, premises and themes can be identified. These, we argue, should serve as starting points for future activities in the inter-related fields of research, teaching and practice, and it is worth briefly reiterating them here.

First of all, vernacular architecture should be explicitly treated as a cultural process rather than as merely a material product. Vernacular traditions are dynamic and generated through a continuous and dialectic interplay of stasis and change, precedent and creativity, stability and innovation. This interplay enables traditions to change while remaining rooted in history and place, and thus to remain meaningful in contemporary times (see Bronner, Marchand, Hoare and Lewcock). Accepting this dynamic and adaptive nature of vernacular traditions allows us to expand the scope of the field of vernacular studies by incorporating the emergence of new traditions (Bronner, Payne) and the way in which existing, 'traditional' ones merge with modern building practices (Vellinga, H. Davis). Furthermore, by accepting the way in which vernacular traditions dynamically respond to the challenges of the present and future, it is possible to envision more clearly the ways in which the vernacular may contribute to the provision of sustainable future built environments (Lawrence, Meir and Roaf, Marchand).

Second, future research and teaching should stress the need of an integrated approach that combines the perspectives and methodologies of different academic disciplines and professions, and that attempts to deal with the dynamic interrelationships between building traditions, cultural identities and environments on an ethnographic and descriptive, as well as a comparative and theoretical level (Rapoport, Lawrence, Asquith). Such an integrated approach, which of course has been called for many times before, will help to identify new perspectives and methodologies that can assist future research in applied fields such as housing, urban design or disaster responsiveness (Asquith, Payne, I. Davis). Incorporating practice based ways of learning and participatory approaches, an inter-disciplinary perspective will be an essential starting-point for a new kind of architectural education and practice that trains architects and others involved in the provision of the built environment to be more culturally sensitive, and to appreciate the input of members of other disciplines and, crucially, local communities (H. Davis, Latter, Payne, Marchand).

Finally, and of vital importance, vernacular research and teaching in the new millennium should be critical and forward looking, paying attention to the achievements and qualities as well as the weaknesses and failures of the vernacular, and shunning any romantic ideas about the suitability or even superiority of the latter (Meir and Roaf). As Oliver (1990: 153) has noted, in nurturing such idealized notions (still a common practice among many in the field), 'we shall learn little, and do little useful service to the advancement of building'. It must be acknowledged that many vernacular technologies, resources or forms are appropriate and sustainable, but it should at the same time not be ignored that there are also those that have failed, or that are currently no longer properly functioning because of changed cultural and ecological circumstances. The challenge is to find out how the accumulated knowledge, skills and experience of the world's vernacular builders may be fruitfully applied in a modern context or development context (Lawrence, Rapoport, Vellinga). A long-established discourse dealing with this issue already exists (Özkan); it now seems time to actively expand and build on its achievements.

At the beginning of the twenty-first century, then, faced with unprecedented cultural and environmental challenges, it is only by adopting a processual, integrative and critical approach that it will be possible to rid the vernacular of its 'thatched cottage and mud hut' image. It is also only by taking such a perspective that it may be possible to provide the vernacular with the more prominent position in architectural research, education and practice that it obviously deserves, and to answer the question 'is there still a place for vernacular architecture in the twenty-first century?' with a confident 'yes'.

Part I

The vernacular as process

Chapter 1

Building tradition
Control and authority in vernacular architecture

Simon J. Bronner

Introduction

Eminently visible, persistent and complex, buildings are artefacts people allow to enclose them rather than to control in their hands. As artefacts, they therefore bring into question control, human ability to create anew and alter the old. Bound up with control is the issue of authority: by whose standards, by what precedents, with whose skills, creation and alteration will occur. Since buildings commonly combine a public façade with a private interior, control and authority are often contested for the physical boundaries between public and private space, and their social borders. After all, buildings rise above human scale, and extend the social interaction that occurs within and around them. Paul Oliver has argued that indeed, they are of a 'scale and complexity that exceeds all man's other artefacts, demanding in many cases considerable investments of labour and resources' (Oliver 1986: 113).

Those buildings that belong to a place, that express the local or regional dialect, are often called vernacular. The linguistic analogy of the vernacular speech of building and dialect is significant because it allows for comparison of grammar and syntax, as well as style or manner of expression in material and verbal forms (Oliver 1997b: xxi). It names a category of expression for the majority of the world's buildings that come into being without schooled architects, and that in fact offer long-standing reminders of the labour and resources in a cultural environment. Drawing on the root of vernacular from the Latin *vernaculus*, or 'native', these buildings tell what is indigenous, common and shared in a community or region. Vernacular identifies buildings as social representations and links them to coherent cultural systems of values and beliefs. This point is evident in Oliver's influential definition of vernacular architecture used in his monumental *Encyclopedia of Vernacular Architecture of the World* (1997a):

> Vernacular architecture comprises the dwellings and other buildings of the people. Related to their environmental contexts and available resources, they are customarily owner- or community-built, utilizing traditional technologies. All forms of vernacular architecture are built to meet specific needs, accommodating the values, economies and ways of living of the cultures that produce them.
> (Oliver 1997b: xxiii)

The above definition apparently is an answer to the question of visibly locating vernacular architecture in the broad global landscape. It emphasizes, after all, the building as a text within environmental contexts and available technologies. The last sentence, however, suggests another question of process, for if buildings represent the cultures that produce them, then how is it and why is it that they are produced? The answer emanates from the ways that a people's values, 'ways of living' and ways of building are transmitted and inherited. In short, we are led to issues of *tradition* to interrogate the visibility, persistence and complexity of architecture as a problem of continuity and change. Considering tradition should allow us to answer not only textual questions of why buildings look the way they do and why they are located where they are, but also processual questions of why they came into being and how they changed along the way.

Central to the explanation of continuity and change is the control and authority by which 'owner' and 'community' conceive the public environments they see and feel, and the ones they allow to enclose them. Those thorny matters often carry a reference to *tradition* since decisions about the shape of the future are based on the influence of precedent. Lest one assume that vernacular architectural texts are static points on the globe, tradition takes into account the balance of individual innovation and social custom in the generation of material culture (Martin 1983; Bronner 1992, 2000a; Glassie 1993). Concern for the traditional in architecture therefore poses not only the question of what tradition is within a society or community, but even more importantly, what it *does* (see Riesman 1961; Shils 1971; Milspaw 1983; Glassie 1985; Rapoport 1989).

Tradition, transmission and creativity

Tradition can refer to both a cultural context as well as a performed text. Oliver offers an analogy from linguistics of *parole* as the 'rule system' (or competence) governing *langue*, the expression or performance (see De Saussure 1972; Hymes 1972; Jakobson and Bogatyrev 1980; Ben-Amos 1984: 121–4; Bauman 1992). Tradition can be both subject and object; tradition shapes building and buildings embody traditions. The common use of 'traditional' to describe building is a reference to the structure as an object within a broader category of vernacular or indigenous, although as a subject all vernacular dwellings embody traditions.

If the vernacular implies a culturally based, generative grammar for material texts, then tradition is a reference to the learning that generates cultural expressions and the authority that precedent holds. This construct of tradition as a process of socially shared knowledge and transmission across time and space is the source of conceptualizing a model of explanation in vernacular architecture. Oliver underscores this direction for explanation by suggesting usage of 'vernacular know-how' or the 'faculty of knowing'. He writes:

> within the context of vernacular architecture it embraces what is known and what is inherited about the dwelling, building, or settlement. It includes the collective wisdom and experience of a society, and the norms that have become accepted by the group as being appropriate to its built environment.
>
> (Oliver 1986: 113)

In making decisions about building, tradition is a constant social reference, and in vernacular building, implies a certain force of authority. It is not equal to 'rule', and in fact, implies unwritten or even unconscious codes of doing things that foster variation, since a single tradition as it has been interpreted (especially in religion) can spawn many versions (see Shils 1971; Glassie 1974; Bronner 1986a). Tradition as a reference to precedent is therefore not fixed, and as a social construction, it is often renegotiated in every generation and in every community. Tradition as an idea invites commentary and interpretation, and negotiation of allowable innovation, which might later become part of the dynamic of tradition. As Henry Glassie has observed, 'Tradition's detractors associate it with stasis and contrast it with change, but it is rooted in volition and it flowers in variation and innovation.' In relation to control and authority, tradition 'opposes the alien and imposed' (Glassie 1993: 9).

Just as the individual (or in reference to control, the 'owner') responds to the perception of tradition belonging to the group or community and works identity into this relationship, so then is creativity a necessary component of tradition because the possibility of change is inexorably linked to continuity of form and process over time (Evans 1982; Kristeller 1983; Santino 1986; Jones 1989; Bronner 1992; Bronner 2000a). In Henry Glassie's study of Turkish traditional artists, for example, he effectively viewed tradition as 'the collective resource, essential to all creativity, and in adjective form it can qualify the products of people who keep faith with their dead teachers and their live companions while shaping their actions responsibly' (Glassie 1993: 9). The linking of creativity and tradition suggests a modern philosophy of the arts that 'the ability to create is not limited to artists or writers but extends to many more, and perhaps to all, areas of human activity and endeavour' (Kristeller 1983: 106). This broadening of artistry suggests creativity as a social ideal. This ideal succeeds the Romantic notion of art and architecture as the sole domain of exceptional cultivated minds, existing free of tradition, and as an expression of originality or genius that can create something where nothing existed previously (summarized as 'creation'). There is not one capitalized Tradition in architecture in a modern philosophy of the arts, but a multiplicity of traditions to explore, for tradition in its multiple, abstract existence does not form a simple contrast with creativity. We may recognize situations and societies in which the pressure to repeat precedent is strong when tradition may be re-introduced or re-established as a creative contribution. In any case, innovation is based on an understanding of precedents, many of which will be perceived as traditional. Creativity and tradition are intertwined, and represent the complex processes of humans expressing themselves to others in ways that carry value and meaning. Tradition, in this view, provides a framework allowing for choice and adaptation. It demands attention to form, fidelity to cultural continuity, while inviting alteration and extension for social needs.

Tradition, choice and expectation

One of the assumptions of the vernacular as part of a modern philosophy of the arts is that in the vernacular, tradition tends to dominate, by which is meant that choices are restricted. By stating that there is *a* vernacular is to imply that there is a shared social understanding of cultural standards, customs and norms. But choices are nonetheless apparent in the performance of customs and enforcement of norms. Geographer Yi-Fu Tuan describes the process of tradition as one of 'constraint' rather than repetition. The question it raises is critical to the reformation of culture: 'Out of all the things that have been handed down to us and that we now possess, what do we choose, and what are we compelled, to pass on?' Tuan reflects that 'perhaps what we must seek to retain are not so much particular artefacts and buildings (though we should try to do so in exceptional instances), but rather the skill to reproduce them' (Tuan 1989: 33). Amos Rapoport picks up this theme in his effort to define the attributes of tradition in relation to the analysis of the built environment (Rapoport 1989). While Rapoport finds tradition associated with 'conservatism' in the sense of accepting the past, continuity and repetition, Tuan offers the less politically loaded (and often negative) sense of 'waiting' in tradition. I prefer the idea of 'expectation' coupled with reliance, implying social connection and trust, and as I shall discuss with some native descriptions of tradition, a notion of security. 'Dependence' would not offer choice, but reliance does, and the sense of reliable connotes the rationality of being time tested. What is significant in the modern concept of tradition is that the past becomes part of the present as a guide to future action. Avoiding the word 'past' as well as its sense of a distant time, Rapoport considers the prevailing question to be 'what is repeated, through what mechanisms it is repeated, and what, if anything, makes it meaningful' (Rapoport 1989: 82).

Using the linguistic model of vernacular, we may think of tradition as the local saying that gains credit by long and frequent use. Structured as a proverb, the saying offers the wisdom of many expressed by one person, but may not need a long precedent (Taylor 1994). Its wisdom is a result of an important aspect of tradition, social acceptance. As an adage, the saying takes on significance because of being transmitted, and tested, through time, and may be used variously influenced by the performer's perception of certain situations and surroundings. Adage implies wisdom that one may choose to follow or at least recognize, in contrast to the 'maxim' which is more of a rule of conduct. The range in these types of sayings shows the way that tradition covers a range of control and authority, and can become contentious for a community as a result. In the modern concept, it is important to take note of the performance of the tradition, and observe that the tradition of the proverb can be customary, as in a practice followed as a matter of course among a people or community, or by an individual as part of the person's experience and way of living. In this analogy, builders can use forms and techniques that they recognize from tradition as socially accepted and time tested, and residents alter and apply their experiences in the house. To be sure, the house is not an utterance, and in its persistence as a form and complexity as a process, it stands boldly on the landscape. It frames experience and custom by providing a basic human need for shelter and symbolizing social existence. Sheltering people as well as symbolizing them, elevated above them and enclosing them, the house can be a constant, longstanding reminder of tradition, and often its standing in a culture.

Tradition as process

Viewing tradition as a process, then, an eye toward tradition in the vernacular landscape takes in several significant implications:

1. As consistencies are apparent in any single time among buildings, the question arises as to the force of tradition as a social construction in dictating the similarities. The understanding is that there is a perception of cultural precedent by which forms are generated. In the interrogation of tradition is revealed the process of learning or direction from 'tradition bearers' to others or the socially shared events in which tradition is invoked.
2. As inconsistencies are apparent in different periods, the question arises to the forces in tradition that allow builders to change and innovate. The understanding is that the vernacular, being rooted in tradition, is less apt to change, and therefore when significant change occurs, it implies major social structural shifts. Implied in this understanding is that communities perpetuate their traditions from one generation to another in various, and often culturally specific ways – in apprenticeships, in rituals and festive events, in family and community institutions, for instance.
3. As variations are apparent in communities or regions, the question arises to the communication of tradition across space. The understanding is that tradition diffuses in traceable patterns and cultural influences. The implication is that various influences affect diffusion, including geographic, social and economic opportunities and barriers, technology and economic connections, linguistic connections from

one group to another, political organization and self-perceptions of insiders (what may be referred to as 'identity') and attitudes toward outsiders.

4   As variations are apparent among the dwellings of different builders and residents who have physically adapted the structure for their use, the question arises about the dynamic of creativity and tradition in individual decisions about the appearance of buildings. Implicit in this question is the role of individuality within a culture. As tradition itself suggests an ethnographic focus on the 'performance' or enactment of a building, the question of variation often adds the behavioural issues (i.e. based on personal, material enactments of ideas and involving communication systems of symbols) of 'technical competence' (culturally developed skills and talents), 'decoration and style', 'use and function', 'arrangement', or 'aesthetics' to the more textual methodology associated with the comparison of form (floor plans and elevations) and materials of construction.

In sum, tradition is performative, and in oral tradition can be traced to other performers with attention to the style and variation of singers, speakers and tellers. Indeed, with songs, speech and tales, tradition as a behavioural process is frequently invoked because the performer is more apparent than in architecture. It appears, in fact, that in uttering, rather than building material texts in situational and cultural contexts, the performer is more in control of tradition in the shaping of a performance (Evans 1982; Bauman 1992; Jones 1997). Part of the reason that tradition as an explanatory concept has not been more applied to building is that scholars typically approach the building as text rather than event or process (see Glassie 1972; Upton 1979; Upton 1985; Herman 1985; Jones 1997). It can be argued, in fact, that vernacular rather than 'traditional' became appealing as an adjective for architecture because it allowed the viewer license to identify cultures through consistencies in building styles without full knowledge of human mediation and customs involving them (see Bronner 1979; Heath 1988). Vernacular allows for distancing people as agents of their own artefacts.

With tradition as a consideration in material culture, it is not just the skill or procedure of construction that is in question, but the way that knowledge of design and values is inherited, adapted and transmitted. From a textual point of view, variation in architecture is assumed to be across space, thus lending itself to an identification of types and the mapping of regions. The impression is given that an organic progression, apart from the volition of human builders, occurs as one surveys patterns across the landscape (see Noble 1984; Ensminger 1992). But with more inquiry into tradition, the relations of self to community, continuity to change and innovation to conformity come to the fore to explain technological and adaptive choices that lead to variation or social perceptions and contexts influencing change.

An important query involving tradition is the way that it relates to modern pluralistic societies in which, as Tuan claims, individual choices are more abundant and innovation is encouraged. Although it may appear that tradition 'is lost', arguably the number of available and emergent traditions greatly multiply. Indeed, a complex issue is the changing nature of tradition within modern industrialized states. Although tradition following Robert Redfield's paradigm of the 'little community' and folk society is associated anthropologically with small homogeneous groups and a limited space, and assumed to be passively received, tradition in modern society is still at work on a mass scale as a cultural reference to a way of doing things or an appeal to the authenticity provided by 'roots' (see Redfield 1947; Foster 1953; Redfield 1960; Bauman 1983; Kirshenblatt-Gimblett 1983; Sider 1986; Kirshenblatt-Gimblett 1995; Bronner 2002a). Because of the perceived need to have tradition as a basis for claiming a cultural identity or marketing authenticity (i.e. cultural tourism), frequently traditions are invented, marketed or constructed (see Hobsbawm and Ranger 1983; Bendix 1989; Cantwell 1993; Becker 1998; Tuleja 1997). Rapoport notes contrasts in the survival of traditional groups and products in centralized states such as France in comparison to the diversity of the US (Rapoport 1989: 99). This survivalist concern is understandable, considering the cultural historian's worry about the preservation of vernacular environments in mass culture, but the query that reveals the dynamic nature of transmitted, inherited tradition, as well as modernity, is one about continuity and change. The focus is on the creative choices made as individuals perceive the external authority of tradition.

Representations of the dynamics of tradition in twenty-first-century America

By way of example of the explanatory project incorporating tradition to analyses of vernacular architecture in complex societies, I offer a comparative summary of three representations of the dynamics of tradition in twenty-first-century America: seasonal construction of the Jewish s*ukkah*, Amish community barn-raising and Houston's 'recycled' houses. I order them in their historical reach: the *sukkah* is an ancient structure codified in a religious text, the barn-raising among the Amish originating in Europe developed in the last two hundred years, and the 'recycled' houses, or homemade environments, date to the late twentieth century. With each, the control that individuals exert over their environments in response to mass society, and therefore the meaning of tradition, are frequently at issue.

The *sukkah* (plural: *sukkot*), translated roughly as booth or tabernacle, is central to the Jewish thanksgiving holiday of *Sukkot* or 'the Festival of Booths'. It drew my attention because it is a holiday revolving around the construction of a primitive dwelling, intentionally meant to be temporary. The building of this structure is an annual reminder for Jews of the exodus from ancient Egypt by the Israelites. Every year, Jews are called to reconstruct the vernacular architecture of the nomadic period of their history in the desert. The Biblical description in *Leviticus* makes plain the purpose to reinforce the transmission of values from one generation to another: 'You shall dwell in *sukkot* seven days . . . in order that future generations may know that I made the Israelite people live in *sukkot* when I brought them out of the land of Egypt, I the Lord your God' (23: 42). In modern

American society given to the individualistic pursuit of wealth and materialism, orientation toward the future and reliance on a service economy, the creation of a simple dwelling and giving thanks for nature's bounty for this time is supposed to have a leveling effect socially and a humbling one personally (United Synagogue 2003). Even if a resident lives in a modern pre-fabricated house, he or she may build the *sukkah* to represent the hand-wrought roots of Jews in the ancient Middle East.

For seven days during autumn, five days after the solemn observance and fast of *Yom Kippur*, many Jews celebrate *Sukkot* as one of the most joyous in the Jewish calendar. As part of the Diaspora, American Jews consider the first two days of the holiday to be *yom tov*, i.e. a major festival on which work is prohibited. The *sukkah* is the major external symbol of the holiday, and inside the booths, symbols of nature's bounty abound. Also evident is its function to encourage close relations within family and community. The holiday is associated with hospitality and the charitable spirit of sharing. Accordingly, it is customary for Jews to invite Jewish neighbours and friends to share a meal in their *sukkah*. Reinforcing this spirit, a symbolic ritual known as *ushpizin* is often enacted, in which a blessing is read to mark the invitation to an ancient Jewish patriarch such as Abraham, Isaac, Jacob, Joseph, Moses, Aaron or David, to sit with the residents in the *sukkah*.

Once considered an individual's obligation in the East European roots of many American Jews, through the twentieth century building the *sukkah* in the US had been increasingly relegated to Jewish community institutions, including synagogues and Jewish community centres. The *sukkot* of the institutions tend to be large, often holding as many as one hundred people. Some orthodox critics consider this trend an undesirable result of assimilation and mass society, as individuals give up their piety and treat the synagogue as a convenient service. During the 1990s there was more effort, especially among the Lubavitch wing of orthodoxy, to publicize among Jews the need to restore the obligation of household *sukkot*.

Few Jews I interviewed who erected *sukkot* consulted the *Talmud*, the written code drawn from rabbinical oral tradition for the design of the building, although it is specific on many aspects of the structure's design. Their most common explanation of their customary 'blueprint' was drawn from two sources; participation in its construction as a youth, and the prevailing design of other structures they have seen in their community. When they established families and single family dwellings, they began to construct their own *sukkot*. Indeed, building the *sukkah* most commonly is a family project; parents place the poles, the father erects the walls and children and wife decorate the interior. Most builders understand the significant features of the structure to be the dimensions and the roof covering. The structure thus invites variation and identification with the family who builds it.

If the exterior has shown more variation because of the use of commercial products adapted for use in the *sukkot*, even more variation is present in the interiors. Families individualize the interiors of the structures to represent

1.1
Sukkah built by family in suburban development in Harrisburg, Pennsylvania, 2002.

their identities. Orthodox rabbis sermonize occasionally on the Talmudic reference to 'make a beautiful *sukkah* in His honour' (Shabbat 133b) based on the Biblical verse 'This is my God and I will adorn Him' (Exodus 15: 2). In a separate section entitled '*Sukkah*', the *Talmud* describes the possibility of decorating with embroidered sheets, hung nuts, almonds, peaches, pomegranates, grapes, corn, wine, oil and fine flour. In the United States, apples, found in abundance in the north-east during the season, are common along with other harvest decorations of gourds, cranberries and dried corn. The interior commonly holds a small table and chairs for eating in close quarters. Parents invite children to help decorate the interior with paper chains and garlands, pictures, sticks and crafted decorations. Sometimes Jewish New Year cards are hung to remind visitors of the previous High Holy Days. Sephardic Jews from Arab countries often hang tapestries. Fruits, nuts and boughs of evergreen often hang from the roof covering. Importance is placed on the activity of decoration, since it engages participants in the process of belonging to the Jewish tradition.

Sukkot stands alone for Jews in late September or early October, whereas *Hanukkah*, a minor holiday, has grown in observance because of its timing around Christmas. Thus Jews are more aware than gentiles whether one celebrates *Sukkot*. The issue of following tradition, and maintaining a Jewish identity, frequently is between 'assimilated Jews' and 'traditional Jews'. Even as demographic studies show that the number of assimilated Jews who have largely abandoned observances of *Sukkot* has grown, they also reveal a concomitant increase in the number of orthodox or traditional Jews (Goldstein 1992; Mayer, Kosmin and Keysar 2003).

Although commonly described as serving the function of thanksgiving and reminder of the exodus experience, the construction of *sukkot* in the US stands more immediately to affirm the ties to tradition itself. In the perception of

Simon J. Bronner

1.2
Sukkah built by synagogue, Harrisburg, Pennsylvania, 2002.

Jews, the building is a primary statement of the resident's fidelity to tradition. As a result of a dramatic rise of inter-marriage, secularization, geographic dispersal and suburbanization in the late twentieth century, American Jews are well aware of divisions concerning the importance of following religious tradition and living in a Jewish community to maintain Jewish identity. In an effort to bolster the

1.3
Interior of *sukkah* with homemade decorations on wood panelling, Harrisburg, Pennsylvania 2002.

holiday, many synagogues and Jewish community centres have offered assistance with construction 'kits' for the *sukkot* and organized children's programs in large community *sukkot*. Yeshivas often take time to train children in the idea of building *sukkot* by involving them in a craft activity creating small booths out of popsicle sticks and decorating community *sukkot*.

Building the *sukkah* in twenty-first-century America has become a chosen signal of involvement. It pronounces the continuity of family ties and the proximity of Jews to one another in the conceptualization of Jewish identity. It states that the family is a key to Jewish continuity in the process of enacting tradition: manually building a primitive structure that stands out in modern suburbia, in willing to work for it and in labouring together rather than individually. In dining outside, often in the cold, the family is stepping out of the comfort of modernity and engaging tradition as vital to their lives. Yet the flexibility of changes to the materials, allowing for modern conveniences of tarp and electric lights, provides integration with the times. Apart from the synagogues, families who elect to build and gather in the *sukkah* as their mode of 'dwelling' are willing to be noticed.

Whereas Jews, at 2 per cent of the population of the US, are concerned about the decline of not only their numbers, but also their fidelity to tradition, the Amish are experiencing growth. Nonetheless, leaders of both groups express concerns for the effects of dispersal and the impact of mass society on their sense of community. Amish barn-raisings today are primary signals of Amish community identity, concentrated in places such as Lancaster County, Pennsylvania; Holmes County, Ohio; and Elkhart County, Indiana. In the three contiguous states, moving from east to west, of Pennsylvania, Ohio and Indiana, the Amish account for around 180,000 individuals, divided into 907 small districts; about 80,000 more are spread out in settlements through the US and Canada, mostly in the Midwest, Upland South, New York and Ontario. The large number of districts insures that the number of members of any community is small, thus encouraging face-to-face interaction in a limited locale. In addition to differences in dress, transportation, language and religion between them and their neighbours, the Amish are also distinguished by their commitment to agriculture. In Lancaster County, 44 per cent of Amish adults are employed in farming whereas only 4 per cent of their non-Amish neighbours are (Kraybill 2001: 81).

Once a common American agricultural practice, barn-raising has been appropriated by the Amish as a special social tradition. Barns throughout nineteenth-century America grew in size as well as importance for the farm, and a social response to meet the challenge of building a large barn was to turn the task of construction into a community event. As American settlements expanded westward through the nineteenth century, the barn in America as a large imposing structure often dwarfing the farmhouse became symbolized as a sign of growth and abundance on the American landscape and the accessibility of individual property (Hubka 1994; Vlach 2003: 1–22). American barns grew in dimensions through the nineteenth century and in Pennsylvania, barns fusing the styles of German bank barns and English entrances on the non-gable end emerged as an expandable hybrid form often called the Pennsylvania barn (see Ensminger 1992; see also Hubka 1994). It featured at least two levels, one below the bank, often to house animals, and one above for storage of grains and vehicles. The tradition of building these large structures could not be perpetuated, however, without substantial use of labour. Many travellers' accounts of the Amish remarked on the palatial dimensions and finish of the barn compared to the simple house (Smith 1960: 263).

The American vernacular term 'raising' referred to the task of lifting into place erect 'bents' or large vertical frames that formed the skeleton of the barn and connect them with cross-girts (Mathews 1966). The task of raising, the visually exciting highlight of barn building, took many men working co-operatively and using ropes and poles (or 'pikes') to erect the heavy wooden bents (Price and Walters 1989). The expectation arose that the community created a festive family atmosphere around the labour-intensive event. Sources can be found for this kind of festival in medieval England and continental Europe, but evidence suggests its association with pre-industrial America because of the increased frequency of barn building on individual farms (see Glassie 1974).

The introduction of individual fire and disaster insurance plans, growth of agri-business, introduction of mechanical devices for raising bents, use of lighter pre-sawn timbers, and professionalization of farm construction in the twentieth century resulted in the decline of barn-raisings as festive communal events. The tradition of barn-raising no longer became necessary because farmers did not need the community for their success or in the event of a disaster, recovery. The farmer, increasingly centred on monetary capital, was not rewarded for providing social capital. The application of technology, and the reliance on professional specialists and services, allowed for more individualism. Among agrarian groups holding on to values of mutual aid such as the Amish and Old Order Mennonites, however, barn-raisings became increasingly important to maintain connection in their communities. They encouraged residents to remain close to one another to take advantage of the labour required for complex tasks such as building the barn. Social capital became even more important to the Amish as modernization brought intrusions of highways and technology between them and their neighbours.

The Amish share with their non-Amish farming neighbours the feature of individual ownership; they are not a communal sect that share property among its members like the Hutterites. But barns carried community functions that distinguish the group from their nineteenth-century English neighbours and help the barn-raisings persist, even as the number of Amish involved in agriculture declines (see Kraybill 2001). Ethnographies of Amish societies showed that the barn was important not only because of its function for farming activity, but also as space for religious services. The Amish do not have churches; instead they use members' barns for services. The religious connection of the barn-raising goes beyond space for services, because the idea of mutual aid is considered a spiritual value of *Bruderschaft* or brotherhood. It is coupled with the spiritual value of manual labour and staying close to the land. The barn-raising is an especially dramatic reminder of everyone in the community involved in handwork. Although the barn-raising is an especially visible example, many activities in Amish life follow this principle, including community care for the sick and aged, help with moving, relief after a flood or drought, and benefit auctions to help families struck with excessive medical bills or injuries. Several analysts of Amish traditions understand the provision of security by the community in exchange for maintaining separation from the world, and it may well be a contributing factor to the high retention rate of young Amish into the church as adults. John Hostetler in his classic sociological study *Amish Society* reflected:

> Security is therefore assured to the Amish individual by the concern of the whole community. If a member is sick, in distress, or is incapacitated, the community knows about it. While the Scriptures admonish the believer to do good to all men, the Amish are especially serious about the advice with respect to their own 'household of faith'.
>
> (Hostetler 1963: 146)

Especially important is that the raising became part of a system of disaster insurance for a group dependent on mutual aid rather than worldly professional services. The Amish believe that church members should be accountable to and responsible for each other, and therefore commercial insurance would undercut aid within the community (Kraybill 2001: 155). After a fire, the community pitches in by restoring the barn or raising a new one. Continuing features of a communal meal, visual excitement for spectators, and festive atmosphere, the Amish barn-raising is a productive integration of pleasure and labour serving the needs, and underscoring the importance, of the community. In sum, the barn-raising is the most dramatic example of 'social capital' among the Amish, which includes face-to-face relationships, extended family, and long-standing traditions and rituals that support them (see Kraybill 2001: 142–60; see also Putnam 2000: 234–5).

Some Amish old-timers note the change in tradition from their time to the present toleration of drink-coolers, electric drills and commercial portable toilets at the barn-raisings, but the process of brotherhood in the large use of labour to build the symbolic working barn remains and will likely continue. John Hostetler, an analyst who grew up in the Amish church, emphasizes: 'Tradition and experience tend to become highly symbolic in structural acts such as a shared style of dress, language, limited education, and mutual-aid practices.' The Amish, he reports, are successful in staying bound by tradition because they have maintained 'substitutionary forms of intense sharing but also by meeting the social needs of the individual' (Hostetler 1963: 147). Symbols such as the Amish barn-raising expresses the group's social unity in material form. In Hostetler's phrase, 'it clarifies the sentiment a society has of itself' (ibid.).

It is a sign of individualistic modernity and mass industrialization that 'traditional' vernacular structures such as the human-built *sukkah* and the Amish barn have drawn attention to themselves because they stand out with their references to the social integration of a community. As a result, in modern journalistic accounts, they are often referred to as 'art' with its implication of being uncommon, even exceptional, rather than a *craft* of everyday living (see Ames 1977). While apparently elevating the status of these traditions, it also implies cultural weakening. Although some critics such as Kosta Mihailović claim that 'the abandonment of traditional production and the traditional way of life is an unavoidable consequence of industrialization and urbanization', the above examples also show that modernization can also result in revitalization of traditions as a response to maintain community and identity (Mihailović 1989: 36). This revitalization raises the question, however, of how conspicuous emergent traditions of homemade built environments, often categorized as vernacular, folk, outsider, informal or grassroots art comment directly on modernization by the use of manufactured materials in striking assemblages (Blasdel 1968; Cardinal 1972; Walker Art Center 1974; Wampler 1976; Rosen 1979; Ward 1984; Santino 1986; Manley 1989; Brackman and Dwigans 1999). Unlike the *sukkah* and Amish barn, the

1.4
Amish barn-raising in early stage showing erection of bent, Lancaster County, Pennsylvania, c.1960.

Building tradition

1.5
Amish barn-raising in later stage showing construction of roof, Lancaster County, Pennsylvania, c.1960.

assemblages are not connected to ethnic and religious roots as much as an approach to the built environment that may have class or occupational connections and a cultural stance of resistance to modern commercial culture. Many vernacular architecture scholars might dismiss the structures as idiosyncratic, marked more by individual oddity than vernacular criteria of a common parlance rooted in place and community (see Marshall 1982; Vlach 1985; Vlach 1986). In another sense, however, as a hand-built environment tolerated within certain locales, are they a sign of the future of the vernacular in complex, industrialized societies?

In Houston, Texas, for example, hand-built environments enjoy special status because of a lack of zoning ordinances. In fact, Houston, partly as a result of its nineteenth-century heritage of an open western frontier inviting to new settlement, is the largest city in the world without zoning restrictions. Its climate is fairly temperate the year round, and usually enables builders to work continuously outdoors. Journalist Michael Ennis refers to 'a state of mind peculiar to Houston' based on the lack of a tie to one culture. He explains it as a distinct local recipe:

> Take a pungent mix of Latino, African American, and Cajun influences, combine with a surprisingly resilient blue-collar ethic, wildcatter individualism, and Southern eccentricity, and stir it all up with Houston's storied absence of zoning regulations; the recipe has encouraged exuberant public displays of art rarely seen elsewhere. *The tradition is a living one* . . .
>
> (Ennis 2002: emphasis added)

37

The reference to tradition as living suggests that there is social toleration for creating the environments and a continued adaptation of practices and forms as builders become aware of one another's work. The process of building homemade environments makes a statement about vernacular production in an industrialized society: in the absence of natural resources, makers employ a type of *assemblage* and *bricolage*, whereby the materials of construction are often manufactured objects which are altered, assembled and arranged, so that their original functions are transformed. One can see many examples of this transformation on front yards where tyres, rain gutters and milk cans become planters and mailbox supports (Nicolaisen 1979; Abernethy 1985; Sewell and Linck, Jr 1985; Bronner 1986b; Sheehy 1998), and in urban settings, where the sidewalk is appropriated for recreational or even religious purposes with recycled furnishings and equipment (Bronner 1985; Sciorra 1989). In the form of the house, the boldest and most total change of the environment, layers may be added by the *bricoleur* so as to essentially refabricate the structure (Bronner 1986a). The items may not only be significant to the *bricoleurs* because of their material, colours, and shapes in an overall design, but also because they represent the transformation of the process of disposing (and obsolescence) intrinsic to commercial culture into a cultural ecology associated with folk cultural use of local resources.

1.6
Section of 'Orange Show', showing construction using wheels, built by Jeff D. McKissack, Houston, Texas, 2002.

The tradition often referred to in this process is one of recycling, akin, some critics observe, to the ecological value in many folk cultures on avoiding excess and waste in the relation of the built environment to nature (Ferris 1974; Greenfield 1986). The transformation of dwelling in these producer cultures occurs from the natural resources in the environment; the argument is that this kind of ecology is still taking place in a consumer society, but the resources are manufactured. Fieldworkers find use of old bathtubs for Madonna shrines, bleach bottles for weathervanes, soda bottles for tree structures, egg crates for herb planters and mailbox supports from welded chains, and interpret them as either replacement for unavailable natural resources in an urbanizing society and an answer to standardization, specialization and disposability of labour and technology (Jencks and Silver 1972; Greenfield 1986). Analysts often find in the *bricolage* structures an artistic outlet for a need to create by hand out of an urge to express a rooted identity in a mass society (Bronner 1986a; Jones 1987; Jones 1995; Sheehy 1998; Jones 2001). Others viewing consumerism as a middle-class movement, sometimes comment on lower- or working-class provision of their needs in emergent forms of industrial folk crafts since natural resources are denied them (Dewhurst 1984; Lockwood 1984). Some studies of the makeshift structures of the homeless, apparently following patterns learned from one another, comment on refabrication of discarded cardboard boxes, corrugated tin, vinyl tarp and plastic milk crates (Daniels 1981). It can also be occupational, as in the notable architectural example of small buildings made out of sewer tiles drawn from a conduit factory (Dewhurst and MacDowell 1983; Dewhurst 1986).

Instead of the outsider, naive or visionary label for such personal environments, Michael Owen Jones urges vernacular art and architecture students to understand the continuities and consistencies in *material behaviour* (Jones 1993; Jones 1994). This approach is a move away from the insistence on a social community for the understanding of tradition, but it nonetheless focuses on the formation of identity and the creative enactment of perceived tradition (Vlach 1985). Instead of displacing tradition as an explanatory tool, this approach can emphasize the ways that individuals consciously shape their own 'traditions' in the sense of their repeated practices that express their values and identities.

The Houston dwelling known as the 'Beer Can House' raises especially provocative questions about material behaviour. Covered in decorations crafted from thousands of beer cans, it appears at first glance to have little precedent as a form, but the process of folk recycling that informed its construction is familiar. Its maker, John M. Milkovisch (1912–88), a Houston native, when asked why he used the cans as his building material, remarked: 'I hate to throw anything away' (Lomax 1985: 19). The beer cans he used to create curtains, fences, windmills and windchimes were ones that he and his wife drank. The result, however, is not an advertisement for commercial labels, because it is only on close inspection that one realizes the source of the metal shell for the house. Milkovisch had been in the habit of saving cans in an act of folk recycling before he got the idea of decorating his house with them. His recycling anticipated some home use for them, and neighbours engaged in similar behaviours such as saving egg crates (planters and jewellery holders), cheese containers (reused as food storage), bleach bottles (liquid storage and scoops), baby food jars (storing nails and small items) and bread wrappers (bags and even tied into rugs). Although the house extended the 'home' use of recycled materials, it did have a relation to house structures familiar in the west using recycled materials, including houses made during the twentieth century of hay bales, railroad ties and stovewood (Welsch 1970; Welsch 1976; Tishler 1982; Graham 1989). Indeed, Milkovisch was aware of a number of houses using recycled bottles for a building material or decoration (McCoy 1974; see also Greenfield 1986: 55–90; Seltzer 2000; Rzadkiewicz and Young 2003). His first decorating of his property, in fact, was with glass marbles and parts of bottles before turning to his beer cans.

Milkovisch sliced off the tops and bottoms and split the seams of the cans with homemade tools. He bundled stacks of fifty flattened cans and filled his attic and garage with them. His occupation was an upholsterer for the Southern Pacific Railroad, but like many working-class residents, he engaged in various handwork hobbies such as woodworking and home maintenance. He lived in the house with his wife Mary Hite, who he married in 1940. The house had belonged to his parents and he bought the modest frame home in 1942. It was not unusual in the neighbourhood for men to build their own patios and loan each other material. As he described the evolution of the project, he began in 1968 with the objective of building a patio and driveway in concrete. Looking to enhance the structure, he used marbles and other small objects such as broken glass, tiles and old keys to decorate the patio in the back yard. As he worked on his project he drank beer and continued to stack the cans he had been saving for about seventeen years. He loathed the constant routine of cutting the grass and decided to pave over his front lawn in concrete as well. Not satisfied with the appearance of the plain cemented ground, he pressed more marbles, tiles and keys into the wet concrete. This made his *bricolage* more public, and he received compliments on his design and even contributions of cans from neighbours.

Positively reinforced in his decorative use of recycled materials, Milkovisch remembered the stacks of flattened cans in the garage. He first hung them as streamers in trees that reflected light much as commercial windchimes he had seen in other yards. Looking to frame the open driveway, he built an arch over the driveway with unflattened cans. Enthused by the malleability of the cans as a construction and decorative material, he applied them to the house itself as a substitute for repainting, creating a ribbon of silver metal around the bottom of the house. Then he kept going, since there seemed to be an inconsistency in his view between the old siding and the cans he was applying. This gave him ideas about the possibilities of the transformation, and he hung streams of cans from the eaves to create a curtain effect as a second layer over the siding; then he made window shades out of them. According to his supportive wife: 'Finally he did it all before I knew what was going on.' Respecting the outside as his work sphere, his wife Mary did not allow him to apply his can decorations on the house's interior.

He added *assemblages* that helped explain some of his values to curious passers-by. One striking sculpture in the garden is a 'ladder of success',

1.7
'Beer Can House' built by John M. Milkovisch, Houston, Texas, 2002.

which he completed after the house was done. As John explained to an interviewer: 'Painted one rung black because most people don't make it. They say every man should always leave something to be remembered by. At least I accomplished that goal' (Lomax 1985: 21). John appreciated being recognized for the skill and creativity that went into the house, but became irritated at interviewers from the media who questioned his drinking or the idiosyncrasy of the house (ibid: 17). To be sure, he wanted to be noticed, but declined to call it art. In his mind, it was his home project after years of unnoticed labour (he refused to upholster at home upon retirement), that gave him and his family satisfaction from the use of productive handwork with recycled materials. His locus was familial and local, as he related with this anecdote:

> Like I told one foreman over there at the railroad, he asked me where was I going on my vacation. We could see downtown Houston from that roundhouse. I says 'You see them tall buildings up there?' That was uptown. 'That's where I'm going.' He says, 'That's where you're going?' I says, 'yeah.'
>
> (Lomax 1985: 20)

Milkovisch was less interested in the expectations of retirement as a time for distant travel and exotic experience as he was in rooting himself as he aged in the familiarity of place. The places he knew had changed dramatically in the time he had been in Houston with high-rise development and urban sprawl, and in addition to building, he would relate his memories of old Houston as a kind of urban village, where neighbours related closely to one another. His building displaced some personal concerns for his suddenly feeling 'out of place' by drawing attention to the kinds of processes he had been familiar with as a long time working-class resident.

Simon J. Bronner

John had passed his love of handiwork on to his sons Ronny and Guy, and after he died, they maintained the house and the various structures on the property for his wife who still lived in the house. They even added some fences out of the cans that line one side of the property. Proud of carrying on John's work, in the 1990s they completed a television commercial in support of recycling (Theis 1998). In November 2001, when Mary, who always expressed pride in the structure, could no longer live without assistance, the Orange Show Foundation (created in 1980 to preserve and promote Houston's vernacular environments) bought the property, with the idea of preserving the house for future generations. The house is one of many that the Foundation features to showcase the working-class creativity in the visible environments of Houston.

The use of cans to create structures is not unique in America. The attraction of commercial cans as a building material is evident in other places such as 'Can City' built by Joshua Samuel in Walterboro, South Carolina (Stanley 1984). The cans are discarded quart oil cans collected by Samuel from gas stations. What began as border trim for Samuel's yard became decorations for trees, and an assortment of rooms built of cans that runs a tenth of a mile long. They, like many folk recyclers, began the endeavours as extensions of their vocations or demonstrations of their hand skills, the work developed late in life when a form of life review occurred and more leisure time became available. To be sure, one often finds religious, visionary or moralistic reasons for the undertaking, and the hope for inspiring others to create. The analysis of tradition in these environments takes in the life story of the creator and the symbolism of the expressions. Unlike the ideal of vernacular as the prevalent common expression of the people, most of these folk recyclers are well aware that they stand apart in their elaborate structure, although many interviews draw out their reference to grassroots skills and processes that were, or should be, part of modern culture (Foster 1984). Their recycling is familiar, although their forms may not be, and sometimes they

1.8
Composition of freestanding patio shelter with bottles and inlaid marbles built by John Milkovisch, Houston, c.1980.

1.9
Wall constructed by Ronnie Milkovisch, the house builder's son, using techniques he learned from his father of cutting tops into different shapes, arranging bottles and applying labels during the 1980s. The son creatively added the use of hanging tabs to the tradition learned from his father.

even draw attention to the Biblical phrase of 'silk purses from sows' ears' to add spiritual force, and perhaps a communal allusion, to the constructions (Greenfield 1986). Their *bricolage* in its transformative powers exerts their control over the very consumer objects they use as raw materials. The resulting construction appears very human rather than industrial, and perhaps a reminder of a lost vernacular.

Tradition as a model for action and commentary on control and authority

Often the inquiry about artistic structures centres on the individuals who created them, but questions persist about the connections of the forms to place and society, and in so doing raise issues of tradition as a model for action. The behaviours that generate various environments as grassroots, hand-built productions in a consumer society invite analysis as a transition from the vernacular of Redfield's 'little community' to the 'mass society' (Bronner 1986b). In the Midwest several religious grottoes made from geodes by devoted Catholic priests suggest patterns of tradition as well as individual motivations (Ohrn 1984; Niles 1997; Stone, Zanzi and Iversen 1999; Brackman 1999); in Los Angeles, the Watts Towers left by Simon Rodia appeared out of place, and without precedent or model, until the connection to Rodia's Italian background of processions with pyramidal Giglio was understood (Posen, Sheldon and Ward 1985); on the Great Plains, gates of horseshoes and structures made from wagon wheels lead to discussions of grassroots efforts to embellish the flat, dreary landscape (Abernethy 1985). I could go on, but suffice is to say that tradition need not be anonymous.

The three examples of living traditions I have given at their surface refer to the continuities they maintain, often with distinct challenges. The continuity, as in the case of the Jewish *sukkah* builders is to an ancient place and form, and their construction has become increasingly symbolized as a revitalization of a community of faith. The instance of the Amish is also about community strengthened by participation in the constructive act, but complicated by the ambivalence toward an outside society which it seeks to distance from them and at the same time upon which it relies for its prosperity. The continuity for the grassroots environments is less to the forms they built than to the processes and values that many of the builders considered displaced by consumer society. All the examples remind us of the constant of change in any tradition. Social discourse is rampant with the *sukkah* and the barn-raising on the tools, materials, and significantly, on the changing functions of their traditions. For the environments, the discourse of change is on the mass society that now views out of place the act of building personal space. Whether the two millennia of the *sukkah*, or the two generations of the beer can house, the builders had to take into account the transmission of the skills and motivations for constructing the structures, lest they became relics rather than continuing experiences or customs.

Less obvious with the examples is the matter of control and authority suggested by engaging tradition, especially in the US where there is imagined a democratic belief in individual liberty, indeed of future orientation, rather than social obligation and fidelity to the past (Bronner 2002a; Bronner 2002b). The examples remind us of the negotiation inherent when traditions are enacted, including the questions of who dictates how or which traditions will be followed, or sometimes how traditions are discouraged and even proscribed by external forces (i.e. authority), and how traditions are adapted and interpreted by their participants (i.e. control). Concerns for maintaining vernacular building in the twenty-first century expressed by Paul Oliver and others as keys to the welfare of migratory and disaster struck populations thus imply not only an accounting of what happens to the forms left behind or swept away, but of what happens to the ideas of traditions themselves. As Oliver has pointed out: 'It's not that the motivations and the skills to build are wanting.'[1]

But one has to wonder whether new conditions allow for the dynamics of tradition to operate adequately. 'Much can be learned from traditional builders, who willingly pass on their know-how and skills as they have in the past', Oliver admonishes, and it is well worth remembering that the physical as well as political environment that supports the process of tradition may also be critical to addressing the 'massive problems of housing in the next century' (Oliver 1999). In so far as many crises of housing result from political campaigns against traditional ecologies, social structures and economies, as well as natural disasters and wars, I would add that these problems are often cultural matters that entail the way that people control, or are prevented from engaging, tradition into their lives. Scholars may assist in responses to crises, then, by questioning in vernacular architecture the tradition that encompasses, vitalises and lends meaning to people. Maybe then we can not only see vernacular architecture for what it is, but also grasp what it does.

Note

1 In northern Niger among the Hausa, the branch and straw shelters built and used by women as 'sleeping huts' and for storage of their possessions were crucial to the social traditions of rural settlements. But the structures were becoming hard to build because the droughts made branches scarce. European advisers who did not have a comparable tradition considered them extraneous, since a family structure already existed, but after consulting native officials realized that despite their concerns for the practicality of the separate buildings strewn around the settlement, the culture considered them essential because they represented social traditions that maintained the culture. Engineers proposed thin-walled domed structures as replacements and the use of woodless construction such as earthen bricks. The replacement of materials was met with approval because the wood was less crucial to the ritual uses of the structure than the existence of the structure itself. An impressive number of traditional builders, both women and men, were trained in how to adapt their traditional structures to earth brick material. Paul Oliver (1999) cites this instance as a prime example of understanding cultural concerns in disaster relief in his essay on vernacular architecture in the twenty-first century. See also Norton (1997). It should be noted that not all groups were able to maintain their traditions in the face of desertification. Reliant on reeds for creating their temporary huts, the Nama in Namibia, for example, have abandoned their nomadic lifestyle. Corrugated iron and western furniture have encouraged a rectangular hut with monopitched roof (see Peters 1997).

Chapter 2

Endorsing indigenous knowledge

The role of masons and apprenticeship in sustaining vernacular architecture – the case of Djenne

Trevor H.J. Marchand

Introduction

This chapter explores the relationship between local knowledge and vernacular architecture.[1] More specifically, it illustrates the gestation of technical learning and socialization that occurs throughout a mason's apprenticeship, and ultimately forges his professional identity, practices and sense of responsibility. Masons, like other craftsmen, gain technical proficiency in conjunction with a social comportment that is publicly recognized and validated as being appropriate to their trade status. An accomplished craftsman must learn to negotiate the boundaries of 'tradition' in his craft and, in some contexts, he may also acquire a repertoire of specialized cultural knowledge and occult secrets related to his trade. Mastery of these various forms of knowledge becomes manifest in his competence to produce and reproduce a built environment that is 'meaningful'. Such a built environment comprises architecture and urban spaces that respond dynamically to, and condition, the changing needs and aspirations of its inhabitants, while remaining rooted in a dialogue with history and place. I will therefore reason that the masons' knowledge (and equally the indigenous knowledge of other building trades) must be central to discussions, studies and projects concerned with the sustainability of vernacular architectures in the twenty-first century and beyond.

My particular case study focuses on the Sahelian mud town of Djenne and its professional association of masons, the *barey ton*. The core of my argument, however, is more generally applicable to the conservation of vernacular architectures worldwide, and aims to instigate a renewed appreciation and engagement with apprenticeship-style learning in western societies. In brief, my past work with masons in Djenne (and Yemen) demonstrates that their pedagogy was not language based, nor was it prescribed in concrete terms by an officiating body. Rather, skilled performance and embodied practices were taught and learned in a participatory forum located 'on-site', and the standards of the apprentice-style training were negotiated and maintained within a hierarchical context of professional interactions between builders. Djenne's distinct architectural tradition, like that of other towns and cities where building crafts flourish, is perpetuated not through a rigid conservation of its surviving buildings and monuments, but instead via the dynamic and responsive transmission of skill-based knowledge from one generation of builders to the next. Effectively, it is this 'tradition as process', as instituted in the apprenticeship-style education, which must be conserved if architectural styles and techniques that respond to, and create, the specificities of 'place' are to survive and potentially proliferate.

The masons and heritage issues in Djenne, Mali: an introduction
'The pronouncement of "*b'ism Alla*" (In the Name of God) opens the sacred space of a benediction, like the start of each *sura* in the Holy Quran', explained a *marabout*. He, like so many other religious scholars of Djenne, was in the business of manufacturing guarantees and protection for clients. Leather pouches concealed Islamic scriptures; natural objects and bodily adornments were sanctified; potent medicines were concocted from leafs, berries and bark; and incantations were uttered. 'The Word of Allah ultimately provides protection', contended the *marabout*, 'and I am merely an agent who possesses secrets for transferring this blessing to people, their things and their plans for the future'. For the construction of new houses, *marabouts* likewise prepared amulets that were buried beneath building foundations or stashed away in hollowed animal horns that protruded above doorways and windows to protect the abodes from wicked *djinn* or malicious visitors. The old gentleman acknowledged that other trades in Djenne also possessed powerful secrets, including the blacksmiths, the fishermen and the masons. A mason's secrets differed, however, from those of the *marabout*. Many masons were illiterate or had received only basic Quranic education, and thus most relied on the power of spoken benedictions that melded Islam with traditional black-African knowledge of the world. These secrets were carefully transmitted from master to apprentice, and from father to son, and were recited privately, and sometimes publicly at barely audible levels to ensure that they were not pilfered by prying ears. The *marabout* considered the masons'

secrets to be potent, citing the old masters' ability to command crumbling walls to stand, or to cause misfortune to fellow tradesmen who betrayed them.

All of the masons I worked with in Djenne proudly heralded their Bozo identity, an ethnic group of the Inland Niger Delta more regularly identified with fishing. Members from other groups of the town's diverse population had also taken up building trades in recent decades and represented a growing minority. In general, masons, like other craftsmen, occupied a middle-class status in Djenne society and, contrary to Prussin's claims (Prussin 1970: 18), they were esteemed for their technical skills, professional occult powers and their coordinated sense of organization. The masons' community was essentially a gerontocracy, embodied in the guild like structure of the *barey ton* and presided over by an elected chief. *Barey* translates to 'builder' or 'mason', and *ton*, a Mande term, means 'association'. Plausibly, the mason's profession is an ancient establishment in this West African region. Archaeological evidence from the nearby site of Djenne-Djeno demonstrates that this strategic location on the Bani River was occupied from as early as the third century BC. The settlement supported a thriving urban civilization with a well-developed division of labour into the twelfth century after which the population shifted to the present town. By AD 1400 Djenne-Djeno was abandoned (McIntosh and Keech McIntosh 1995). The remains of rectilinear mud-brick buildings on the site date to between the twelfth and thirteenth centuries AD (Keech McIntosh 1995: 392), and terracotta figures were discovered buried in wall foundations and on a platform next to the entrance of a house (De Grunne 1995: 77), likely providing protection for inhabitants in similar manner to the buried and concealed amulets in contemporary constructions.

Detailed accounts about the masons did not transpire until the arrival of the French in 1893 (see especially Monteil 1932; Ligers 1964; and the later study of Djenne's architecture by Maas and Mommersteeg 1992). Patronized by the town's wealthy merchant class, religious leaders and nobles, the *barey ton* firmly controlled the reproduction of Djenne's architectural heritage during the first three quarters of the twentieth century. During the period of colonial rule lasting until 1960, they profited additionally from French commissions for public and administrative building throughout the region and beyond. Masons reported that their corporate organization at that time was robust: a mason's wages were fixed and clients supplemented their pay with rewards of *kola* nuts and supplied the builders with daytime meals. During the two decades of severe drought that afflicted Mali in the 1970s (1972–4) and 1980s (1984–7), however, a significant proportion of the town's young males, including many masons, left to find work in larger West African cities. Many never returned. This period was characterized by severe hardship, witnessing a serious degradation of the town's built environment and a demise of the *barey ton*'s traditional authority. Economic instability rendered wages negotiable in favour of clients, and payments allegedly consisted only partially of cash and included grains, food and articles of clothing. The builders, many of whose fishing and agriculture pursuits had also been detrimentally affected by drought, competed for the scant work available and, through necessity, accepted what was on offer. The patron's practice of supplying meals all but ceased, masons' incomes plummeted, and the structure and authority of the *barey ton* largely collapsed.

More abundant rainfall in the late 1980s improved economic prospects throughout the Sahel. New commissions and building activity gradually increased and the *barey ton* slowly reconsolidated. The professional association nevertheless faced an uncertain recovery of its former power and influence as growing demand for new building materials and technologies threatened the masons' traditional practices, hierarchical organization and relations with local suppliers and manufacturers. A growing availability of corrugated iron sheeting, concrete breezeblocks, cement, steel reinforcing bars and the use of formwork in new constructions spread throughout Mali. Some masons, especially those returning from abroad with newly acquired skills, were capable of working with the 'modern' materials that some clients now identified with status and prestige, and that they could now afford. The neglect and decay of many mud-brick buildings, as well as the progressive abandonment of the distinctive 'Moroccan' and 'Tukolor' styles that graced the town's grand residences, provoked local and international concern that Djenne's distinct architecture was under siege. Likewise, pillaging of the adjacent archaeological site to supply foreign collections was sabotaging Mali's ties to its ancient material culture. As a result, the town and nearby Djenne-Djeno were added to UNESCO's roster of World Heritage Sites in 1988. A Cultural Mission was established to lead efforts in safeguarding the archaeological remains and character of the city, and to educate the local population regarding heritage issues.

Evolving concerns with heritage have sparked a myriad of positions within the circulating discourses on building practices, architectural styles, authenticity and sustainable conservation. Government officials, funding bodies, architects, conservationists, historians, archaeologists, anthropologists and, importantly, members of the resident population have been engaged in co-operative efforts and competitive struggles over the town's building and planning practices. Idealized images of Djenne's monumental mud edifices, largely informed by the Mosque and noble two storey houses, have remained significant in constructing both regional and national identity. Since the earliest years of French administration, the so called '*style-Soudanaise*' has been promoted in photography, on postcards and postage stamps, and in a wide spectrum of literature (from academic research to travel writing) to emphasize the existence of a highly developed urban culture in this pocket of West Africa (Gardi, Maas and Mommersteeg 1995). The French erected splendid pavilions at their turn of the century exhibitions, eclectically fusing features from the buildings of Djenne and Timbuktu (Prussin 1986: 18). With further adaptations borrowed from North Africa such as verandas and arches, the French re-made the *style-Soudanaise* into the hallmark of their regional colonial style. For contemporary Malians including government leaders, the tourist industry and craftspeople dependent on foreign patrons, Djenne's architectural heritage, along with Dogon villages, the allure of Timbuktu and Mali's renowned music industry, represents a vital resource for drawing much needed international investment and foreign capital.

2.1
Traditional 'Tukolor'-style home in Djenne.

A co-operative venture between Dutch and Malian experts produced a credible plan for the rehabilitation and conservation of Djenne's architecture (Bedaux *et al*. 2000). The Dutch-funded project spanned a period from 1996 to 2002, and successfully restored one hundred monumental houses deemed to be of historic and architectural merit. A subsequent phase has been planned to uphold conservation efforts with annual plastering and maintenance; to propagate grass-roots interests and local expertise in heritage issues; and to publicize project results (Bedaux *et al*. 1995: 24). From the start, the scheme recognized the masons and *barey ton* as pivotal agents necessary for its implementation and long term success. Respect for the *barey ton*'s autonomy and regular consultation with masons about project aims and scheduling consequently strengthened the internal ties and coordinated efforts of the professional association. The project provided masons with valuable opportunity to acquire practical experience in restoring (and reconstructing) historic Moroccan- and Tukolor-style buildings, and to foster trade skills that will hopefully be inculcated in successive generations of builders. In the following discussion, I suggest that the practical immersion and honing of expertise also fostered the potential for innovative and well informed interventions that ultimately give rise to a 'dynamic' building tradition in Djenne.

The following section provides an overview of the mason's apprenticeship based on ethnographic examples from the sites where I worked and studied as a labourer and apprentice. The progression of the apprentice's training will be presented with excerpts from various stages in the construction of a Moroccan-style house, from its foundations to the sculpting of its decorative roofline crenellations. Apprentices, as differentiated from labourers, normally worked closely with a single mason (though sometimes spending periods of varying duration under the direction of other masons), preparing his tools and assisting directly in the building processes. In cases where a mason was training someone other than his own sons, this special relationship of participatory learning was typically founded on a contractual agreement initiated by one, or both of the boy's parents. It was also the case that some masons had more than one apprentice at a time under their tutelage. The mason could be expected to provide the necessary vocational training over a period of several years in exchange for the apprentice's obedience, respect and provision of free labour. In this active context of close collaboration and often intense social relations, not only were performances perfected, but personalities and worldviews were also formed. I will ultimately assert that the maintenance of an apprenticeship system that endows young men with not only technical skills but a sense of social identity and professional responsibility is the most effective way to guarantee a sustainable reproduction of a distinct architecture and an urban landscape imbued with changing and dynamic meaning for the Djennenké population that live in it.

Building and apprenticeship: learning skill and performing identity

Beginnings and blessings
The building period in Djenne was short, with peak activity occurring during the dry and dusty *harmattan* season lasting from late December through March. Average daytime temperatures at this time of year dropped to tolerable levels of between 28° and 38° Celsius, and receding water levels along the river banks and in the marshes at the edge of town exposed rich deposits of silt and ideal sites for brick making. During the remainder of the year, if not employed by patrons to execute necessary repairs and maintenance, masons worked in their fields, fished the rivers, or engaged in petty trading from their homes or at Djenne's vibrant Monday market. When building, they worked a six-day week with a rest on Mondays, and Friday afternoons were reserved for congregational prayers at the mosque. The typical workday commenced early in the morning and finished at three in the afternoon, with sporadic breaks to share a glass of tea or ladle of millet porridge. Start and finish times were somewhat weather dependent since masons abhorred handling the cold, moist mud on chilly mornings, and conditions were occasionally made unbearable by the blistering heat of late afternoon and by the incessant, virulent *harmattan* winds.

Trevor H.J. Marchand

Masons took regular precautions to ensure the safety of their entire work force and well-being of the future inhabitants who would occupy their buildings. They made benedictions that warded off the harmful spells cast by other masons and guaranteed the structural integrity of their works. New projects, like any journey in life, were commenced on auspicious days. Such dates were calculated according to the lunar calendar and constellations (among other things), and conferred upon by local *marabouts* and contingencies of respected elders from the mason's neighbourhood. Both masons and *marabouts* blessed stones, bones and other artefacts, and they prepared amulets containing written verses or entire *suras* from the Holy Quran. The masons buried these along property boundaries and below foundation lines to protect the terrain and future edifice from malice. Prior to commencing daily work, masons made benedictions that guarded the team against accidents. Their verses typically comprised a blend of *bai quaré* and *bai bibi*, epistemological classifications referring respectively to Islamic-based and traditional black-African forms of secret knowledge. Such verses can obstruct wicked intentions, alter destinies and change the state of the world. Proper memorization and correct performance of the secret verses was part and parcel of the young builder's training. Non-kin apprentices were normally granted only limited access to their master's repertoire of secrets, and they had to supplement these with incantations passed on from their own fathers and powerful verses they memorized from the Quran. Invocations of the Prophet Abraham (builder of the Kaba) and recitations of the 112th *sura* from the Holy Book were considered to be especially potent. Builders, like most residents of Djenne, also donned protective devices on their bodies. Amulets, or *gris gris*, manufactured by *marabouts* and medicine men, were concealed in leather pouches worn around the neck, upper arm or abdomen. Blessed rings of copper and silver came in a variety of plain, twisted or inscribed bands, and some were set with large semi-precious stones such as agate and carnelian. In addition, respect for the masons' totem, the yellow headed *unbaka* lizard that scurried over wall surfaces and basked in the hot sunshine, endowed the builders with balance and agility when working high above the ground.

At the start of a new construction, two site masons sprinkled blessed grains, nuts, seeds and cowry shells into shallow foundation trenches. The mixture of grains, including fonio, millet, sorghum, maize, rice and peanuts, represented the staples of the local diet and its placement in the foundations secured plentiful nourishment for the inhabitants. Cotton seeds procured clothing, and the cowry shells, once used as money, attracted wealth and prosperity. The foundations, like the walls to be built above, were constructed of sun-dried mud bricks. The bricks were manufactured along the banks of the Bani River and by the water at the edge of town. The local laterite contained ideal quantities of clay and, when mixed with chopped straw, provided a resilient, thermally adaptive building material. The brick makers used standard wooden moulds with interior dimensions measuring 36 cm long × 18 cm wide × roughly 8 cm high to produce rectangular bricks called *toubabu ferey*, or 'white-man's brick'. The employment of these uniform modular bricks replaced the hand-shaped, cylindrical *djenné ferey* during the first half of the twentieth century, thereby rendering newer constructions with crisper lines and fewer molten-like contours. Simple handles nailed to either end made it easy to slip the wooden mould up and off the freshly formed brick, and accelerated the pace of production. In 2001, bricks were sold at a unit price of ten CFA (approximately 0.015 Euros), and the drivers that transported them to the building sites by horse-drawn cart levied an additional five CFA per brick. Heaps of iron-red laterite were likewise delivered to site. The soil was pulverized with hoes, water added and bare legs trampled the mixture to produce mortar. Bricks and woven basket loads of oozing mortar were passed in assembly line fashion from the labourers to the apprentices who stockpiled supplies and handed them on cue to the masons.

Tasks and tests of an apprentice
Though some apprentices only began training in adolescence, most were initiated as children who followed fathers, uncles and older siblings to the work site. For Bozos, socialization within their ethnic community reinforced historical ties to both fishing and building. The boys, sometimes as young as six or seven years old, were given the small tasks of cleaning tools, passing bricks and fetching glasses of sugar saturated tea for the masons. Very often they were simply left to play, taking intermittent naps in the shade. As they grew older they were delegated increasing responsibility and they began to work more closely at the mason's side learning the skills of the trade through participation. The apprentice's privileged position introduced him to the range of tools that included the crowbar, plumb line, string level, French level, tape measure and two types of trowels; the large and sturdy *courou bibi* (black skin) trowel produced by local blacksmiths, and the delicate, sharply pointed French trowel used for more detailed work. By assisting, observing, mimicking and practising, the trainee acquired basic techniques and understanding. Through listening to negotiations and disputes with clients, suppliers, team members and other masons, he became further immersed in the concerns, worldview and social performance of his mentor. These factors, in combination with the secret knowledge he gained, forged the young man's identity as a member of Djenne's building community, and ultimately made possible his appropriation of a publicly recognized social position and fulfilment of related professional responsibilities. It should also be noted that the intense engagement between master and pupil equally educated the mason in becoming a mentor. Mentoring processes coerced a critical reflection upon the mason's own skilled knowledge, and upon the control that he exercised over the reproduction of his trade.

The two masons marked out where the new section of wall should begin and end by slicing grooves into the top face of the foundations with their *courou bibi* trowels. For each successive course, the two bricks at the opposite ends of the wall were first positioned and the plumb bob was lowered along the vertical faces to verify their alignment with the courses below. Next, a blue nylon cord with a long nail tied to either end was tautly drawn between the outer faces of the two end bricks and fixed in place by wrapping it around and inserting the nails into the mortar beneath the bricks. The cord served to guide the remainder of the course that could now be laid in rapid succession. Dollops of mortar delivered in baskets were spread thickly over the top of the wall in patches large enough to lay four to six bricks at a time. It was normally the task of an apprentice to toss the bricks one by one to the mason who, in a continuous sweeping motion, caught and set them. Once positioned, the mason guided his palms and fingertips over the surface of the bricks, checking for irregularities and wedding the bricks to the mortar with a gentle twisting motion. Each brick was tapped into exact place with the butt end of his trowel to meet the horizontal stretch of nylon cord. Verticality was periodically checked with the plumb line. Mortar spilling out between courses was fingered along crevices to fill gaps and the excess was smoothed with the trowel along the outer planes of the wall.

The apprentice's typically mundane duties of catching, stacking and passing importantly integrated him with the mechanics of the labour force. Participating with the labourers better familiarized him with the physical properties, preparation and efficient handling of materials. It also expanded his understanding of the division of labour on a construction site, and coerced him to think and act as a team member. In these circumstances he learned to issue and receive directives, to temper his emotional displays of anger and frustration, and to earn respect at all levels of the hierarchy. His aspirations to partake in actual building activities were typically communicated by attempts to lay bricks on his own, carefully miming the procedures, practices and gestures of his mentor. Progressively, the mason would permit his apprentice to build alongside him, correcting the young man's errors and instructing almost entirely by means of demonstration. Choice verbal cues pointed the novice's attention to salient aspects of the mason's actions, and reprimands and a serious tone of voice drew boundaries. The main test arrived when the mason directed his apprentice to erect a flawless brick wall by himself. Repeated displays of his capacity to do so were rewarded with a new plumb bob. The presentation of this first key instrument augmented his status as an apprentice, signalling his promising progress. He would not be declared a mason, however, until mastering other essential tasks such as setting level palm-wood lintels over windows and doorways, rendering wall surfaces and ceilings with evenly applied coats of mud plaster, and constructing flat roofs with their characteristic angular geometries of quartered palm-trunk beams.

Sculpting decoration: tradition and innovation
One afternoon, the senior apprentice was delegated the responsibility to assist his master in positioning ceiling beams. The mason selected several straight and sturdy lengths of palm timbers from the pile and instructed a labourer to chop one to a specified length. Palm timbers (*sébé sen*) were the most costly component in Djenne's mud-building tradition and, because they had to be imported from outside the region (and often from Burkina Faso), they were not always readily available. The mason and apprentice straddled adjacent walls, positioning themselves diagonally across from one another over the open space of the room. Two

Endorsing indigenous knowledge

labourers hauled the heavy timber, wary of the dense, needle-like fibres of the palm wood, and lifted one end to the mason who stretched downward to take the load, and hoisted the second to his apprentice. With both ends resting securely on top of the walls, the mason directed his assistant in their effort to position the beam at a 45° angle across the room. The procedure involved an intensely co-operative concentration to rotate and shift the quartered palm trunk into a precise and stable position. The mason requested successively shorter lengths of palm wood that were similarly cut, carried and hoisted up to the two men who placed them until the triangular space above the corner was completely covered with timbers. The three remaining corners of the room were spanned in similar fashion and generous coats of mud were spread over the top of the beams to produce a more homogenous assembly. Two further layers of palm beams were placed running parallel to the walls and between the four triangular corner assemblies, producing a typical *fata taki* ceiling. This highly developed and aesthetically pleasing triangulation method enabled the builders to create rooms of greater linear dimension than would be possible if they restricted widths to the available lengths of palm wood (i.e. three metres). The *fata taki* also created a shallow dome-like effect of geometric angles in the interior, and its slightly humped exterior profile, covered in thick layers of mud, efficiently drained rainwater to the roof's edge where it was siphoned off through clay spouts.

With the roof completed, the next stage involved constructing the exterior embellishments that would lend the house its distinctly 'Moroccan-style' appearance. Four slender mud-brick pilasters that stretched the full height of the two-storey street façade had been integrated into the wall construction, and these would serve to organize the arrangement of sculptural elements above the parapet. Single pilasters (*sarafar har*) terminated either end of the long, planar street façade and two others (*sarafar woy*) framed the somewhat asymmetrically positioned entrance to the house. The clear span between the two pilasters

2.2
An apprentice assisting a mason to set palmwood lintels above a doorway.

framing the entrance measured 3 metres and the width of each pilaster was approximately 0.4 metres (the length of a standard *toubabu* brick) totalling 3.8 linear metres. Based on this measurement and on his existing knowledge of the dimensions of the mitres (*sarafar idye*), the mason determined that there should be a total of seven spiky projections crowning the crest of roofline above the door. He insisted during his patient explanation that the spacing between each element must be carefully calculated, warning that a small mistake could spoil the whole effect of the *potigé* (the decorative assembly of the façade). The mason squatted down on the road in front of the house where he was instructing his senior apprentice and me, and proceeded to etch two parallel lines in the dirt representing the two pilasters. He capped these by a perpendicular line indicating the top of the parapet, and then made seven evenly spaced impressions above this line with his thumb evincing the overall majestic effect of the mitres.

Six tiny rectangular apertures (*funey*), three bricks high, pierced the parapet wall above the level of the carefully arranged bundles of projecting palm wood sticks (*toron*), and each aperture was framed by alternating tiny pilasters and half columns. The pilasters (*gaga*), numbering four in total, terminated with bold rectilinear capitals that supported a span of long palm timbers running the length of the *potigé*. The three half columns centred between the pilasters were composed of semi-circular drums carved from the brittle mud bricks with the mason's *courou bibi* trowel. The half-column heights equalled that of the six apertures, and each was capped with two superimposed spherical projections made from bricks and sculpted with plaster, and referred to as the *musi bumo*, or 'cat's head'. In the narrow space between the top of the *musi bumo* and the plastered span of palm trunks above, the masons introduced an additional tiny round aperture that they called the *musi motondi*, or 'cat's eye'. One mason described how builders of the past had no knowledge of making *musi motondi*, but this feature was invented nearly two decades ago by a now deceased and celebrated master mason. Since then, he claimed, most of Djenne's accomplished builders have learned to produce this decorative feature that flaunts their skill and expertise. The second mason agreed with this account of the *musi motondi*, and added that this same aforementioned master was also the author of other decorative and architectural features that have now been incorporated into the local architectural language.

As in other similar instances when masons attributed innovative forms and ideas to particular masters of the *barey ton*, it was not so much the historical veracity that interested me as an anthropologist, but rather the degree of esteem they accorded to creativity and signature in the building trade. Though Djenne's architecture has, for the most part, remained remarkably consistent over the past century or longer (and this is arguably due, at least in part, to the hegemony of colonial and post-colonial discourses on heritage and authenticity), the building tradition has by no means been static. An apprentice's long, first hand acquaintance with the restricted palette of tools and materials served to inculcate a practical knowledge in him about structural possibilities. His introduction to negotiations that took place between masons and clients concerning design and budgets, as well as his growing familiarity with the practices of other masons in the community, shaped his aesthetic sensibilities and his judgement of quality, proportions and composition. This combination of structural and aesthetic principles, together with an accretion of practical experience, has enabled masons to creatively expand the existing repertoire of built forms and decorations in a way that is deemed both traditional *and* innovative by their fellow colleagues and public.

Becoming a mason

A raised wooden plank resting on stacks of bricks was set up on the roof adjacent the section of paràpet wall framed by the *sarafar woy*. The two site masons and the senior apprentice stood balanced in a line along the plank ready to build up the seven mitres (*sarafar idye*) in perfect tandem. First, the bases of the mitres (each composed of three whole bricks and measuring 40 cm wide by 58 cm deep) were carefully arranged along the top of the wall leaving equal spacing of approximately 16 cm between each. In total, the top of the mitres reached eight bricks high,

2.3
A labourer passing a woven basket of mud mortar to a mason.

Trevor H.J. Marchand

2.4
A mason sculpting the tip of a *sarafar idye* – the mitre-like projections that crown the front facades of traditional Djenne houses.

and the builders tapered the width and depth of each successive course, achieving a slight convexity on all four sides that ultimately terminated in a fine point. The eighth course comprised only a small, pyramidal piece of brick. A tautly drawn horizontal cord was raised from course level to course level to ensure that the masons and apprentice were producing the same diminished front and back faces of their respective mitres. The tapering of the inside faces was judged for symmetry by the masons' eyes, and corrections were made regularly to ensure that all the *sarafar idye* matched perfectly. The *sarafar idye* were thickly covered with mud and their identical finished forms were sculpted and smoothed with French trowels. The apprentice worked on the central mitre and the masons each built up the three on either side. The apprentice was performing well and his master paused to exclaim loudly to the nearby labourers who were passing bricks and mortar, 'See! My assistant is a mason today!' This sort of praise encouraged the young man and instilled a growing sense of confidence. It also publicly confirmed his rising status and authority in relation to the other team members.

When a mason judged his apprentice to be fully qualified, he announced his decision to make the boy a mason at the following congregation of the *barey ton*. Apprentices were not present since, customarily, only fully-fledged masons attended the meetings. Afterwards the mason conveyed the good news to the young man's family, and he and the boy's guardian made benedictions for his prosperous future as a craftsman. The mason presented his pupil with a *courou bibi* trowel and a flared-end crowbar (*sasiré*), effectively completing his kit of basic trade tools. The gifts, issued by the master, officially validated the transformation of the boy's many years of toil and determination into a new professional status. Young builders recently declared masons continued to work with their masters or with family members in the trade, and they received a mason's daily wage (2,500 CFA in 2001, or roughly equal to 3.75 Euros). Without an established clental base of their own, they remained dependent on senior masons in the community for work. During the early years in his new role, a young mason continued to be mentored, but he now assisted more directly in the actual building processes than he did as an apprentice. This arrangement enabled him to hone skills and learn new tasks with the benefit of continued support and limited liability. By way of example, he also learned to effectively negotiate with clients and translate their needs into spatial planning and buildings, devise contracts, estimate quantities, calculate costs, work with budgets, schedule the construction phases and manage a team of (often unskilled) labourers. In time, the junior mason would be allotted small projects to complete on his own. Though senior masons would monitor these, his new autonomy would be an important stride toward eventually having his own clients, and one day his own apprentices.

Conclusions: endorsing indigenous knowledge

In my anthropological study of the built environment and material culture of Djenne, I have emphasized the role of masons and their apprenticeship style of education. As illustrated in this summarized version, the duration of a builder's apprenticeship was lengthy, lasting four and more years depending on the age of the initiate. The young mason's formal learning continued under the tutelage of senior trade members until they had established their own client base and came to be regarded as 'masters' by their colleagues and public. The master–apprentice relationship and continued mentoring process were fundamental not only to learning technical skills but also to the social development of both men involved. The mason learned to be a mentor within the participatory context of the building site and in his social interactions with the apprentice outside the official work place. He gradually took responsibility to introduce his trainee to tools, materials, secret benedictions, decision-making processes and procedures, and, importantly, to the negotiated nature of relations with clients, suppliers, team members and fellow craftsmen. The mason's own dialogical immersion and active engagement in these various social and professional relations informed his worldview and produced his sense of identity. An intense involvement with his apprentice provoked recognition and critical reflection of his role, power and responsibilities in reproducing the profession and Djenne's built surroundings.

The mason's knowledge, like any other, was not privately concealed in his mind, but was embodied in his skilled activities, social performance and the things that he made. All three were publicly manifested and thus accessible to the apprentice's scrutiny and appropriation. The mason's manifest knowledge and performed identity, in conjunction with the tools, materials and physical setting of the building site, structured the apprentice's learning environment. The apprentice's committed investment of unpaid physical labour, obedience, heightened attentiveness, strategic social manoeuvring and carefully composed mimicry resulted in a technically and socially competent performance. His practised skills were hopefully recognized by his master and rewarded with a gift of basic trade tools and, ultimately, the transformation of his status to mason. It must be reinforced, however, that the reproduction of qualified masons and trade skills was by no means static or one of generic conformity. I would argue rather that the lengthy gestation of practical and social knowledge during the apprenticeship and the subsequent mentoring period fostered an intimate understanding of material, structural and aesthetic possibilities that, in turn, enabled a degree of creative innovation. These innovations, typically ascribed to honoured master masons, expanded the discursive boundaries of 'tradition' and re-inscribed Djenne's architecture with contemporary meaning and value for its modern inhabitants.

This focus on the masons' performance based knowledge collapses the classic dichotomies between mind and body, between theory and practice, and between tradition and modernity. From an anthropological vantage point, Djenne's building tradition is understood primarily as a set of meaning-making practices rather than a landscape of physical objects to be conserved for their unique forms or some inherent historic value. For the town residents, living with the historic styles has been salient to their everyday, phenomenological makings of 'place' (Casey 1996). Djenne's architecture has been as much a vehicle for social and cultural expression as it has shaped people's social and cultural understandings of themselves. As discussed earlier, the *style-Soudanaise* architecture continues to be pivotal in the construction of local, Malian and even pan-Sahelian identity. International interest in this monumental, urbanized form of cultural expression, beginning with colonial administrators and more recently attracting UNESCO officials, conservationists, foreign scholars and travellers, has introduced a host of new and shifting positions that compete for control over the meaning of Djenne's architectural heritage. However, as I have argued thus far, it is the knowledge produced and reproduced by the masons and their professional association, the *barey ton*, that is central to the perpetuation of this discourse. Any proposed intervention of conservation or development works to the urban fabric must therefore acknowledge the historically negotiated expert status of these craftsmen, and the decision-making power that their position entails. Djenne's masons not only provide dwelling: they erect buildings that denote (and construct) the socio-economic prestige of their patrons; they guarantee prosperity and well being for the inhabitants with their secret benedictions, blessed objects and amulets; and crucially, they produce the successive generations of qualified agents endowed with a complex combination of technical, social and occult expertise.

If Djenne's building tradition, defined as the masons' practices and their system of education, is to endure in the twenty-first century and beyond, then a number of key issues must be considered and taken on board by players with a stake in the town's heritage. By no means an exhaustive list, some of these issues are as follows:

- First, the apprenticeship system must be left to operate unhindered. Attempts to control the reproduction of the historic styles by introducing formalized trade schools will defuse the dynamic quality of Djenne's mud-building tradition. The formulaic learning environment of trade schools cannot replace the rigour of the master–apprentice relationship that instils a moral constitution of character in both teacher and pupil in relation to craft production. The long period of direct engagement in the practical context progressively imbues the builder with a broad knowledge of existing social and technical parameters. A creative capacity balanced with reason enables the master mason to strategically manipulate these parameters and to negotiate and transform existing traditions while keeping them meaningful.

- 'Museum' approaches to the conservation of the city should be avoided. As a living urban site, Djenne's mud-brick houses have always been in a state of flux and process rather than finished products or completed architectural statements. Second stories were added, the form and function of rooms altered, decorations sculpted and re-made, all in response to generational changes to family size, household economics, personal needs and tastes. The building materials allowed for unparalleled flexibility and could be reused, recycled or left to decompose back to soil, and should therefore be celebrated and promoted in light of these economic and ecological attributes.

- Development initiatives should aim to promote and establish sustainable supplies of locally available building materials. Plantation and forestation projects are necessary to maintain a supply of mature, healthy palms for *toron* and ceiling construction, and various hardwoods for the manufacture of doors and 'Moroccan-style' lattice windows. Water management schemes must strive to keep the river and system of rivulets and marshes around the city clean in order to supply a hygienic source of mud for bricks and plaster. The introduction of piped water two decades earlier failed to implement a sewage system for the evacuation and filtration of water-borne waste. Open gutters and pools of stagnant water have since posed a serious health menace and are detrimental to the structural stability of the earth structures. Garbage and toxic waste embedded in the sedimentations of mud are equally hazardous to brick makers and builders. Perhaps the greatest threat posed by a large scale water management venture is the Talo Dam project planned for the Bani River, and situated upstream from Djenne. Not only might the dam dramatically alter fishing and agricultural practices, the local economy, the climate and the natural environment of the Inland Niger Delta, but it might also substantially deplete the annual alluvial deposits of silt from which the region's houses and mosques are built (Gallier 2002).

- Local appreciation for Djenne's architecture and building tradition must be bolstered, and its social, economic and ecological value recognized. In order to revive and sustain popular associations of the *style-Soudanaise* with status and prestige, the post-colonial dichotomy between tradition and modernity must be challenged. The popular association of tradition with stasis and 'backwardness', and the conceptual affiliation of modernity with concrete, corrugated iron and all things Western must be debunked. Changing attitudes can only be achieved through educational processes that promote scholarly investigation, publications, public displays and open discussion. As long as Mali's elites continue to conceive of mud architecture as the property of their poverty-stricken rural brethren, Djenne's building tradition, as well as the diversity of other building traditions throughout the country, will be progressively denigrated and may one day cease to exist.

In conclusion, conservationists, architects, planners, development experts and academics should strive to empower Djenne's masons (and other craftspeople) not by paternalistically protecting them from the fragility of their economy or from the changing tastes and demands of their public, but by granting them greater autonomy and centrality in their projects and studies. The masons must be brought in from the peripheries in discussions concerning their town's architectural heritage and its future as an urban environment. Academics and professionals must actively engage them in an interdisciplinary dialogue that addresses the issue of 'knowledge' in relation to all building traditions. An expanded regard for 'knowledge' should go beyond conceptual knowledge expressed in the propositional statements of spoken language, and include the skilled technical and social performance of the masons. This embodied knowledge is learned in the situated context of the building site (Lave and Wenger 1991), and therefore the role and maintenance of the apprenticeship-style training must figure prominently in the conservation of Djenne's building tradition. Recognition of the importance of apprenticeship for sustaining a dynamic, meaningful architecture will hopefully bring about a re-evaluation of this 'traditional' mode of education and prompt serious consideration of how it might be revitalized in our own modern societies.

Note

1   I would like to thank the British Academy and the School of Oriental and African Studies for their fieldwork support in 2001 and 2002, and Professor Rogier Bedaux for his invaluable assistance and encouragement. I would also like to thank the masons of Djenne who made my research there so thoroughly enjoyable and rewarding.

Chapter 3

Forms and meanings of mobility
The dwellings and settlements of sedentarized Irish Travellers

Anna Hoare

Introduction

The concept of nomadism has been much criticized in recent decades. As a corollary of ecological adaptation to physical environmental constraints, spatial mobility has been described as merely 'a central pastoral technique' compatible 'with technologically advanced and profit-oriented economic activity', which should not be associated with nomadism (Humphrey and Sneath 1999: 1). The line generally drawn in the literature between pastoral and 'peripatetic' or 'service' nomads (the 'other' nomads (Rao 1987)) implies a typological distinction in the minds of many scholars between the rationales and functions of mobility among nomadic pastoralists and non-pastoral nomads. Discussions of non-pastoral and pastoral nomadic societies are separated in the distinct literatures on the basis of their economic systems, leading the reader to infer that the nomadism of pastoralists must be understood as 'a trait of cultural ecology' (Spooner 1972: 130), as distinct from the movements of 'non-ecological' nomads (Weissleder 1978: xviii). In general, spatial mobility has received little attention outside its role in economic systems, where the ecology of pastoralism, in terms mainly of physical environmental factors, dominates. More usefully, the analytic separation of spatial mobility and pastoralism advocated by some scholars (e.g. Dyson-Hudson 1972; Tapper 1979) recognizes that the implications of each for social, political and economic systems may be distinct, and motivated by different factors. This allows for the possibility of preserving the concept of nomadism in order to describe and analyse forms of spatial mobility within a range of cultural and political, as well as economic contexts, and to consider the internal dynamics, as well as the partibility of different mobile systems. Gulliver's (1975: 369) emphasis on nomadic movement 'as a response to the internal context of the socio-cultural system' of a people, calling for greater specification and analysis of the 'sociological implications of recurrent movement' suggested the potential for more varied and subtle readings, although the study of movement as a central principle of culture and society remains, nevertheless, underdeveloped.

In this chapter I explore ways in which mobility may continue to be socially, politically and culturally significant in the lives of sedentarized nomads, manifested in the social organization of sedentarized communities, and in the material culture of dwellings and settlements. Detached from the regular practice of spatial movement, do 'nomadic' cultural understandings and invention influence the practical and symbolic organization and experience of social life, and if so, how? This approach seeks to shift the emphases which have dominated descriptions of mobile societies and groups, and to question the common assumptions that economy, ecology or the facts of spatial movement play determinant roles in shaping the internal social and political structures of nomadic societies (e.g. Salzman 1967). Movement and mobility, which, of course, are not the same thing, may serve multiple, overlapping purposes for nomadic peoples. As a central mechanism of culture and social organization, mobility shapes perceptions and experience, and generates and sustains fluid networks of relationships both within a mobile society and with its sedentist neighbours (Barth 1961; Asad 1970). In relation to wider economic systems, movement and mobility can enable a nomadic group to negotiate multiple positions in different tiers or at nodal points of an overall regional economy (Salzman 1972; Chatty 1980). And at the level of the changing residential community and its shifting alliances and residence patterns, they are at the heart of inter-family relations, marriage and domestic arrangements, connecting and rearranging groups with changing needs and capacities; reinforcing or reducing mutual dependence and co-operation between the well-off and the poor, siblings and their conjugal families, and the old and the young. Movement offers ways to manage conflict and competition in politically uncentralized societies (Irons 1974); to renew, maintain and transfer alliances between kin (Asad 1970); and to maximize opportunities, skills and knowledge (Hoare 2002). The manipulation and interplay of relatedness through kinship and residential closeness, in flexible, affective and functional collectivities, means individuals and conjugal families can mobilize social support from different sources, for diverse purposes.

Studies of the role of mobility in economic adaptations influenced by ecological approaches (e.g. Johnson 1969) have added little to the understanding of social practice, identity and ideation stemming from, attached to, and generating, mobility itself. If a radical decline in mobility is likely to become the experience of larger numbers of nomads in the twenty-first century (e.g. Casimir,

Lancaster and Rao 1999), the question of how post-nomadic cultural, social and political systems, modes of dwelling and settlement patterns may emerge and develop, is of considerable importance. If mobility is culturally generative and socially dynamic, as suggested above, rather than simply economically or environmentally determined, what are the implications for post-nomadic societies, and for the complex and fluid networks which many writers have identified as essential features of their social organization? How will post-nomadic societies integrate their political systems and social values within state structures, or will they remain independent and peripheral? These concerns suggest that the concept of nomadic mobility as a complex cultural phenomenon is not ready to be relegated, even if, as has been argued, it requires refinement and perhaps expansion.

Tapper (1979: 45) remarks that, for some Middle Eastern peoples, 'movement, tent-dwelling, and stock-rearing are not imbued with central meanings at all' and that, 'in these cases nomadism is an economic rather than an ecological, cultural or political adaptation'. Such 'uncommitted nomads', he suggests, typically form part of social and political structures composed of both nomadic and sedentary elements. My concern here, however, is not to suggest that movement and mobility generate the same types of meaning, or serve identical functions in socially, culturally and politically varied societies. Rather, as Burnham has argued, 'the contingent quality of inter-personal and inter-group relations in spatially mobile societies is a fact of their political organization and must be considered as an independently significant phenomenon whatever the environmental conditions' (Burnham 1979: 350).

In the first half of the chapter I argue that mobility, being meaningful, practical and adaptable, can be multidimensional for nomads; that movement and the social and economic systems it supports are not necessarily coterminous, but flexible and partible; and that 'nomadic' flexibility can therefore, paradoxically, include mixed modes of dwelling, including sedentary forms. This may include the division of interdependent groups into sedentary and mobile formations; the sequential use of sedentary and mobile dwellings; or extended periods of sedentary dwelling followed by a resumption of nomadic migration. The observation of mixed modes of dwelling among nomads has frequently been made, adding to the problems of definition and distinction between nomadic and sedentary cultures. While these are, therefore, by no means original insights, I wish to emphasize both the distinction and relationship between movement and mobility in 'mobile' cultural systems, and to suggest that the 'structures' of spatial movement and moveable dwelling conventionally understood as nomadism are only one possible manifestation of mobility. I conclude this section with a discussion of the sense and experience of place that may exist for nomads, which I take to be distinct from that of sedentists, with further potentially important implications for the cultural trajectories of post-nomadic societies.

In using the term 'nomadic' in the course of this chapter I seek to describe a cultural disposition and forms of social organization in which movement and/or mobility shape understandings, influence living systems and structure practical, strategic and fluid systems of social and political relations, which may, or may not, involve actual spatial movement. This usage both distorts and stretches the original, problematic term. My purpose is not to reinstate a word for its own sake, but to explore mobility as a cultural phenomenon, and to differentiate 'nomadic' mobility from forms of practical or contingent movement and mobility among sedentists. Recent uses of the term 'nomadic' for describing the movements of mobile workers, transnational migrants or refugees, for example, draw upon sedentist presumptions that nomadism is a category of movement by defined degree, and are invested in the sedentist association between the idea of home and spatial fixity. Those who are delineated as 'homeless' according to sedentist understandings are thus classed as 'nomads', and vice versa.

In the second half of the chapter I shall describe aspects of the way of life of co-resident groups of sedentary Irish Traveller families on 'official' Traveller residential sites in the Republic of Ireland and the UK, for whom mobility, as part of a structured cultural disposition or 'habitus' (Bourdieu 1977) continues to inform the social composition of the settlement; to shape relationships and identities; to influence the production and built forms of dwellings; and to frame the experience of a larger world, and responses to the events of daily life.[1] Although either one or two generations have been born into a sedentary way of life and reached adulthood among the families involved in the research, close relationships, marriages, and cultural understandings continue to reflect and generate a distinctive 'Traveller' style and approach to life.

Multidimensionality and partibility in nomadic systems

Discussions of movement among nomadic peoples have focused predominantly on the rationale for spatial movement in relation to particular internal or external aspects of the life of a nomadic group or people, such as economic systems, ecological adaptation and herd management; resistance to state control; the social, political or economic organization of the camping unit or larger group; or processes of fusion and fission, avoidance and alliance, within wider nomadic society. While describing the impetus for movement in one area of life, many writers have noted its capacity to serve diverse purposes in different contexts (e.g. Gulliver 1975; Burnham 1979), and so spatial mobility has been described in a variety of terms, as a strategy, mechanism, technology or mode of life. In particular, the capacity for dispersal and change in the social composition of residential groups, as well as strategic aggregation for the purposes of 'contingent' political organization has frequently been observed (e.g. Salzman 1967), suggesting that social and political fluidity are primary characteristics of mobile, or nomadic societies (e.g. Burnham 1975). At the level of inter-family relations, the movements of women between camps represent a further important social and political function of movement (Stenning 1994; Marx 1978), although receiving little attention. The range of ways in which spatial movement and mobility may be understood as integral aspects of several, intersecting or layered social, domestic, political, economic and ecological systems within an overall cultural frame is therefore considerable, although the implications of this multidimensionality have been neglected.

Formal, traditional, strategic, adaptable, improvised, ordinary, ritual, public and private, symbolic and practical – in its extensive functional, intuitive and expressive possibilities, nomadic movement may be said to possess many of the characteristics of the vernacular dwelling. Spatial movement is, in some senses, flexibly 'inhabited', made to fit varied, changing, yet broadly predictable circumstances. This suggests that the motivations for movement may not always be what they seem, or be acknowledged to be what they are (Gulliver 1975). Individual members of a residential group or conjugal family may be mobile in different ways, in different social spaces or time frames, and for different reasons. The many 'etic' approaches to the phenomenon of mobility have neglected the ideas of nomads themselves about mobility, and largely ignored the varied characteristics of individual or gendered movements, focusing instead on an overall 'rational' pattern of migration, conceptualized spatially and geographically, rather than socially and symbolically.

The dominance of centralized sedentary states at the expense of the former political power of nomads in many parts of the world led Gellner (1973: 7) to the conclusion that nomadism had become 'an economic rather than a political adaptation'. But the political 'peripheralism' of nomads in modern state systems clearly does not preclude possible political benefits; mobility may be an effective way of avoiding the reach of the state, as well as an important political mechanism within nomadic, and nomadic-sedentary, group relations. The Yomut Turkmen of northern Iran used mobility in order to avoid conscription and taxation (Irons 1974), and for the Pukhtuns, '*azadi*' expresses a determination to retain 'political freedom': to 'escape or, where possible, exploit the administrative structures' of the state (Ahmed 1982: 1101). In spite of an overt political ideology expressed in both cases, these are surely unexceptional uses of mobility, and both Turkmen and Pukhtun nomadic systems were also inevitably integrated within local and regional economies, and simultaneously served the needs both of domestic systems and wider social relations. The problem then, is not whether nomadism should be understood as either 'political' or 'economic', but how multiple cultural, social and political aims, livelihoods, ways of thinking and being are motivated, embodied and expressed by patterns of mobility for a particular society within its wider contexts.

Burnham observes that, 'once established as a principle of political organization, fluidity of local grouping proves to be a remarkably resistant feature of any society' (1979: 351). The return of the Gbaya to their former patterns of mobility, dispersal and fission, following the relaxation of enforced village consolidation in Cameroon by the French, leads Burnham to focus on the role of mobility as a mechanism of political adjustment, resisting the tendency to stratification induced by sedentary institutions and fixed social organization. Absence of authoritative political hierarchies, competitive alliances of subordinate and dominant groups (clan segments), clientship and protection, are features of both Gbaya and MBororo societies, and these political relations are activated and expressed through marriage and patterns of residency. Marriage and residency practices can be considered as independent, functional and symbolic elements within a system of mobility, understood as both cultural and political. The partibility of mechanisms which generate structural fluidity is an important point, in considering mobility within sedentarized Traveller culture. The transformation and modification of networks of relationships by means of marriage and residency are crucial features of mobile political and social systems, which can operate in conditions of minimal spatial mobility.

Among the Basseri, although years of enforced sedentarization under Reza Shah had dispossessed annual migration of its former economic rationale, the collapse of Persian administration in 1941 was marked by an immediate resumption of the migration cycle: families without viable herds went on the move.

> All the Basseri expressed their reaction as one of resuming migration – not as 'becoming pastoralists again'. As a matter of fact, most of them had very few animals, and some appear to have resumed migration entirely without stock – the supreme value to them lay in the freedom to migrate, not in the circumstances that make it economically viable.
>
> (Barth 1961: 149)

Among actively mobile nomads Barth acknowledges the 'methodological problem to demonstrate the value that is placed on migration, when this value is not, in fact, expressed by means of technically unnecessary symbolic acts and exotic paraphernalia' (1961: 153). Nonetheless, the 'whole basic system of activities involved in the economic adaptation of the Basseri, of camping and herding and travelling, are pregnant with such meanings, and ... the context within which they take place, that of the great migration, is vested with extreme value' (1961: 147).

Aesthetic, social and ethical significance may lend migration, and the idea of migration, dense emotional potency for nomadic agents. An urge and desire for movement is sometimes said to be 'in the blood' among people who have adopted sedentary modes of life (Hoare 2002: 58), and symbols of nomadic life are generated dramatising 'difference' by former nomads, the display of material culture associated with nomadic life forming markers of identity, desire and allegiance. Cartwheels, images of horses and wagons adorn the dwelling and garden, announcing the identity of sedentary Irish Travellers on Irish housing estates. Sedentarized Basseri continue to camp in tents in the gardens of their houses 'with a continuing emotional interest in and identification with nomad life and ways' (Barth 1961: 106). Former nomads in Mauritanian Sahara, traders in gasoline, entertain guests in tents pitched in courtyards behind the earth dwelling, where 'the traditional symbol of the original shelter is alive' (Morgenthaler 1977: 152); and 'Somali sailors in London, Toucoulour sweepers in Paris, Tuareg workers in Benghazi, Baluchi labourers in Oman, Gambian senior civil servants, and Sudanese cotton farmers ... continue to invest their savings in herds at home' (Aronson 1980: 184).

To the array of practical and political purposes, symbolic and social meanings of nomadic mobility, the frequently noted 'flexibility' of many nomadic peoples must be added: a readiness to adopt new environments, technologies and economic opportunities, and the evident willingness of nomads to take up sedentary life as an elected strategy in some circumstances (Barth 1961; Salzman 1980; Berland and Salo 1986; Lockwood 1986; Salo 1986). This flexibility, which may be more typical than exceptional, leads to evident difficulties in developing a unified theoretical framework for nomadic mobility: the range of issues at stake for any particular group at a given moment will be highly specific and forms of response equally varied.

I suggest that nomadic livelihoods and the changing patterns of movement and sedentary dwelling with which they are commonly associated may be understood as aspects of complex, partible systems; active and overlapping forms of social, economic, political and cultural aims, whose elements, however inter-related, are neither necessarily coterminous nor interdependent. 'Mobility' as part of an overall cultural ethos and related social and political system, is multi-dimensional, within which spatial movement itself is a negotiable element. This suggests why the Basseri could passionately resume movement after years of sedentary dwelling: the social structures, meanings and articulations of mobility are multi-layered.

Structured invention and the *habitus*

A conjunction of economic adaptation, ecological opportunity and historical and political conditions, which have so often been seen as causative factors, might instead be regarded as facilitative circumstances for forms of social, political and cultural life which generate their own dynamic momentum. In relation to Xoraxané mobile economy, Lockwood (1986: 66–7) observes: '"Niche" . . . really carries the wrong connotations. The term implies a more or less stable situation, a structured symbiotic relationship with people occupying other niches. The reality is much more fluid than this'; each new economic strategy can best be understood as a 'holding pattern'. This concept, which Lockwood applies to the flexible livelihoods of the Xoraxané, is suggestive of the capacity for 'holding patterns' within other aspects of nomadic life: residential groups, sedentary and mobile dwelling systems, migration patterns – as well as livelihoods. In any particular situation, what is put on hold, so to speak, or adopted as a holding pattern must conform to an intuitive practical sense that is particular to the people in question, a skill for improvisation, arising from a unique history, a body of knowledge and an essential flexibility.

Salzman (1980: 6) describes 'latent institutionalized alternatives' among nomads, which make possible far-reaching reconfigurations in economic systems, social groups and modes of dwelling. Bourdieu's concept of the '*habitus*' lacks the connotations of a menu of preconceived options, conveying instead the fluidity of structured invention and creativity: 'a feel for the game.' The *habitus* consists of:

> systems of durable, transposable dispositions, structured structures predisposed to function as structuring structures, that is, as principles

which generate and organize practices and representations that can be objectively adapted to their outcomes without presupposing a conscious aiming at ends.

(Bourdieu 1990: 53)

Both nomadic and post-nomadic societies might then be expected to develop highly varied cultural, social or economic trajectories under a range of different possible circumstances, some elements of which nevertheless bear hallmarks of continuity. And conversely, this emphasis on the structured agency of nomadic peoples suggests that groups that migrate between different ecological and political environments might nevertheless retain similar forms of economic and social organization. The empirical evidence of both pastoral and non-pastoral nomadic migrations between different environments supports both of these divergent theoretical possibilities (Johnson 1969; Frantz 1978; Salo 1986). Such instances run counter to the trend of ecological interpretation that focuses on the micro-adaptations of economic and social systems to local environments, and understates or ignores wider social and political contexts, as well as the overarching cultural frame or *habitus*, which shapes particular adaptations.

Terms such as 'response', 'adaptation' and 'flexibility', frequently employed in descriptions of nomadic regimes, suggest passivity on the part of people in their relations towards external environments and changing circumstances, to which they apparently merely react (Frantz 1978: 114). I concur with Gulliver (1975) that active choice and decision-making informs the modes of living of nomadic and post-nomadic groups, rather than passive and inevitable response to 'ecological compulsion' (Johnson 1969: 2), even where the constraints of physical environments or political situations may seem overwhelming. Adaptation and flexibility among nomads imply unusual creativity in finding multiple ways to divide and re-combine groups based on changing interests, assets and skills, and to mobilize networks and alliances for varied purposes, or in order to exploit a wide range of resources. The 'structural conservatism' (Burnham 1979: 351) of highly dispersed and mobile societies therefore requires mechanisms for the stable affirmation of cultural commitments in spite of movement and geographical separations, as well as the capacity to sustain a way of life which feels 'true to self', and which can be reproduced across generations and in widely different environments, through either nomadic movement or sedentary dwelling. The comparatively greater risks of social disintegration require proportionately stronger affective mechanisms with which to forestall them. Why should the claim be made that nomadic peoples display these characteristics to any greater extent than sedentists – who, after all, also migrate, adopt new technologies and livelihoods, move from small to large scale communities and so forth, frequently retaining loyalties, languages, social and economic connections with a place and community of origin? All this is true, but let us look at the comparison from another perspective.

Nomads, I suggest, do not *move* from one 'place' to another, as described from the sedentist position (e.g. Ingold 1986: 175), they simply *live* in much larger places than sedentists; places that, as a consequence of their geography, are often characterized by enormous internal variation, which may be physical, demographic, political, technological, cultural, religious, social, economic and linguistic. And these variations are experienced, not in a single moment of change between one environment and another, as in sedentist migration, but as a constant dynamic feature of the migratory sequence. Nomadic conception of place, therefore, is not of one here and one there, each being a fixed and located set of references to which attachments are formed, identity linked and between which acts of adjustment must take place; it is, rather, of passage and variation as perennial concomitants to the experience of the physical and social world. 'Place' is therefore often reckoned in terms of time and season, arrival and departure, not as a separable object in 'space', as Barth (1961) and others have observed. Individual flexibility, initiative, inter-personal co-operation, powerful group loyalties, as well as a taste for variety and capacity for change are therefore the stable qualities indispensable for managing life in large and complex places; and the idea of 'place' to which nomadic people may be attached (in the sedentist sense), is movement itself.[2]

Casimir, Lancaster and Rao (1999: 3) recently suggested that, in Africa, where 'perhaps, pastoral nomadism has outlived its usefulness . . . so the way of life and social behaviour that is associated with it becomes redundant and inappropriate'. The question must be: redundant and inappropriate for whom? If nomadic perception, morality, social life and mobility are creative, affective factors in themselves, rather than dependent corollaries of economic and ecological adaptation, the distinctive skills, understandings and values of nomadic cultures are likely to find new forms of social expression, and distinctive ways of living.

Irish Travellers and sedentarization

Throughout the twentieth century Traveller families have adopted a sedentary way of life from time to time as part of an economic strategy, where occupational specialisms could be practised to greater advantage within a sedentary setting (Hoare 2002). In such cases, Traveller identity was retained by the family concerned in the eyes of both settled and Traveller populations, and children raised in sedentary settings might return to a nomadic way of life as adults, regarded as being 'in the blood'. Ethnicity was therefore symbolized by the nomadic lifestyle, but the suspension or relinquishment of nomadic life did not compromise Travellers' distinct identity. In such circumstances, 'towns Traveller' children might grow up to marry settled or 'country' people, and, regardless of whether the couple adopted a sedentary or nomadic way of life, their individual settled and Traveller identities remained socially fixed on the basis of their perceived ethnic inheritance. Individuals did not 'pass' from one identity to another by virtue of either marriage or lifestyle, as has sometimes been imagined (Hoare 2002). Since the late 1960s and 1970s, however, the causes and pervasive pattern of state intervention in Traveller sedentarization have differed considerably from that described above, and the settlement types and forms of resident social groups created by Travellers have varied accordingly. Traveller sites, once regarded by the state as intermediate stages in the inevitable process of assimilation, are now frequently the only form of sedentary dwelling available, and in many cases, the type preferred by Travellers themselves.

Demographic, economic and technological factors combined to bring about the demise of small farm life and culture in Ireland between the 1950s and 1980s (Hannan 1979; Ó Tuathaigh 1998), and as part of the same processes of change, Travellers experienced the loss of their livelihoods and of familiar roles, relationships and patterns of movement among diverse sections of the settled population (Hoare 2002). Full agricultural mechanization of the industrial farming sector and of the peat industry took place throughout the 1960s and 1970s, both of which had employed Travellers in considerable numbers into these decades. The Irish sugar beet industry, founded in the early 1930s, was at the height of its production in the post-war decades, and as a mobile population with a diverse economic base, Travellers formed a workforce with a unique capacity to meet the seasonal demands of the sugar beet industry (Hoare 2002). The beet harvest was dependent on a seasonal labour force composed of a declining population of under-employed settled farm labourers, growing numbers of whom were opting for migration to work in the UK construction sector, and of Travellers. Between 1950 and 1970, around 68,000 acres of sugar beet were grown annually in the Republic of Ireland, and in 1967, only 50 per cent of this production was harvested mechanically (Foy 1976). Mechanization of the beet harvest reached 98 per cent by 1976, and its spread was a critical factor in the widespread sedentarization of Travellers.

From the 1960s Travellers began to be affected by determined state policies to 'assimilate' them within the wider population, the first of these being to inhibit and, if possible, prevent nomadic movement and camping. In these efforts, the state was assisted by technological change. Tractors enabled farmers to plough into the wide headlands of their fields, formerly created by the turning distance of the plough team, and thereby reduce to narrow strips the wide roadside verges where Travellers camped and grazed their horses. Roads were gradually widened to accommodate a growing volume of motor traffic, the two moves effectively destroying many former camp sites. Kerbs were introduced along miles of country roads, stopping wagons or trailers from leaving the road, and a 'boulder policy' was eventually introduced, preventing access to lay-bys or other off-road camps. The legal, physical and political reconstruction of landscape represents an important aspect of Ireland's construction of its national identity, adding to the significance of the exclusion and, ultimately, containment of Travellers. Legislation prohibited the use of the tent, which continued to be an integral part of the nomadic dwelling into the 1970s, and a succession of ever-tighter legal measures has continued to impose a relentless pressure against the nomadic way of life in Ireland.

Together these circumstances produced a growing crisis in which a long and painful process of sedentarization began, accompanied by increasing social isolation, and the hardening of ethnic boundaries. The 'difference' between Irish nomadic and settled populations, undoubtedly long recognized and formulated in ethnic ideology by Travellers themselves, was nevertheless an unremarkable feature of Irish social life for much of the twentieth century, particularly among those with whom many Travellers had most interaction: the small farmers of the west of Ireland and the midlands.[3] Social identities and relationships shifted in Ireland in the course of the twentieth century, leading to the profound stigmatization of Irish Traveller identity, both in Ireland and in the UK. A full account of these important subjects must await separate discussion.

The following account describes features of the residential communities, built settlements, and contemporary life of sedentary Irish Travellers in the UK and Republic of Ireland. Mobility, as an organizational feature of social relations, continues to shape the formation of residential groups, to inform the values and community life of residents, and to be reflected in the built forms of the settlement. Consciousness of living in a 'large place' influences the experience of time and place, shapes the perceptions of daily life and reinforces shared understandings.

Irish Traveller sites

The settlements of sedentarized Travellers are popularly known as 'Traveller sites', of which there are several different kinds in the UK and the Republic of Ireland. While a few residential sites in the UK are privately owned (Okely 1983), others were established by County Councils to meet the legal requirements of the Caravans Act 1968 in the UK, or the Housing (Traveller Accommodation) Act 1998, in the Republic of Ireland.[4] Sites are awarded under licence agreements to 'compatible' groups of Traveller families, publicly represented by a senior male figure, following what are invariably tense negotiations over the type of infrastructure or accommodation to be provided, and the conditions of occupancy. Travellers are at a considerable disadvantage in these processes, and are frequently forced to accept terms, locations or built structures they find repugnant.

The decision by a local authority to create a residential settlement for the tenure of an existing residential group of allied or related families may result from a court ruling in favour of the group, following attempts to eject them from an informal settlement where they have become established and where the land occupied is subsequently earmarked for development. Processes similar to this resulted in the creation of the three sites on which the following descriptions are partly based, two in Ireland, the other in the UK, although a court ruling was involved in only one of the Irish cases.

Travellers provide their own living accommodation on some sites, in the form of trailers or chalets. In other cases local authorities construct 'service units', consisting of a small brick structure per family 'bay', containing a kitchen, bathroom and toilet, and each family uses one or two trailers for sleeping. Some sites have houses for each family, and the keeping of trailers is then strictly prohibited. Such conditions confirm the official policy to impose 'permanent' sedentarization on site residents. In both the Republic of Ireland and the UK, sites are situated in isolated, peri-urban or rural situations that would be considered unsuitable for standard local authority housing, because, for instance, of their proximity to busy roads or rubbish dumps, and the absence of nearby shops, pavements or street lighting. All the settlements I have visited are situated in such settings.

It is important to emphasize the degree of control and restriction to which residents are subject, which considerably exceeds that imposed on settled local authority tenants. Traveller residential sites thus represent places of exclusion as well as of containment. They are invariably bounded by walls, banks or fences to inhibit expansion. Sites are provided with a single entrance (i.e. no means of escape) and are always subject to special monitoring and surveillance. In the Republic of Ireland this takes the form of commercial security firms or retired *Gardai* as daily visiting wardens. This brief outline of site conditions conveys the essential physical features of typical Traveller settlements, and the pronounced political control to which many Travellers' daily lives are subject.

'Coming and going' and the mobility of the sedentary group

The phrases 'on the road' or 'travelling' are used by Travellers to refer to any particular instance of movement as part of a migratory cycle. 'Coming and going', however, describes the particular use of movement to maintain the important links between a married couple and both sets of their parents and living grandparents, and with the wider group of first and second cousins. 'Coming and going' described the crucial practice of maintaining practical kinship links through joint travel, in the course of which a married couple mobilized their extended family ties, and embedded their own children within the daily lives, identities and histories of their parents, siblings and cousins on both sides. It was a long-term, strategic and intensely practical activity, enabling a young, inexperienced family to practice co-operation with different travelling companions, peers and elders from related families; to observe and learn from others; and to share skills and resources. Through coming and going, a young married couple gradually widened their geographical knowledge of different travelling circuits, and of their varied opportunities or limitations. The closeness of camping relationships helped to nurture familiarity and affection between their young children and those of their siblings, cousins and friends, with a view towards the marriages of the next generation. 'Coming and going' lasted until all the children were themselves married, and the couple would then travel by turns with their own adult children.

The restrictions of sedentary life have severed the scope for the important social, strategic and practical functions of 'coming and going' through movement. Contacts with the wider group of family and friends are nowadays limited to family visits and larger gatherings where marriages or christenings are celebrated, or at funerals. These occasions now loom much larger in scale and importance than they did before sedentarization, although they lack the multiple functions and flexibility of 'coming and going', and hardly generate the same bonds of mutual dependence and intimacy.

It is the residential community of the settlement, as a functional collectivity, which continues to generate and embody co-operative and affective bonds, concentrating some of the most important relationships of growing families within a single locus, and seeking to distil some of the social activities and values of 'coming and going', which took place through time and movement, into the spatially fixed community of the site. There is, of course, a compromise if not a contradiction in this. In all the sedentary communities I have visited, a group of adult siblings and their conjugal families forms the core of the residential group of between ten and fifteen households. The 'breed' name of this dominant group (a segment of an agnatic kin group) gives the settlement its identity. Together with their parents, children and affines, some of whom are patrilateral cousins, this social group offers the security of shared histories and identities, an expectation of mutual defence and co-operation, and the hoped-for satisfaction of future marriages between the cousin group of the next generation. Although most settlements appear to include one or two households of 'friends' of different breeds with whom intermarriages are probably intended, only one adult sibling group and their parents can normally be represented in such a structure.

The undoubted strengths of such communities, who are deeply committed to their own continuity and reproduction, are achieved at a cost. Close alliances with the families of husbands and wives of other breeds are sacrificed, and relationships with dispersed segments of the breed may be subordinated. The children of sites may associate with the majority identity of the residential community, sometimes their mother's rather than their father's name, risking the status of agnatic identity, and weakening the social position of a husband and father who is from another, under-represented breed. Nevertheless, the fluidity of multiple, shifting identities and layered relationships helps to moderate the ascendancy of the dominant breed, through the mobility of its social structure. A mother-in-law may also be an aunt, or an aunt a sister-in-law, and their assistance invoked in different ways for different purposes. Marriage links through sisters and mothers complexify identities, and multiple bonds tend to engender trust and warmth, and to turn kinship into friendship. The marriage of first cousins turns siblings into affines in the same process. Such shifts have the potential to subtly reconfigure a wide constellation of affections, obligations and commitments, and so the residential community is deeply engrossed in all matters pertaining to the marriages of their children: their arrangement, celebration and on-going support, as each new marriage both regenerates and transforms a family and its group relationships. The daily lives, histories and futures of families are thus closely intertwined but never fixed, residents having multiple ties and layered relationships with others both within and beyond the settlement, in structures which are made dynamic and mobile through the marriages of the younger generation.

Dwellings and material culture

By virtue of their functions, their materials and their creation, the shelter and bed tents were formerly the most intimate and important objects in nomadic life. The tent parts, made and assembled by a husband and wife, were regularly moved, reconstructed, renewed and remade. A set of hazel wattles, forming the curved ribs of the tent frames, lasted up to two years, and the cover or 'piece' between three and four months. The 'rigging pole', into which the wattles were inserted, into drilled or burnt holes, would last many years. The 'piece', or tent cover, was sewn with carefully folded double seams, and waterproofed with gas tar and margarine, the latter added to 'keep it limber', that is, to prevent the piece cracking when it dried, a process which took several days. Descriptions of making the piece, and of tent construction in general, evoked associations with female sexuality. A story was recounted of a wife who, walking through the camp, tripped and sat squarely on a sticky piece shortly after it had been spread out on the grass to dry. Her husband's effort to remove the tar, requiring the liberal application of margarine, left her 'as limber as anything for weeks after that'.[5]

The tent, its materials and its construction, carried multiple symbolic resonances for a married couple, and discussing the making and use of tents, and the ritual burning of the old winter piece of the shelter tent in spring, when the winter camp was vacated, evoked deep feelings of nostalgia in older Travellers. An association between the material culture of the sedentary Traveller dwelling and the marital relationship persists, its renewal and transformation continuing to enact a changing, renewing partnership between a husband and wife. The dwelling grows with the enlargement of the family and frequently regenerates itself in new forms through acts of repainting, the addition of an extra chalet or trailer, the tarring of a drive, external lights, internal redecorations and so forth. New ornaments or pieces of 'delft' mark birthdays and anniversaries, growing collections forming a record of the growth of the marriage. Religious icons memorialize pilgrimages and significant journeys. Material forms are embodiments of time and movement. Front yards or gardens fill with displays of plaster horses' heads, cartwheels or religious statues, and when no more room can be found, new colour schemes are dreamed up. Production and reproduction are irresistibly linked in the mobile materiality and symbolism of the dwelling as it enacts the characteristics of marriage: the living artefact of a woman and man, always on the move.

Different dwellings on a site may articulate varying attachments to the former spatial organization of the camp. A senior couple keeps separate chalets spaced some yards apart: a bed chalet, kitchen chalet and store chalet (the latter like the wagon), reminiscent of the physical forms of the camp and its separated activities. Conversely, the attached chalets of younger families, more house-like in appearance, emphasize the cohesion and autonomy of younger households within the wider group. The spatial organization of the settlement always includes solid boundaries around each dwelling, affirming the autonomy of the conjugal family and its separability from the residential group.

The performance of rituals of renewal and transformation asserts the independence and mobility of the conjugal family within and through the dwelling, forming visible, public expressions to the rest of the community, who smile approvingly as each new endeavour gets underway. A household is seized with a desire to reorganize, expand or transform, and in doing so, becomes a discrete, self-contained entity, temporarily removed from the rest of the group for days or weeks. Such impulses express the self-determination once asserted through dividing the camp and striking out on a new route, asserting one's independence of the community: of 'coming and going'.

Confirming this interpretation of the value of autonomy through the essential mobility of separate conjugal families within the residential community, it is notable that the marriages of young couples are not, as might be expected, symbolically embodied by the establishment of a new dwelling, but by their departure from their respective extended families and settlements to go travelling for at least a year, in order to forge a durable bond away from the solicitous eyes of parents, siblings and cousins. This behaviour directly echoes the former pattern of nomadic life, when young couples went alone on the road together, before they took up the social practice of coming and going, marking adulthood. Irish Travellers say that solitary travel by a husband and wife, today just as previously, is indispensable for strengthening the conjugal bond in the early, vulnerable years of marriage, and parents born to sedentary life often continue to send their children onto the road after marriage.

The Traveller site and the world of others

The *habitus* makes a virtue of necessity (Bourdieu 1977: 77). Containment and the restrictions of the site are converted into defence and protection of its internal world, its boundaries put into service from within. The dwelling of a senior figure is invariably situated in eye line of its entrance, which is almost constantly kept in view by children playing, or women passing between houses or trailers. Unfamiliar visitors are questioned; news passes rapidly and invisibly through the site; the appearance of an uninvited official at the trailer door is greeted with feigned surprise.

But penetration of the settlement by unwelcome others can take different forms; the unpredictability, complexity and essential 'largeness' of the nomadic world are always present. Ghosts are sometimes heard at night, whispering outside windows, scratching at doors, tormenting people, 'it was *himself*', who fearfully resort to prayer. Their intention, unquestionably, is to harm those within. In an account of her community's *novena* (nine day prayer cycle), Kathie described how an unknown child appeared among them one day as they prayed. Seen by several of those present, his appearance was accompanied by the scent of roses, signifying the presence of Mary, mother of Christ. The vision followed hard on the heels of a previous manifestation of the devil as an unknown dog at night, whining and scratching at doors, in an attempt to scare them from their religious endeavour. This remarkable demonstration of the power of their prayer 'for peace in the world' transformed the small settlement into a site of cosmological conflict, in which their joint spiritual strength ultimately triumphed, and they overcame their own fear as well as the supernatural forces that threatened them.

Such real, physical encounters with spiritual forces (their malevolence, trickery and disguises, or their power to heal and renew) form a familiar aspect of experience in relation to the visible and invisible world. The meanings of manifestations are often ambiguous, and people seek the interpretations of others: should I be reassured or take it as a warning? Has it happened to you or anyone you know? 'There was a robin in my shed when I opened it this morning. What does it mean?' The uncertainties, dangers and contradictions experienced in nomadic life came at any moment and seemed to emanate from the most unexpected source. Although small farm communities looked the same, there were 'very different kinds of people' (Hoare 2002: 63). Some would welcome you in to a seat by the fire; at other times a gang of local people might stone a woman and her children asleep in the tent (ibid.). The deceptiveness of people is linked to their power to harm in invisible ways: 'whenever she's been here, nothing else goes right for the whole day.' The settlement community is alert to unexpected dangers because they are *always* expected, and protects itself with the known physical efficacy of mobile phones, van keys at the ready, traditional wisdom, faith and prayer.

Keeping up with the movement of events

While the settlement community may be seen as a distilled, spatial compression of the structured time and movement of nomadic social organization, concentrating key agents of 'coming and going', time as it is experienced in established settlements is also distinctive. Involvement in the regular daily lives and unexpected events of other families creates an ever-shifting focus of interest, and women in particular often describe a pace of life in which they are swept along, barely in control. A birth, a series of visits, an accident, someone falling sick, preparations for a marriage, a trip to Ireland: all demand response and absorb attention, and life is experienced as constantly changing, moving at its own tremendous speed, filled with substance and feeling. The pace of events continually invokes Traveller initiative, skills, pleasure and morality, and in doing so, repeatedly demonstrates their value and relevance. Rising to the occasion, making the best of a bad job, resolving a problem – all call for ingenuity, commitment to the group, optimism and co-operation, and daily life presents numerous instances which affirm the rightness of the Traveller way of interpreting and doing things.

A young boy drives a car through the site and, reversing into a wall, demolishes it. Women rush to the scene to check the children. They're all in one piece, 'Thank God'. Luck was on their side: it was never *meant* to be a disaster. In an instant, a mother organizes the group of toddlers into a work gang, constructing a safe pile of bricks ready for the repairs to be done. Everyone is delighted with this spontaneous transformation of the accident, in which essential Traveller ideals are elegantly embodied: co-operation, practical ability, finding the good in the bad and making the best of things. The children are admired as they proudly stagger back and forth with their bricks; the women are pleased, observing their enthusiasm and new ability; and later, a certain boy has his first lesson in bricklaying. What about the car? 'Ah well, it was written off, but it was no good anyway.'

Individual members of the community embrace their rapidly changing roles in the regular flow of both minor and more serious events. As exemplars of change, of life moving on, they offer something of the emotional satisfaction and a sense of the fullness of life associated with movement itself: you never know what is round the corner. Even disturbing events (the breakdown of a marriage, sickness of a parent or child, hostility from other Travellers or intrusion by settled authorities) represent testing grounds for group stability and mettle, occasions for mutual support and personal initiative. Every crisis is discussed at length within and between the families, appropriate responses and strategies are debated, and, although not exposed to the detailed substance of their parents' concerns, children observe the emotional tone and demeanour of adults, and absorb and emulate the seriousness, humour, anger or excitement which attends them.

Conclusion

The organization of relationships and social life in Irish Traveller settlements has many complex resonances of the mobile systems and structures of Travellers' nomadic culture. Does this imply that Travellers are engaged in a fruitless effort to preserve a way of life that no longer fits the world they inhabit? That can only be answered through trying to recognize the characteristics of that world. Movement, change and unpredictability are, for Travellers, the self-evident features of the natural and social world: a world understood and experienced as processual and mobile, rather than stable or fixed. Human lives, changing relationships and families, the flow of events, all endorse this understanding of an unstable natural order. The multidimensional mobility of Traveller culture reflects responsiveness to these unavoidable realities – the strategic and creative endeavour to keep ahead of the game. Gulliver (1975: 373) states: 'the paradox is that although movement is an inescapable necessity, yet this affords a degree of freedom of action.'

Social identities and alliances operate at different scales: at the public level of the agnatic breed and its dispersed network of human and material resources; within the widespread cognatic group of a conjugal family; and in the close, multiple relationships of settlement residents. The interplay of strategic relationships and identities at different social levels, through both spatial and social reconfigurations and rituals, continues to express an essentially mobile, 'nomadic' social structure. Layered relationships reinforce mutual commitments, so that they can be called upon to meet the unpredictable needs of changing circumstances. The arrangement of the marriages of the young constantly shifts and dislodges the established order, offering new dimensions to some relationships and potentially reducing dependence on others. Rearrangement, renewal and transformation are ways of constantly redressing the balance in a complex and ever-shifting social world. The size and dispersal of settlement communities across Ireland and the UK, consisting of multiple segments of the breeds in varied relationships of interdependence, both within and between families, can also be understood as an essentially fluid and resistant form of political organization, an uncentralized society, which, in spite of sedentarization, continues to generate structures according to its own needs.

The fluidity of strategic social relations involves the conjugal family in a demanding process of balancing its own needs, autonomy and unity against the varied interests of the residential group of which it is part, within which a husband and wife may have different roles, status and levels of influence. The dwelling occupies an important position in this endeavour, binding a couple together in acts of physical creation and renewal, which belong to them alone. Prussin (1996: 93) observes that the 'repetitive, hands-on dialogue with the materials and process of construction generates a closer identity with the material artefact than is normally experienced through other art forms'. The ritually repeated construction and renewal of the nomadic dwelling in the course of migration formed a symbol of marriage for Irish Travellers. Amid the many pressures of social life within the sedentary settlement, the ability of the dwelling to become a site of creation, separation, autonomy and mobility reinforces its symbolic value in sedentary life.

Nomadic mobility therefore implies far more than the practice of migration as part of an economic adaptation. Among Irish Travellers it suggests a way of understanding the coterminous physical and spiritual world; of organizing changing relationships and multiple identities within and between families; and of creating and regenerating the dwelling, settlement and community. Mobility is a simultaneously moral, strategic and inventive way of being and of living with others, in which 'no move . . . is . . . decisive' (Gulliver 1975: 379). The Traveller site is a place of fixity, limitation and constraint, but its physical boundaries, guarded in multiple ways, are open to penetration at any moment, in ways that confirm the largeness, ambiguity and instability of place in nomadic imagination and experience. The life of the settlement, in many ways insecure, constrained and vulnerable, can nevertheless be constructed as a 'holding pattern' by Travellers themselves, in ways that conform to the skills for improvisation and creative optimism. Sedentary life among Irish Travellers continues to represent a distinctive cultural trajectory, its 'generative schemes' neither a 'simple mechanical reproduction' (Bourdieu 1977: 95) of habitual cultural forms, nor a novel departure from the structures of nomadic life. Like many other forms of nomadic culture, it exhibits complexity, durability and flexibility.

Notes

1. The data used here forms part of my ongoing research with Irish Traveller communities who have been sedentarized for between fifteen and over fifty years. The different communities have been resident on official sites for between three and fifteen years, and before that, in informal settlements for between ten and more than twenty years. This research was started in 2000 as part of a dissertation for the MA in International Studies in Vernacular Architecture at Oxford Brookes University, and is continuing within an MRes in Anthropology at University College, London.
2. Different types of movement may influence the nomadic sense of place posited here.
3. The use of the term 'settled' to denote the sedentary population in Ireland is common among both nomadic and sedentary sectors of the population, and is therefore employed here. For Travellers, the term 'settled', like 'country people', denotes ethnic status. Although press reports occasionally use the term 'settled' Traveller, meaning 'sedentarized', it is a misnomer: Travellers themselves maintain there is no such status. The term 'towns Traveller' formerly denoted Travellers who adopted sedentary life as an elected strategy.
4. For a detailed discussion of the policy ethos of the 1960s to 1980, and impact of implementation of the 1968 Caravans Act, see Okely (1983: 105–24). This gives a description of varied forms of site provision in the UK, and of the 'temporary' and 'permanent' dichotomy which influenced official thinking on what was appropriate for 'different kinds' of Travellers. Okely (1983: 113) states: 'site provision was equated with settlement, and in turn equated with assimilation'.
5. This story was pronounced completely apocryphal by the lady herself, who angrily berated her husband as 'a liary oul man' during the telling of it, although suppressing amusement at his shamelessness.

Chapter 4

Engaging the future
Vernacular architecture studies in the twenty-first century

Marcel Vellinga

> *There's something cosy about vernacular architecture; it's a sheltered retreat for many who fondle the adze-marks, feel the fit of the ashlar or marvel at the assembly of post, wall-plate and tie-beam. Somehow, there's not the craftsmanship anymore; all that honest workmanship with simple tools and muscle – it's gone.*
>
> (Oliver 1984: 17)

> *But it's happening, here and in scores of other estates around the country, new messages are being uttered in the vernacular but, as far as I'm aware, no one is devoting much attention to finding out what they mean.*
>
> (Oliver 1984: 19)

Introduction

Recently, in his *Village Buildings of Britain* (2003), Matthew Rice lamented over the state in which vernacular traditions in Britain find themselves at the beginning of the twenty-first century. Largely ignored by conservationists and those involved in the provision of housing, the future of the British vernacular is far from bright, so Rice tells us. From the nineteenth century onwards rapid cultural and economic change, spurred on by processes of industrialization and urbanization, has

Marcel Vellinga

dramatically altered the face of the British countryside. As rural people left to work and live in the cities and, more recently, city dwellers moved back to the country, the social infrastructure of many village communities has, in Rice's words, been 'destroyed'. Concomitantly the vernacular heritage of Britain 'is now in the hands of those who do not understand it', the retired couples and white-collar commuters who have bought the houses left by the agricultural workers who went away or were forced to move into council estates. As a result many old vernacular buildings in Britain are today converted in an 'insensitive' way, having modernist views of 'a perfect country cottage' imposed upon them that paradoxically turn large parts of the villages into the suburbia that the new owners tried to leave behind when they moved to 'the country'. Local councils and planning departments have not done enough to stop these 'damaging' developments by allowing property developers to built estates filled with bungalows and semi-detached houses, while not fully recognizing, let alone enforcing the importance of maintaining architectural connections with the local vernacular (Rice 2003: 8–14).

Clearly, Rice has a less than positive perspective on the contemporary status of the British vernacular and, more particularly, on the way in which it is made to respond to the demands of the time and the wishes of its current owners and inhabitants. Essentially based, it seems, on aesthetic and emotional judgements, his view of the vernacular is a static and conservative one. For Rice, British vernacular architecture consists of historical rural buildings that were built before the Industrial Revolution, in a time when villages in Britain were agriculturally based and supposedly self-contained, and the construction of the railways had not yet facilitated the replacement of traditional materials and crafts with imported modern ones. Although he recognizes that architectural development and expansion is unavoidable in many places, it is a prime responsibility of the householders and developers of today to carefully preserve this 'pristine' vernacular of half-timbered houses, limestone cottages and granite farmhouses so that the much celebrated British vernacular landscape will not be 'spoilt' any further. Thus, for example, he laments over the way in which 'overzealous' conversion has turned too many Cotswold barns into 'awkward hybrids' that have been 'stained' by the use of 'particularly nasty treacly brown' colours and the addition of 'horrid car ports', arguing instead that such buildings should be maintained and used in as original a manner as possible (Rice 2003: 95).

Rice's purist and static perspective by no means represents the views of all those involved in the field of vernacular architecture studies. Nonetheless it is fair to say that his point of view is not unique or isolated either. Ever since the vernacular became an area of academic and professional interest in the late nineteenth century, a predisposition towards the study and untouched preservation of the oldest and therefore supposedly most 'authentic' or 'traditional' buildings has been strong. Today this tendency is still prevalent among many scholars in the field, regardless of whether they work in western or in so called developing countries, and despite repeated calls for new and more dynamic approaches, and the growing number of studies that have tried to answer them (e.g. Oliver 1969; Bourdier and Alsayyad 1989; Abu-Lughod 1992; Upton 1993).

Up to a point, this conservative and defensive approach towards the vernacular is understandable and justified. Numerous vernacular traditions today face challenges that seriously threaten their survival into the twenty-first century. The vulnerability of vernacular traditions in the face of forceful processes of modernization and globalization makes it desirable to document, study and preserve historical and traditional buildings before they may be lost or become irreversibly changed. Yet ultimately, I believe, it is an approach that is too narrow and restricted, as it results in representations of vernacular traditions that are frozen in time, incomplete and, quite often, romanticized. Besides, and equally importantly, it effectively hinders the development and survival of those traditions by reconfirming, unintentionally no doubt, the persistent stereotypes that represent vernacular architecture as picturesque and charming, yet out of date and irrelevant.

In my opinion, the continued tendency of scholars and conservationists like Rice to approach the vernacular as comprising of pre-modern historical and traditional buildings that have to be studied and appreciated in their 'pristine' state, and that accordingly need to be safeguarded from the onslaughts of modernization and change, has restricted the scope and development of the field of vernacular architecture studies and continues to hamper the recognition of the vernacular as an architectural category worthy of full academic and professional attention. Rather than helping vernacular traditions develop and endure by pointing out their dynamic character and their potential relevance to the provision of sustainable architecture in the future, it relegates them to the past by emphasizing either their historical or traditional, but in any case unchanging and outdated status.

In this chapter I will suggest that what is needed instead of this static and essentially historical perspective, or in any case alongside it, is an approach that explicitly focuses on the dynamic nature of vernacular traditions, one that attempts to show and understand how vernacular traditions, here and now, at the beginning of the twenty-first century, will change and adapt to the cultural and environmental challenges and circumstances of the present and future. By no means the first to have called for such a perspective (e.g. Abu-Lughod 1992; Upton 1993), I believe that a more dynamic approach that views tradition as a conscious and creative adaptation of past experience to the needs and circumstances of the present will significantly broaden the scope of the field of vernacular architecture studies, allowing for studies that focus on new and emerging traditions as well as on enduring ones and, crucially, on the ways in which they interact and relate to one another. In doing so, it will help to rid the discourse of the persistent stereotypes about 'disappearing worlds', underdevelopment and irrelevance that are so common among members of the academy, architectural professionals and the general public. What is more, it will pave the way for a more action oriented approach that perceives the vernacular as a source of architectural knowledge and that critically examines the way in which this know-how may be integrated with new forms, resources and technologies so as to develop culturally and environmentally sustainable architecture for the future.

Marcel Vellinga

Before discussing the possibilities and challenges of such a more dynamic and active approach, it is first necessary to critically consider the concept of the vernacular so as to understand the current predisposition towards stasis and the past.

Narratives of loss and decline

Some ten years ago, Dell Upton (1993) argued that the field of vernacular architecture studies, though more or less established and increasingly recognized within the academy, had so far been held back by the limitations of its own assumptions and definitions. Regarding the vernacular as an enduring, but essentially static and passive category that is defined in opposition to the more dynamic categories of the modern and the formal, most scholars of vernacular architecture have tended to concentrate their work on a small number of buildings only. These are the rural and pre-industrial log cabins, farmhouses and barns that, in the British context, also constitute the focus of Rice's attention. Meeting the ideal characteristics of the vernacular, they are perceived to be vernacularly 'authentic', having been built by their owners in pre-modern times, in keeping with the values and needs of their local communities, and using local resources and technologies. As a result of this restricted focus on so-called 'pristine' buildings, most work on the contemporary use and meaning of such vernacular traditions has, as in Rice's lament, tended to emphasize processes of loss and decline. In focusing on the pre-industrial rural building heritage, Upton writes, the discourse on vernacular architecture has committed itself to models of 'acculturation, contamination, and decline, models of impaired authenticity and reduced difference'. Rather than acknowledging and trying to understand the transformations of the buildings in an era of post-modernism and globalization, 'our tales are tales of woe or tales of heroic resistance (which are simply their complement)' (1993: 12).

Though Upton's discussion is mainly directed at the discourse on Anglo-American vernacular architecture, he rightly notes that his remarks are equally applicable to much of the work done in a non-western context. Though perhaps not as dominant a tendency as in the discourse on the western vernacular, seeing that analyses that focus on the way in which non-western traditions change in response to the process of globalization have become more common in the last fifteen years or so, many studies of African, Asian or Latin American vernacular architecture have tended to focus on those building traditions that are regarded as 'traditional', in the sense that they are or have directly evolved out of indigenous building traditions that existed in the period just before or during the colonial encounter. In comparison to the studies that document and analyse such traditional patterns of space use, construction, design and symbolism (e.g. Blier 1987; Waterson 1990; Prussin 1995; Bourdier and Minh-ha Trinh 1996), studies that pay attention to more recent and modern (or modernized) indigenous building traditions are relative rare; despite the fact that these traditions, though admittedly not always as exotic or distinctive, arguably meet the definitions of what vernacular architecture is and, quite often, constitute the majority of buildings in the societies concerned. Again, as in the discourse on the western vernacular, what is identified as vernacular architecture are the 'authentic' traditional buildings that, still today, are seen to form part of some indefinable ethnographic present that preceded the modern present. If the influences of new and modern traditions are dealt with, it is often in negative or, not infrequently, derogative terms (e.g. Bourgeois 1989). In general the perception is that the advance of the one inevitably leads to the contamination, destruction and disappearance of the other.

As Upton notes, this persistent predisposition of scholars of the vernacular to focus on the past and tradition and, more particularly, the emphasis on narratives of cultural decline and damage that tends to accompany it, does not stand on its own. Daniel Miller (1995: 264) has observed how, during the mid-twentieth century, anthropological monographs often ended with a chapter on social change. Typically these chapters reported on the introduction of modern (usually western) practices, ideologies and, especially, consumption goods, and discussed the negative impacts that their arrival had on the perceived integrity and authenticity of the cultures concerned. Like Rice, the authors of these monographs presumed the bygone existence of a traditional era in which indigenous cultural traditions were still pure and authentic; an era, that is, against which the advance of modern western influences could be juxtaposed. Marshall Sahlins (1999) has recently shown how this perspective on cultural authenticity and change is intricately bound up with what he calls the despondency theory. Particularly dominant in anthropology during the mid-twentieth century, this theory posed that traditional cultures, once they had been brought into contact with the west, would irrevocably fall into despondence, as a result of which indigenous social structures, practices, values and beliefs would soon and inevitably decline and be lost. As a manifestation of western modernist attitudes towards history, culture and tradition, and as a logical successor to the social evolutionist theories of the late nineteenth century, it posed that, ultimately, traditional cultures would have to become modern and 'just like us – if they survived' (Sahlins 1999: iii).

And of course, no one can deny that such processes of cultural assimilation, decline, conflict and loss have all too often taken place in the time of western expansion and colonization. Nor, indeed, can anyone deny that they have been common in the post-colonial period of modernization or that they are still widespread in the current era of globalization. Yet, Sahlins notes (1999: ix–x), as real and widespread as they may be, they do not necessarily make up the whole story, for next to the unmistakable tales of woe there are also tales of cultural persistence and vibrancy. Thus he gives the example of the Siberian Yupik on St Lawrence Island who, despite an increased incorporation into the world capitalist system and the introduction of modern means of production, transportation and communication still maintain their hunter-gatherer culture and basically 'are still there – and still Eskimo [sic]' (1999: vii). Rather than having succumbed to the pressures of modernization and development, as early ethnographies predicted, the Yupik have adapted their culture by incorporating certain modern elements (mainly technological ones), while maintaining or even strengthening traditional others; as such further developing a culture that is, ultimately, still distinctly Yupik. They have, in Sahlins' words, indigenized modernity by creating their own cultural niche in the global scheme of things. And, as Sahlins notes, the Yupik case does not stand on its own. Nor, indeed, is it an example that is characteristic of or restricted to the late twentieth century only. As Eric Wolff (1982) has so convincingly shown, contacts and interconnections between cultures have taken place in the past as much as in the present, and most, if not all cultures in the world are the result of processes of encounter and confrontation, as well as of cultural borrowing and merging.

A major shortcoming of much of the current vernacular discourse, I believe (especially that dealing with western traditions), is that it does not really acknowledge this processual, heterogeneous and adaptive character of cultural traditions. Overall it still tends to regard vernacular traditions as homogeneous, passive and rather static entities that can be classified into bounded geographical, chronological and typological categories (Cotswold, colonial, yurt), and that may consequently be lost in the encounter with other, more active, modern traditions. In doing so, the processes of cultural interrelation, merging, change and indigenization that have been increasingly acknowledged in disciplines such as anthropology, cultural geography and history are largely ignored. Effectively, much of the discourse still focuses on the study of building traditions in particular regions or, especially in the case of the western vernacular, in specific time periods. When discussions of change are entered into (as, admittedly, they frequently are), these are often restricted to analyses of typological transformations in time or space or, in the case of changes resulting from encounters with other (usually 'modern') traditions, they tend to be set off against an ideal and dehistoricized past or ethnographic present. Like the cultural traditions in the anthropological writings of the 1950s and 1960s, in most of these cases the vernacular is defined as a separate category consisting of traditional buildings that may be opposed to modern ones and that are in danger of losing their authenticity and integrity when confronted with the impacts of modernity. Many are the authors that, like Rice, still describe the arrival and incorporation of new technologies, materials, uses and meanings in confrontational terms, instantaneously viewing them as the beginning of the end of a distinctive (vernacular) era rather than as an active adaptation and continuation of a living tradition. The vernacular and the modern, it seems, cannot go together.

Yet they *do* go together and merge, right at this moment and all around the world, in all kinds of different and sometimes surprising ways; just like cultural transfers and exchanges have always taken place in the past and will undoubtedly continue to do so in the future. And the result is the contemporary emergence of all kinds of new and adapted traditions that, though different from the ones that preceded them, are authentic in their own right and that, I will argue below, can still be seen as vernacular. After all, many traditions that are now seen as vernacularly authentic in fact evolved out of the amalgamation of different traditions. North American log cabins, Nootka (Nuu-chah-nulth) plank houses, Maori *whare hui*, fired brick vaults in Afghanistan and Pennsylvania barns, to name but a few, all, in one way or another, result from the encounter of different cultural traditions (Weslager 1969; Marshall 2000; Austin 1997; Szabo and Barfield 1991; Ensminger 1992). Many other examples of architectural adaptation, borrowing, hybridization and amalgamation can be given. In all cases one now speaks of truly vernacular traditions, yet none of the traditions concerned has a static, isolated or homogeneous history. As in the case of the Yupik, over time outside influences in terms of technology, use, resources or form have been incorporated into existing building traditions, as such further developing the latter or creating new hybrid or creolized forms. The outside influences have, in Sahlins' terms, been indigenized or, perhaps more appropriately in this context, 'vernacularized'. Of course, I do not want to deny that in the course of time, in the process of such cultural encounters, many distinctive building traditions have been lost, or that they still continue to do so at present; once more, however, such examples do not necessarily make up the whole story.

De-reifying the vernacular

As Upton (1993: 10–12) has argued, the narratives of decline and loss that characterize much of the discourse on vernacular architecture are traceable to and reflected in the way in which the concept of the vernacular has generally been interpreted and defined. Though the diversity of traditions makes it difficult to draw up a single definition (see Oliver 1997b: xxii), there are a number of elements that seem common to most, if not all, current understandings of the concept. At the heart of these, Upton notes, are a number of 'us' and 'them' dichotomies that serve to define the vernacular in opposition to categories like the formal and, especially, the modern, and that essentially relegate vernacular traditions to a time and space that is distinctly different from the latter. Thus the vernacular is generally said to be the culture of 'true' communities that, like Rice's pre-industrial British villages, are supposedly largely homogeneous, socially cohesive and, not infrequently, agriculturally based. Besides it is often understood to embrace the building traditions of the people rather than those of the elite and is generally, as noted, seen as stable, passive and instinctive rather than as changing, active and conscious (Upton 1993: 10–11). Added to this may be the widespread notion, recently reiterated by Steen, Steen and Komatsu (2003), that the vernacular is generally built using natural and ecologically sustainable resources and technologies rather than with the manufactured materials and mechanized means that characterize much modern architecture. In essence, then, the vernacular is seen to belong to a distinct time and space in which, in Henry Glassie's (2000: 49) words, there is an active engagement with nature and other people; it is part of 'a discrete, pre-lapsarian arena of social experience that lies just beyond our own experience and can never be directly accessible to us' (Upton 1993: 12).

Glassie (2000: 20) recently rightly noted that, in giving the vernacular a name, it has been given an existence. Having isolated vernacular architecture as a field of study, the importance to architectural history of long neglected non-monumental and non-western building traditions has been increasingly recognized and acknowledged. Yet, following Upton, I would argue that in the

Marcel Vellinga

process of naming and defining, the vernacular, as a category, has become reified. To speak with Wolff (1982: 3), a name (a residual category of buildings that, ultimately, are not fundamentally different from other forms of architecture) has become a thing. In the worthy pursuit of recognition for building traditions other than the so-called Great Ones, a distinctive and bounded category has been created that can be opposed to other categories such as the formal, the modern, the popular and the informal; categories that are themselves in fact as much reifications as the vernacular. Unfortunately, in doing so, what those involved in the field set out to achieve was partly lost. By interpreting and defining the vernacular in terms that oppose it to the modern, the category has essentially been referred back to a pre-modern past, notwithstanding the repeated reminders that, at the beginning of the twenty-first century, the vernacular still comprises the vast majority of buildings in the world. Consequently, contemporary building traditions largely neglected by architectural history (for example, suburban houses, squatter settlements, self-built 'counter culture' architecture) have also been ignored in the field of vernacular architecture studies. Besides the interconnections between those traditions identified as vernacular and those that are modern, or popular or informal, and the new traditions that emerge from their creative amalgamation, are not really incorporated into the discourse.

Because of the reified nature of the definition, all changes that take place to the vernacular in the present will automatically be seen as cultural decline and a loss of authenticity. If a building is to be truly vernacular, it will have to be part of a cultural context (pre-industrial, rural, socially homogenous and self-contained) that, in contemporary times, will be ever harder to find. In the process of definition then, though crucial in terms of the recognition of vernacular traditions as forms of architecture, the vernacular has effectively been banished to the pre-modern past by those who championed it, while simultaneously, by not really allowing for change, it has been denied both a history and, indeed, a future. In doing so, the dynamic indigenization or vernacularization of outside (modern, formal, global) cultural influences has largely been ignored. Furthermore the common and persistent stereotypes about vernacular architecture are confirmed, which in turn further affirms the ambivalent status of the vernacular and strengthens the perception that it is irrelevant to the future. With the unstoppable advance of modernity, the vernacular field of study finds itself in a serious predicament, getting smaller and smaller every year. A disappearing world, indeed.

Yet, as noted, vernacular traditions have not all vanished, but in many cases have merged (just like they have always done) with modern ones to create new manifestations of tradition or localized hybrid forms that better suit current circumstances and requirements. What is needed therefore, I believe, to evade the predicament of the vernacular, is to break free from the limitations of the current conceptualizations by adopting a more dynamic interpretation that more explicitly recognizes the ways in which old and new building traditions merge, adapt, combine and, in the process, become vernacularized. The vernacular, in other words, needs to be de-reified. Rather than treating it as a category that consists of buildings that, as static objects, can be categorized in neat types and periods and that, concomitantly, may be more or less real or authentic depending on which type or period they belong to, the dynamic and processual nature of buildings, and of the building traditions of which they form part, should be the starting point of analysis. As has in more general terms been observed by Kopytoff (1986), all buildings (whether vernacular or modern, bicycle shed or cathedral) are constructed, modified, renovated, updated and, ultimately, demolished. Throughout this process, the function, use and meaning of the buildings changes continuously. Each building therefore has its own biography or life history, which is written in line with the changing circumstances, expectations, insights and possibilities of their owners.

Similarly the building traditions that comprise individual buildings constitute processes of continuous change and adaptation. Despite persistent popular conceptualizations of a tradition as a package of ideas or practices that is handed down from generation to generation (and the associations of stasis that continue to accompany the concept), it is difficult to maintain that a tradition constitutes a bounded, de-personalized and unchanging body of knowledge or customs. Rather than as a fixed entity that exists independent of the people that transmit and live by it, a tradition is best regarded as a continuous creative process through which people, as active agents, negotiate, interpret and adapt knowledge and experiences gained in the past within the context of the challenges, wishes and requirements of the present. In this process (which Abu-Lughod (1992), following John Turner, referred to as 'traditioning'), existing ('traditional') skills, know-how and practices may be straightforwardly applied to deal with contemporary challenges, but they may also be adapted to better suit current needs, or indeed be rejected because they are no longer perceived as being relevant or useful. In the latter instances, it is not so much that the tradition is lost, but rather that is has been adjusted to comply with current circumstances. Tradition, then, is a process of active regeneration and transformation of know-how and practices within a contemporary local context, that does not exist on its own or apart from the people that transmit it.

Given this fundamental processual nature of buildings and building traditions, it is of course difficult, if not impossible, to identify and classify buildings as truly vernacular, informal, modern or popular. For what may be regarded as vernacular from an outsider's point of view (a west African palace, say, or a Japanese shrine), may be monumental or formal from an insider's perspective, while particular elements of traditions (for example, technologies, resources or forms) may be shared by popular and formal buildings, or by vernacular and informal ones. Conversely, what is generally seen as one of the quintessential popular house types in the United States and western Europe (the bungalow, a building type that, incidentally, evolved out of the conjunction of Bengal and British colonial traditions) can just as well be described as a 'suburban vernacular' (King 1995: 152). In a similar way, it is difficult to compartmentalize individual buildings, as what may initially be considered a vernacular building, can in the course of time be subjected to so many changes in its construction and use that it no longer fits the current definition (a sixteenth-century timber frame house in an English high street, for example, repeatedly renovated and now used as a mobile phone shop). Again, the same could well be said for a popular house type such as the British suburban semi, many of which in the course of time have been subjected to so much individual adaptation that they may arguably be said to have been vernacularized (Oliver, Davis and Bentley 1981; Oliver 1984).

Engaging the future

As noted, the current pre-disposition of many scholars in the field is to regard as inauthentic, damaged or contaminated those vernacular buildings and traditions that show influences of modern, popular or formal traditions. Yet to do so is to deny the processual nature of buildings and building traditions, as well as that of the application of meaning, and indicates that many of those involved in the vernacular discourse are still unable to deal with the conjunction and transformation of traditional and modern elements that characterizes much of the world's architecture at the beginning of the twenty-first century. The dominant tendency remains to keep the vernacular from development and change, whereas what we should be doing, I believe, is focusing on the way in which those traditions that are now called vernacular actively and creatively combine with those called modern, popular and informal to create new buildings that suit contemporary and future requirements, needs and expectations. All buildings, whether traditional, modern or modernized hybrids, are authentic cultural expressions in themselves. Thus a Minangkabau house in West Sumatra (Indonesia) that is built of concrete and provided with a modern bathroom and garage as well as a traditional spired roof is no example of the contamination and decline of a vernacular building tradition, nor is it a 'fake', 'replica' or an 'imitation' of an older timber building. It represents a new phase in the living Minangkabau building tradition and as such it deserves as much attention and admiration as its older and supposedly more 'authentic' counterparts (Vellinga 2004a and 2004b). The same, I would argue, goes for an urban Mongolian yurt provided with a concrete base and electric lights, a Lakota sweat lodge used by both Native Americans and whites or, indeed, a Cotswold barn that is now used as a luxury weekend retreat by a successful stockbroker from London (Evans and Humphrey 2002; Bucko 1998; see also Jolly 1992).

Besides all such buildings should be regarded as vernacular in the sense that they are distinctive cultural expressions of people who live in or feel attached to a particular place or locality, and as such they form part of, or indeed help to constitute the local architectural dialect. A modernized Minangkabau house built by an emigrant living in Java is still distinctly Minangkabau, and a gentrified Cotswold barn owned by a London insurance advisor still distinctly Cotswold, despite the fact that they may serve similar purposes (as tourist accommodation, or holiday retreat) and may have been built or renovated using similar (modern) materials and technologies. The one can only be found in West Sumatra, and the other only in the Cotswolds. Both constitute distinctive cultural artefacts that relate to localized cultural needs, economies and values and that are, in their own and unique way, intimately related to the identities of their owners, builders and inhabitants, using available technologies and materials.

Besides, both have developed out of the amalgamation and hybridization of traditional and modern elements, each having vernacularized the manifold manifestations of modernity in its own distinctive way. It is the way in which these processes of vernacularization take place, not just in West Sumatra and England, but in Mali, Mongolia, the Pacific, Ecuador, Canada and elsewhere that, I would argue, deserve much more attention from scholars of the vernacular. At the beginning of the twenty-first century, it is only by taking seriously the ways in which the historic and the modern, the informal and the formal, the urban and the rural, the indigenous and the migrant, the traditional and the contemporary, and the popular and the monumental combine, interrelate and in the process become vernacularized that, to borrow Upton's (1990: 211) words, a 'more genuine architectural history' can be written.

A focus on the way in which old and new building traditions today merge and become vernacularized opens up a wide and largely unexplored field of research; a field in which current architectural categories, types and periods interrelate and overlap, and which is increasingly inhabited by contemporary buildings that, though unique and authentic in themselves, have still not received the academic and professional attention that they deserve. Writing about the anthropology of material culture and consumption, Miller (1995: 269) called for studies that would expunge the latent primitivism of the discipline by taking seriously New Age Californians, Singaporean puritanism and 'West Africans in suits playing video games', so as to end 'that romantic nostalgia which had the effect of allowing only "western" peoples to be the true inheritors of the industrial revolution'. In a similar way, I believe that the field of vernacular architecture studies will need to break out of the bounds of its own static and, indeed, latent primitivist definitions by acknowledging and taking seriously new and contemporary hybrid buildings such as Native American casinos, Japanese bio-regional eco-houses, contemporary North American log houses, Indonesian cultural heritage centres and lightweight concrete fantasy homes in Mexico, to name but a few (Krinsky 1996; Skinner 2002; Cohen, Yamaguchi and Spengler 2003; Vellinga 2004; Kahn 2004). Integrating such buildings and the emerging traditions to which they belong in the vernacular discourse will increase our understanding of the varied ways in which people build and live, providing better insight into how architecture is fundamentally involved in the constitution of cultural identities and how in time, and interdependently linked to such identities, traditions become established, change, adapt and ultimately endure or disappear. Besides, it will help to discard the persistent images of the past and irrelevance that currently surround the vernacular and, as such, will allow for a more serious incorporation of vernacular know-how in modern and future building projects.

Of course, expanding the scope of the field of vernacular studies by incorporating the dynamic amalgamation of those traditions now called vernacular, modern, popular or formal should not inhibit us from studying or conserving the traditional and historical buildings that have so far received most attention. Such work constitutes a legitimate and invaluable part of the architectural history discourse and by all means should continue to help document and understand the

Marcel Vellinga

rich and diverse architectural heritage of the world; especially since so many distinctive traditions and buildings are today faced with disappearance and destruction. But there is a need to complement these historical and traditional studies with an approach that more explicitly engages the future by looking at the ways in which contemporary building traditions from all around the world creatively adapt to current cultural and environmental contexts and processes. Not to want vernacular buildings and traditions to develop by proposing, like Rice, rigid preservation, and by condemning as contamination or adulteration any change or adaptation is to deny history and to make vernacular traditions all the more out-of-date, irrelevant and destined to disappear. On the other hand, unconditionally assuming that all change is good and to be regarded as a positive achievement is to ignore the manifold negative and, not infrequently, destructive consequences of a lot of modern development, and would inevitably result in a similar sad fate for the vernacular. What should be done to sidestep both of these extremes, is to develop an approach that acknowledges the existence of change and development, but that tries to understand how and why it takes place and attempts to ensure, through critical assessment and engagement, that the changes that are effected are sensible, appropriate and, most of all, sustainable.

Of course, such a dynamic, critical and more applied approach to the vernacular, taking its place alongside the current academic discourse, already exists, and in time has resulted in many important studies and projects (Fathy 1973; Cain, Afshar and Norton 1975; Afshar and Norton 1997). It now seems more important than ever to elaborate on this work, and to try and make it a major component of the vernacular discourse. Roland Barthes once spoke of 'a moment of gentle apocalypse' (Sontag 1982: xxii). Although referring to the literary climate of his time, his phrase seems to encapsulate the era of the early twenty-first century, which to a large extent is characterized by a worldwide, slow and gradual accumulation of both cultural and environmental transformations and problems. The process of globalization, typified by rapid developments in the field of ICT, increased mass consumption, continued urbanization and the growing internationalization of capital, business and power, around the world has led to profound cultural changes and dislocations, to new patterns of ethnic relationships and to the emergence of new hybrid cultures. At the same time it has contributed to increased environmental problems on a global scale, which may well result in a true ecological disaster and are exemplified by a rapid loss of natural resources and species, high levels of energy consumption, and increasing amounts of waste and pollution. It goes without saying that, as a prominent cultural category and a major consumer of energy and resources, the built environment, including the vernacular, is seriously implicated in both processes. As such, there has been a slowly growing interest among architects, planners and engineers in the design of architecture that can address the many environmental, economic and social problems in a sustainable way (e.g. Edwards and Turrent 2000; Williamson, Radford and Bennett 2002).

As Oliver (2003: 17) recently noted, as a source of 'much accumulated wisdom', vernacular traditions have the potential to contribute much to the development of such sustainable architecture. Like indigenous medicinal and agricultural knowledge (Ellen, Parkes and Bicker 2000), vernacular practices, skills and ideas, which have often developed over time as part of a continuous process of trial and error and are as such often well adapted to local climatic and cultural contexts, may offer many valuable lessons and precedents to the scholars and professionals involved in the development of sustainable buildings. Indeed, as a category the vernacular will *have* to be taken serious in this respect, as the vast majority of people in the world currently lives in vernacular buildings and is likely to continue doing so throughout the twenty-first century. Yet, as Oliver notes (2003: 14), in order for such recognition and integration of vernacular know-how in a modern or development context to materialize, much more applied research is needed. An essential premise of such research, I suggest, should be the explicit acknowledgement that all vernacular traditions constitute dynamic and creative processes that result from cultural encounters, borrowings and conjunctions, and that, as such, should be allowed to change and develop. For if the current static and conservative interpretations of the concept that confirm the vernacular stereotypes of a backward past and underdevelopment are maintained, the incorporation of vernacular traditions in new and modern projects is not likely to succeed or, indeed, to be taken seriously.

As is shown by the work that already has been done (e.g. Fathy 1973), an approach that focuses on the active application of vernacular technologies, forms and resources in a modern and development context will not be without its problems, challenges and setbacks, and will have to address themes and issues that so far have been largely disregarded in the field of vernacular studies. For instance, as it will have to engage with, or indeed be part of, the so called 'development discourse' (Grillo 1997), there will be a need for critical discussions of the political and ethical dimensions of key concepts like sustainability, development, intervention and participation. At the same time there is an urgent need for research into more pragmatic issues that have so far not received much attention in the field of vernacular architecture studies, but which are nonetheless crucial in terms of the sustainable development of the vernacular. Forming part of the global 'ecumene' (Hannerz 1989), many of the vernacular builders of today are increasingly confronted with 'modern' issues like planning regulations, climate change, resources depletion, building performance standards, population growth, technology transfer and even, in some cases, mortgage restrictions and insurance criteria. So far, the implications of such practical issues on the way in which vernacular traditions are transmitted have hardly been the subject of research. Yet it is only by critically examining the way in which the vernacular will be able to deal with these contemporary challenges that it will be possible to show that vernacular traditions are not necessarily just anachronistic survivals of an era destined to disappear, but may well have an opportunity to endure and contribute to a sustainable built environment of the future.

The end of the vernacular
Ultimately, like Upton (1990: 210–11) and Glassie (2000: 21), I believe that the category of vernacular architecture will be obsolete. This does not mean that we

should discard of it as yet, because as an analytical concept it is useful and, as long as architectural history remains narrowly concerned with the study of great buildings and architects and does not recognize the importance of all building traditions in the world, it is much needed. Nor should it be taken to imply that the building traditions that are currently referred to as vernacular will no longer be useful or relevant in the future. Indeed, as noted, there is every reason to acknowledge that, as a source of know-how, the vernacular will still have an invaluable part to play in the creation of a sustainable built environment for all.

What is important in order to assess such a future role, however, is the explicit recognition that vernacular traditions, like their modern, informal, environmentally responsive or popular counterparts, constitute dynamic and creative processes of development and change. A limitation of much of the current field of vernacular architecture studies is that this active and processual nature is still not really acknowledged, as a result of which disruptive processes of decline and loss are emphasized, while contemporary examples of architectural continuity, revival and amalgamation tend to be disregarded. Although the buildings that are the result of these latter processes undoubtedly differ from the ones that preceded them, often combining traditional elements with modern ones, they are nonetheless distinctive cultural artefacts which, as authentic expressions in their own right, are uniquely related to the particular cultural and environmental context in which they are found. In that sense they are still vernacular, or in any case the outcome of the local vernacularization of modernity.

Recognizing the existence of these contemporary vernacular or vernacularized buildings alongside the historic and traditional ones that so far have been the main focus of the vernacular discourse, opens up a wide field of research; a contemporary, varied and exciting field in which new and enduring building traditions continue to come together in creative and new ways. Paying attention to this dynamic field will help to change the current 'thatched cottage and mud hut' image of vernacular studies and is a necessary first step on the road that needs to be taken if the incorporation of vernacular know-how in modern and development practice is to become a reality.

Ultimately, engaging the creative cross-fertilization of architectural periods, categories, ideas and practices, it may also enable a more comprehensive architectural history to be written. Such an architectural history, when it emerges, will have to accept that all building traditions in the world, regardless of their prefix or location, come into being, develop, adapt, combine, endure and disappear, and will no longer need to identify rigidly bounded and static categories to distinguish buildings that, in the end, are not fundamentally different from one another. It is only when the vernacular, like the modern, the popular, the colonial or the informal, has become analytically obsolete, that its existence and importance will truly have been recognized.

Part II

Learning from the vernacular

Chapter 5

Traditionalism and vernacular architecture in the twenty-first century

Suha Özkan

Introduction

The study of architecture and its theory has posed a continuous dilemma over the centuries. Especially after positivism in science and philosophy prevailed, architecture remained rather crippled, placing itself neither as an objective enquiry as in science, nor as a speculative and personal form of expression as in the arts. As far as its theory is concerned, architecture as a field of knowledge and as an area of practice occupies perhaps the most obscure place among both the sciences and the arts. It is a field of enquiry where the end product, the building, is formed by a process of design and is based on experiential values. Although the latter may be considered a reflection of the 'soft' values of art, they are implemented through the hard and verifiable knowledge of applied sciences, i.e. engineering. Therefore, in the tree of knowledge of the philosophy of science, architecture falls between the hard and soft realms of scientific enquiry. To be objective and therefore scientific is an ambition of positivist thinking in the design process. However, when the so-called scientific pursuit is over-exercised, the meaning of the process, one that relies so heavily on creativity and psychological factors, is reduced.

Throughout history, architecture has informed its practice simultaneously in terms of values and aesthetics, which have been recognized as the mission of the profession, and in terms of safe and correct building methods. Therefore the theory of architecture is a collection of disparate contributions that combine the ideas, missions, assertions and approaches of many individuals. In the end, the theory is actually the literature that expresses and externalizes findings, convictions and manifestoes in literary form. When there is a prevalent agreement on what is 'good', or 'beautiful' for that matter, the theory becomes instructive and valid. When these agreements are challenged, alternative movements arise and spawn change.

Throughout history, we have witnessed distinct eras when these common agreements were effective among architects. The discourse of Ancient Greek and Roman as well as Renaissance architecture was based on these openly expressed theoretical assertions. However, this open transmission of knowledge was challenged by Gothic architecture, when the original theoretical premises being developed were retained by master masons as trade secrets, and only discreetly transmitted orally to following generations, in order to safeguard the know-how of the epoch (Rykwert 1988: 31–48). Subsequently, Neo-Classicism and Modernism returned to what had previously prevailed, enjoying open agreements to guide the profession. The theoretical discourse consisted mainly of proliferations and of variations on agreed valid principles of architecture.

Since the beginning of the twentieth century, Modernism has encapsulated and declared itself within two slogan-like assertions: those of Adolf Loos, 'Ornament is crime' and Louis Sullivan, 'Form follows function'. Modernism, addressing the aesthetics of industry and mass production, developed in leaps and bounds and became the uncontested *lingua franca* of architecture. It not only flourished in many other expressions of art, but also became the political manifestation of 'progressive' thought. Although this 'progressive' tide coincided with the worldwide spread of uniform, tedious and uninteresting buildings and urban environments, it would be unfair to blame Modernism for this, as these undesirable consequences were mainly the result of profit driven enterprises. Nevertheless, the distorted use or abuse of Modernism has been regarded by many as a set of values and premises that lacks respect for cultural identity, historical continuity and climatic relevance.

The reaction to Modernism took many forms, starting with post-Modernism, a short lived movement, inspired by shallow ethics. Unlike Modernism, which found expression in other fields like painting, sculpture, music, dance and industry, post-Modernism emerged within architecture and spread to other fields rather thinly. The notable exception to this is in the field of literature, where post-Modern structure was employed by many authors to great effect, significantly enhancing the quality and breadth of contemporary writing. The post-Modern movement was tirelessly preoccupied with reflections on meaning, historical continuity and the expression of identity.

From another angle, a 'revolutionary' movement known as the 'Architecture of Freedom' became a widespread repudiation of any control. Here Modernism was not revered. This was a denial of the need for political and planning control over building practices by the people themselves, who took the initiative to solve their own, primarily housing, problems. This movement based its discourse on *de facto* construction, generating vast settlements like the *favellas* and *barrios* in South America, the *basti* in the Indian subcontinent, the *prosphika* in Greece, the *bidon-villes* in North Africa, the *kampung* in Indonesia, and the *gecekondu* in Turkey, all of which took off in the aftermath of the Second World War as 'people's solutions' for housing in rapidly urbanising countries. Recognition of this form of building as a solution that challenged the incapable institutional (or, for that matter, Modern) architecture, one that did not cope with the dynamism of the vast post-war demand, found its catch phrase in John F.C. Turner's declaration: 'Freedom to build' (Turner and Fichter 1972; Turner 1976).

The third and perhaps most influential reaction is traditionalism. Traditionalism's focus on research in vernacular architecture and the revitalization of traditional building practices placed it critically in the centre of architectural theory.

Vernacular architecture research

Many have contributed to the research on vernacular architecture, including Bernard Rudofsky (1910–87) and Paul Oliver – two pioneers in the subject's recent history.[1] Rudofsky did not have an academic interest or similar pursuit when, in 1964, he put together an exhibition of traditional architecture at New York's coveted Museum of Modern Art. The title itself was a challenge to the profession: 'Architecture without Architects' (Rudofsky 1964). Indeed, Rudofsky did not address the familiar discussion about whether or not builders who create spectacular architecture ought to be considered architects. Instead, he provokingly brought to the fore and introduced onto the agenda of world architecture an area of architecture that had, as yet, gone largely unnoticed and which used to belong solely to a rather hidden field of academic architectural research. Suddenly, edifices that had been kept within the field of interest of human geographers, folklorists, anthropologists and architectural scholars became a subject of wider architectural interest.[2]

Paul Oliver, an artist whose main research interest was the origins of Blues music, discovered an enormous wealth of architectural expressions. From among many qualifications for this form of building, he borrowed the term 'vernacular' from linguistics and opened a vast field of exploration. In his first book, *Shelter and Society* (Oliver 1969), he quite eloquently and thoroughly places vernacular architecture within the discursive history of architectural theory. He orchestrated the subject by generously voicing the research of a wide spectrum of, particularly young, researchers worldwide, including myself, allowing me to present one of my first research papers to an international readership. Oliver crowned his dedication to the subject with a three-volume *magnum opus*: *The Encyclopedia of Vernacular Architecture of the World* (Oliver 1997a).

Since the early 1970s, thanks to Oliver and his followers, research in vernacular architecture has developed into a respectable academic field and has yielded extraordinary findings. Vernacular architecture research filled the biggest vacuum within architectural theory: the lack of laboratory conditions within the theory of architecture, which previously had prevented the discipline from deriving valid and verified knowledge from cases and field studies. Research done in the realm of vernacular architecture embraces social, economic, cultural and technological aspects of architecture and describes and analyses how buildings emerge and are sustained through cultural processes. The research carried out on this premise also seeks to understand the historical depth behind architectural formation and development, together with the cultural and environmental factors that are involved. As one might expect, vernacular architecture research, a truly multi-disciplinary field, not only expanded the scope of architecture but also provided niches of academic interaction, offering faculty members the opportunity to work with people from departments outside their own. This led to the emergence of a new field of academic enquiry, which defined its scope as architectural anthropology, using the research techniques of anthropology, enriched with those of architecture (e.g. Amerlinck 2001). In short, vernacular architecture research became a mainstream academic activity, using processes of objective analysis and evaluation for architecture by displaying its determinants in their contextual entirety and historical continuity.

Until Paul Oliver's appropriation of the term 'vernacular', architecture that evolved from within communities and perfected itself with the test of time in conformity with societal, climatic and technological conditions was referred to by many different terms. The term 'traditional architecture' emphasized a process that had culminated in built form, one that is sustained by tradition as the binding tissue of that particular society. For those who named it 'primitive architecture' (Guidoni 1975), it meant that the architecture in question contained the basic necessities of society in their simplest form. When it was referred to as 'folk', it signified that the architecture formed part of the ethnographic premises. The term 'indigenous' architecture regarded this form of building as a particular and original aspect of building in a definable geographical setting. This type of architecture was also referred to as 'anonymous', given that the buildings did not have any significantly determinable architectural authorship. Finally, in the same vein, the term 'un-institutionalized architecture' was used in some academic discourse to define the same phenomenon. Thus, the term vernacular defined the subject by embracing its entirety, including the complexities of societal and cultural processes (Oliver 1969).

In the relatively short period of the three decades since then, hundreds of academics have explored vernacular architecture within their reach and presented their research in international fora. One of the leading media for sharing the outcomes of vernacular architecture research was established by Jean Paul Bourdier and Nezar AlSayyad as a part of the Centre for Environmental Design Research at the University of California, Berkeley. Their first international symposium, entitled 'Traditional Dwellings and Settlements in a Comparative Perspective', in 1988, brought together more than one hundred papers and as many scholars, sharing a similar interest. Their seriousness of purpose was made clear as these symposia continued to take place periodically in many different countries and on a range of different themes, making this newly founded institution an important medium for interaction and the exchange of findings on this particular subject.

Vernacular architecture in the history of theory

Throughout the centuries, those who have theorized on architecture have attempted to define the genesis of architecture or the point zero condition from which architecture originated and developed. The earliest known architectural treatise that deals with this issue is by Antonio Filarete di Averlino (Filarete 1965). In his well-illustrated treatise, he focuses on that episode in human history in which the need for shelter first emerges. Although his work concentrates on the origins of architecture, Filarete's initial aim was not to write a book on the subject, but to foster interest and try to persuade Lord Sforza of Florence to build a new ideal town named Sforzinda, by relating his vision of architecture and urbanism in the form of evening stories. In concordance with the monotheistic faiths, human life on earth is explained by the Biblical legend of Adam and Eve and their eventual expulsion from Paradise, where perfect conditions precluded any need for shelter. Only on expulsion from Eden did the prophetic couple and, thus, humankind, encounter the harshness of earthly conditions and climatic realities, and thus the need for protection. A simple shelter as a means of protection for survival during this period is Filarete's explanation of the genesis of architecture.

Filarete does not take a ready-made set of circumstances as the generic mode for his theory. Instead he returns to the basic necessities of refuge for human beings in order to discover the origins of architecture. From a religious perspective, the point of genesis comes from Adam and Eve's need to protect themselves as they are driven out of Paradise. At this point, Filarete refers to the sticks and leaves as the most original and pragmatic aspects of building. Filarete's point of departure, from the essential as opposed to the formal conditions, aims to have a sound, generic and unquestionable beginning for architecture and its theory. Unfortunately, others did not share this line of thinking until as late as the mid-eighteenth century, when a similar understanding was defended by the French architectural theorist Marc Antoine Laugier using the concept of *beauté primitif*. Laugier took the same primeval existence and named it 'primitive beauty', prior to his elaborate discourse on the 'high' architecture of his time.

In his most important and probably most influential works, the eighteenth-century French theorist Laugier bases his discourse on the rudiments of architecture. His two editions of *Essai sur l'Architecture* (1966), which were released in 1753 and 1755, and his *Observation sur l'Architecture* (Herrmann 1966), twelve years later, reflect the mainstream theory of the period.[3] This is not surprising given that Laugier essentially adhered to the line of classical theory, only rarely diverging from it. Nevertheless, his written work remains a valuable contribution to the most up to date establishment of classical, normative, canonic aspects of architectural theory.

This judgement does not aim to underrate his various efforts. Laugier, for instance, derived the basic rules or reasons for the existence of architecture by referring to the primitive hut, a popular reasoning even in today's architectural discourse. Laugier's point of departure for architectural principles, which is reflected in the frontispiece of *Essai*, is probably the same generic pattern followed by those who search for the governing and generic patterns of architecture in the vernacular mode of building. Laugier asserts that the first step in architecture begins with four sticks and beams to connect them, i.e. the rustic hut. The hut becomes a generic, iconic model for architecture from which many issues have been derived. In that respect he reminds us of Filarete's story of Adam and Eve's expulsion from Paradise.

By taking the 'rustic hut' as the generic model for architecture, Laugier makes a revolutionary leap from the Renaissance analogy that deduced all its laws and principles from 'Man'. Laugier's contribution is extremely important, because it brings reason as opposed to dogma into architectural thinking. Primarily, the Vitruvian model in which a creation of God, i.e. man, is analogous to man's own creation, i.e. architecture, is replaced by a point of departure where ideas and valid results are generated from an archetypal solution to man's necessity for shelter. Laugier takes an analytical view of the subject, as opposed to an 'unquestionable dogma' of an analogy, which would be impossible to validate.

When we compare Laugier's 'rustic hut' with Filarete's archetypal shelter with 'four posts', we may conclude that Laugier takes up a pre-Renaissance position on the genesis of architecture. But he does so with a difference in purpose: he intends to derive the basic principles of architecture analytically instead of pandering to the scholastic dogma of the Renaissance, a dogma that arose from prescribed analogical paradigms.

Contemporary architectural discourse and traditional building practices

As previously mentioned, since the beginning of the twentieth century, the prevalent paradigm in architecture has been Modernism. Even though it dominated both educational theory and the practice of architecture, reactions to this prevailing 'ideology' have never subsided. Over time, Modernism, with its strong roots, has become diversified and has developed a plurality of its own. We may categorize this plurality within the following seven groups.

The first group comprises those who unquestionably adopt the minimalist architecture of Ludvig Mies van der Rohe and the Modernist principles of Walter Gropius as they were initially set out and declared. Minimalist modernism as an expression of progress claimed to be valid for any geographical or cultural context. Gradually it became not only a conviction but a lifestyle, and even a political attitude towards the built environment. Architects who committed themselves to this approach have adopted it as their mission. Minimalist internationalism in time became the most challenged and criticised aspect of Modernism, as it not only ignored the cultural and climatic aspects of life, but also ventured to reform them.

The second group of architects consists of the followers of Le Corbusier and his simple and sublime expressions, explained by his affinity for the Mediterranean. This opened up avenues to explore a valid Modernism for different cultural and specific climate settings. Perhaps the most prominent architect in this respect was Alvar Aalto, who developed a new Modernism specific to Finland, without making any compromises to Modernist principles. Luis Barragan, Geoffrey Bawa, Tadao Ando, Charles Correa, Balkrishna Doshi, Rafael Muneo, Ricardo Legoretta, Alvaro Sisa and Sedad Eldem can be mentioned among the hundreds who are committed to culture- and climate-specific modern architecture, which has been referred to as 'modern regionalism'.

The third group of architects can be called 'new moderns'. They do not diverge from most of the principles of Modernism, yet they do not take 'function' as the basic determinant of form. On the contrary, they believe that when function is underplayed, a huge area opens up for free expression by engaging many means and techniques of contemporary design. Architects such as Frank Gehry, Zaha Hadid, Peter Eisenman, Wolf Prix, Renzo Piano, Daniel Liebeskind and Santiago Calatrava are generally held to belong to this group.

The fourth group embraces those who made the visions and the ambitions of the Archigram Group a contemporary reality: the use of the most sophisticated contemporary technologies and the highest possible precision for building. Jean Nouvel, Norman Foster and Richard Rogers are the pioneers who sustained this line of work and designed notable buildings of our times.

The fifth group is made up of those who from the outset regarded Modernism as a rigid canon that would limit original and creative expression. In their opinion, architectural design should not obey any limitation of expressiveness. Talents like Antonio Gaudi, Hans Scharoun, Paolo Soleri and Bruce Goff took risks by not taking part in the mainstream. They remained on the margins and maintained a vivid line of opposition to all, obliterating modernity.

The sixth group of architects blindly obeyed Modernism, but did so lightly and with creative cynicism. They regarded the realities of life, as they manifested themselves in building, very seriously. Their main opposition to Modernism can be seen principally by their lack of symbolic references to identify buildings within their context, their past and with the aspirations of their clients. Theoretically led and enriched by Charles Jencks, this reaction became a short-lived movement spanning the two decades from the 1970s to the 1990s, and gave rise to many novel expressions. Charles Moore, Robert Venturi, Paolo Portoghezi, Aldo Rossi, Rob Krier, Rifat Chadirji and Richard England are mentioned as the proponents of post-Modern architecture.

The seventh group are the conservatives who believe in the wisdom and accomplishments of the past and commit their careers to perpetuating history. Even though this group shares a similar discourse, social aspects divide them into two fundamentally different sub-groups. On the one hand, classicists who enjoy the royal support of Prince Charles, and who have been represented by Quinlan Terry and Leon Krier, believe that whatever was built in the past is good enough for the urbanized world to repeat in most loyal form. By doing so, they believe, we respect our architectural heritage and enjoy more culturally relevant urban environments. On the other hand, traditionalists, who in essence hold similar aspirations, have a mission that is more geared to the rural environments and the use of appropriate technologies.

Suha Özkan

Traditionalism and Hassan Fathy

Traditionalism in architecture cannot be discussed without taking a close look at the Egyptian architect and activist Hassan Fathy (1900–86), who single handedly challenged Modernism.[4] Through his discourse and influence, Hassan Fathy has come to occupy a 'saintly' position in the world of architecture, even though he has not enjoyed a similar degree of success or recognition with regards to his own architecture. His honesty and determination have made him a hero among generations of architects who, like him, respect the social concerns present in architecture, and its mission.

Hassan Fathy's strife and commitment to architecture was an endless battle and has endured long after his death. He symbolized his own convictions by keeping a statue of Miguel de Cervantes's timeless character Don Quixote in his bedroom as a continuous reminder of man's aspiration to help those less fortunate. In literature, Don Quixote has become an icon of honesty, conviction, perseverance and a continuous struggle against power and those who possess it. The symbolic association between Fathy and the protagonist was very clear, and was understood and acknowledged by his visitors over the years. Fathy also dedicated his life to an uncompromising struggle against the prevalent forces of internationalism, which he felt were becoming the architectural reflection of 'modern society'. He regarded internationalism as a forceful intrusion that obliterates the meaning and social consciousness of architecture. He was cognisant of the fact that he was engaged in an uphill battle, but took little notice. He persevered throughout and in the end left us with an enormous legacy. In time, he had radicalized his position and turned it into a battle against the totality of institutionalized architecture. After seeing a building with a curtain wall and mirrored glazing, he once remarked to me: 'Look at the architect. He is so much in shame with his own design that he only dares to reflect the architecture around it.'

The intrinsic creativity, modesty and dedication of Fathy and his architecture have never been denied, even by those who do not share his vision. However, his failure to garner support from the social forces in which he had confidence has been widely used as proof that the traditionalist approach used and pioneered by Fathy does not work. This was clearly illustrated when two of his rural resettlement projects, New Gourna Village and Bariz Village, failed to succeed, even though this disappointing outcome was due in large part to pre-existing social and economic realities, rather than to the settlements themselves. The inhabitants of Gourna had originally based their livelihood on local clandestine archaeological digs, and so when prompted to move, they refused to do so as they did not want to risk losing their only source of income. This refusal to resettle scarred Fathy's work methods, which relied heavily on communal participation. In fact, Fathy had ingeniously re-mastered the traditional Nubian vaulting system of mud bricks without using any wooden formwork, instead stacking bricks at a slight angle and resting them against a wall to form vaults. Although this was a very simple construction technique, it yielded well-insulated, comfortable spaces and impressive, robust architectonics for architects and non-architects to admire.

'Building with people', Fathy's main thesis, did not hold for Bariz Village either.[5] In the end, Fathy was criticized not for his architecture per se, but for the negative reaction of the communities involved. In short, the very people to whom he had dedicated his mission very sadly rejected him. Naturally this was not a deliberate reaction on their part, or one targeted specifically at him, but the indirect result was a betrayal on behalf of those whom he loved and wanted to help the most. Fathy's priority had been to improve the living conditions of the poor by using appropriate architecture, but the outcomes of his endeavours led him instead to the design of exquisite stone masonry villas for the well-to-do intelligentsia of Cairo. Cynics, envious of Fathy's international reputation, voiced this contradiction with slogans such as: 'He is writing about the poor and building for the rich.' However, they deliberately ignored Fathy's underlying belief that if the leaders of Egyptian society would appropriate and make good use of his architecture for their high aspirations and lifestyles, the underprivileged would follow suit. But this dream did not materialize, at least not in his lifetime.

In the end, Fathy divorced his beloved wife over a trivial argument (the music of Johannes Brahms) and dedicated his life to his powerful discourse and intellectual existence. His residence, which he shared with more than thirty cats, was a portion of an old Mamluk house next to Saladin's Citadel on Darb el-Labbana, and soon became frequented by many who admired his mission and wished to benefit from his wisdom. His permanent appeal to those interested in talking to him, or at least listening to what he professed, took the form of an invitation to join him for tea every afternoon. From the 1960s to the late 1980s, having tea with Fathy was considered a deed, a responsibility or perhaps a ritual not to be missed. He was possibly the only internationally acclaimed person for centuries whose door was open to everyone. His outward reach and hospitality linked him with people all over the world – which is how I met him for the first time in 1969, and then many times after. Fathy had clearly become an undeclared *guru* for alternative architectural discourse.

Fathy's followers

Many who deliberately chose his discourse and benefited from his talks became important architects and activists, perpetuating his mission in their work. Among these, many have received the prestigious Aga Khan Award for Architecture including: André Ravereau (Mopti Medical Centre, Mali, 1980), Abdelwahed el-Wakil (Halawa House, Egypt, 1980, and Corniche Mosque, Jeddah, Saudi Arabia, 1989), Jak Vauthrin (Pan African Institute for Development, Ouagadougou, Burkina Faso, 1992) and Fabrisio Carola with Vauthrin (Regional Hospital, Kaedi, Mauritania, 1995). When classicism dominated architectural discourse in the late 1980s, the Egyptian architect Abdelwahed el-Wakil always referred to Fathy's prophesies with pride. He was also among the pioneers of the classicist movement, along side Quinlan Terry and Leon Krier.

As important as these individual accomplishments in architecture may be, groups have been formed that have institutionalized Fathy's teachings and refined his mission in order to use architecture for human, social and economic development. One of Fathy's dedicated followers, John Norton, together with Alan Cain and Farokh Afshar, established the Development Workshop (DW). This group coupled international consciousness and economic support with local needs and technologies, blending all with architectural wisdom. They have since become one of the outstanding forces in socially responsible architecture, a helping hand working to bring architecture into the social realm.

In addition to its social and technological concerns, DW has distinguished itself by commitment to the protection of nature, by developing materials and methods of construction that take into consideration the scarcity of natural resources. DW started work in the mid 1970s in Iran and later became involved in Angola, Vietnam, Mali and Mauritania. In each of its projects, DW developed innovative and appropriate technologies based on abundant and readily available natural materials. Moreover, in order for the projects to be successful in the long run, they regularly implemented very thorough training programmes for local builders. Over the years, DW has received funding from many international organizations and continues to enjoy the support of the international community in general, who wholeheartedly share the wider environmental concerns.

Among the DW projects is 'Woodless Construction', an initiative not to use scarce wood in geographical areas of sub-Saharan Africa that are susceptible to or already suffering from desertification. Supported by the United Nations Organization for the Conservation of Nature, 'Woodless Construction' in essence is faithful to Fathy's legacy and teachings of Nubian vaulting and doming without wooden formwork construction systems.

Hugo Houben is another follower of Fathy's who frequented his tea meetings for many years. With the help of Patrice Doat, he established CRATerre (Centre for Research in Earthen Architecture), a research centre based at the University of Grenoble that immediately gained the support of Jean Dethier. Dethier was responsible for organizing the legendary earthen architecture exhibition entitled 'Down to Earth', at the Centre Pompidou in Paris, and became the director of CRATerre's architecture section. This exhibit tirelessly travelled the world for two decades and was viewed by millions; it has been as influential as was Rudofsky's exhibition.

Houben, a civil engineer, had a very pragmatic and in many ways scientific approach to earthen construction. CRATerre established laboratories and workshops at the University of Grenoble to test various technologies of earth construction. Mud bricks, compacted earth, earthen building blocks with reinforcing additives and *pisé* were among the materials they investigated. In the field, they demonstrated fast building techniques and displayed prototypes. Their prototype of a primary school building in Somalia and their Exhibition Centre in Riyadh attracted professional interest.

Perhaps the most interesting exercise that CRATerre conducted was in collaboration with the local authority of the new town of Ile d'Abeau in southern France. It involved the commissioning of five architects to design, and five contractors to build, a series of houses. This experimental project aimed to encourage architects to use mud and mud bricks as the principal building material, and to allow them to explore possibilities in mud construction techniques. These houses have now been in use for more than fifteen years and convincingly prove that the material, mud, indigenous to the Lyon region, can be put back into use. The serious commitment of the local authority in Ile d'Abeau to make use of local resources, an approach long espoused by Fathy, was rewarded by naming one of the streets after him, 'Rue Hassan Fathy', as a gesture to recognize and eternalize the great architect. It is worth noting that no such recognition has yet been made in Fathy's city of origin, Alexandria, or in Cairo where he spent most of his life.

CRATerre also conducted major housing projects using appropriate local materials other than earth. Their involvement in mass housing in Mayotte spanned more than two decades and provided hundreds of houses. In various types of compounds, different house types, architectural solutions and site plans were all executed in respectful appreciation of the plurality of expressions among this remote island population.

Jak Vauthrin is another of the leading disciples of Fathy. He established ADAUA (Association for the Development of an African Urbanism and Architecture), an institution that works primarily in Africa. The mission resembles that of DW, although perhaps with stronger emphasis on the development of local capabilities. ADAUA established branches in Mali, Mauritania, Burkina Faso and Senegal. Vauthrin remained faithful to his mission and persevered, despite the change in name of his central organization to Mirhas and then FISA (International Foundation for Architectural Syntheses), and moves from Geneva to Seville.

ADAUA's most substantial urban intervention was to supply housing for a refugee population that moved out of their original settlements in Mauritania, due to drought, to Senegal's coastal town of Rosso. In the early 1980s, more than three hundred houses were built in brick, with particularly developed and very simple vaulting technology. Simple load-bearing brick walls formed a single square module as the basic room. The roof was derived from a construction technology of spiralling equidistant layers of bricks to form a dome. They based their solution on the abundant earth available in the area, transforming it into building materials and eventually forming expressive domed dwellings. Although outsiders admired the project for its simplicity and for local people's participation, the residents themselves did not like the houses. It has been suggested that the domed appearance reminded them of funerary structures, repelled them from living in them. Part of the resistance also stemmed from the fact that this innovative solution had been brought in by foreigners, even though it employed local resources. The similarity of this experience with Fathy's Gourna is compelling. Both are well-intentioned and good-will gestures that were prevented from fulfilling their goals because of pre-existing, complex cultural forces that emanated from within these local communities.

Under such circumstances, it is not easy to be successful. First of all, it is impossible to objectively select the criteria that define the term 'success'. Is the measure of success the number of houses provided for people? Is it people's acceptance of what has been offered to them? Or is it their own definition or expectation of a lodging? Whatever the criteria may be, the critical issue at hand is the lack of understanding by local people for those who came to their assistance with good intentions and an open heart, bringing what they thought would be 'good' for them. Unfortunately, local suspicions and continual questioning of the innocent nature of this 'volunteer spirit' have led to the defeat of these projects.

Sustainability

Among the new environmental ethics of the twenty-first century, sustainability has emerged as one of the most important and internationally endorsed principles, especially in the world of architecture and in terms of appropriate building practices. Its international adoption means that it has now become a major criterion in the judgement of any architectural or planning practice, and has placed itself within the expectations of new policies set out by international organizations. Needless to say, vernacular architecture is the highest form of sustainable building, as it not only uses the most accessible materials, but also employs the widest available technologies.

Architectural theory, which encompasses all the factors that surround the art of building, is embedded within society and is passed on from one generation to the next by means of tradition. This is a cyclical period of sustention. It applies to dwellings as well as to religious or communal monuments. It is when these cycles of transmission of information or technology are broken by outside forces that tradition ceases to be active. Unfortunately, changes that ignore the complex nature of social and environmental forces yield inappropriate architecture.

At the beginning of the twenty-first century, globalization has a forceful impact on every aspect of our lives. From music to food, and from lifestyles to architecture; there are no areas of our existence that have not been affected by global forces and values. While bringing convenience in living and communication, this globalization has a homogenising effect and threatens to reduce the meaning of the architecture and built environments we live in. Meaning naturally comes with cultural awareness and historical continuity. In order to combat the threat of homogenization, the issue of cultural appropriateness, advocated by Oliver and others, will have to be taken more seriously.

Awareness of the importance of the conservation of architectural heritage, especially in areas where haphazard urban and rural building in inexpensive materials has taken place, has been raised. In the last ten years or so, many decision-makers have realized that the cultural identity that they are proud of is clearly related to the architectural heritage that they have been losing. The carelessly condoned heritage, when destroyed, becomes an important priority for conservation. Unfortunately, in most cases, what has been selected for conservation are the physical shells of the traditions, i.e. the buildings, rather than the cultural values and practices underlying them.

In time, the followers of Hassan Fathy and Paul Oliver are destined to be successful. In a world where the scarcity of energy resources and synthetic materials is only likely to increase, their determination to make use of abundant local resources and their desire to respect and engage with the complexities of cultures, historical contexts and the pressing needs of habitat, will most certainly give rise to the impressive, durable and socially conscious vernacular architecture that Fathy, Oliver and their followers hoped to realize.

Notes

1. Later, in 1977, Rudofsky revisited the subject as 'notes toward a natural history of architecture with special regard to those species that are traditionally neglected or downright ignored' (Rudofsky 1977).
2. Prior to Rudofsky and Oliver, there had been a vivid line of research on vernacular architecture. Among this work was a series of theses written at the Istanbul Technical University in the 1950s. However these works primarily focused on traditional, regional architecture in various cities such as Ankara, Diyarbakır, Konya, Erzurum and Kastamonu.
3. I only had access to the 1755 edition of *Essai* (1966). The references made to the 1753 edition of *Essai* are based on Herrmann (1962).
4. In 1972, Paul and Valerie Oliver, and Hassan Fathy visited the Middle East Technical University in Ankara as my guests. They enjoyed a field trip together, looking at the extraordinary rock cut vernacular architecture of the Central Anatolian region of Capadoccia.
5. Hassan Fathy's *magnum opus* is his *Architecture for the Poor* (1973), which was originally published as *Gourna: a tale of two villages*, Cairo, Ministry of Culture, 19 9. This book found its real meaning when translated into French (Fathy 1970).

Chapter 6

Learning from the vernacular
Basic principles for sustaining human habitats

Roderick J. Lawrence

Introduction
Vernacular buildings are human constructs that result from the interrelations between ecological, economic, material, political and social factors. Human habitats in the Alpine region of Switzerland are interesting examples of the intersection between these sets of factors, which have been modified over at least a thousand years. Given the changing nature of these factors, and the interrelations between them, it is unrealistic to consider an optimal sustainable state or condition of vernacular buildings, or any larger human settlement. Instead, it is more appropriate to discuss ways and means of sustaining this heritage in constantly changing human ecosystems. In essence, all human societies regulate their relation to the biosphere and the local environment by using a range of codes, practices and principles based on scientific knowledge and community know-how. Societies can use legislation, surveillance, monetary incentives and taxes, as well as behavioural rules and socially agreed conventions that define practices and processes in order to ensure their sustenance over many generations.

All living organisms, individuals and species aspire to survival. The mechanisms used to sustain humans depend on their capacity to adapt to changing local conditions, such as climate and the availability of resources, especially food. A human ecology perspective stresses that adaptive processes for sustaining human settlement processes are based on both ecological principles and cultural practices. These principles and practices stem from the fact that specific localities or sites provide intrinsic opportunities and constraints for all living organisms, including human individuals and groups, to sustain themselves. The site of any human habitat is also a small part of a much larger region that has interrelated sets of indigenous, ecological, biological and cultural characteristics. Therefore, no site of an existing or future construction should be interpreted in isolation from these interrelated sets of characteristics.

This chapter challenges the common viewpoint that sustainable human settlements are ideal visions, plans or physical models that can be used to construct an optimal kind of habitat. The approach advocated here acknowledges that sustaining human settlement involves a range of human practices and processes that ought to adapt to the dynamic circumstances of a constantly changing world at both local and global levels (Lawrence 2000). This contribution presents some basic principles that social and natural scientists, architects and urban planners can apply at the beginning of the twenty-first century in order to improve living conditions and sustain human settlements for current and future generations. The principles are derived from a synthesis of the accumulated knowledge and know-how of the benefits and shortcomings of vernacular building layout and construction. These can be illustrated by vernacular buildings, hamlets and villages in the Alpine region of Switzerland (Oliver 1997a: 1239–54). Alpine communities had to adapt to living with the constraints provided by the ecosystems of their habitat. Indigenous knowledge about the ecosystem has evolved in an adaptive way over many generations. Therefore, the vernacular buildings and rural hamlets are not simply curiosities that deserve a place in a folklore museum. Instead, they ought to be considered as part of a large warehouse of natural and cultural heritage that shows how humans have adapted to extreme conditions of ecosystems over long periods of time. The mechanisms used for these adaptive processes can provide lessons for future generations.

What is sustainability: an objective or a constraint?

The word sustainable is derived from the Latin *sustenere* meaning to uphold, or capable of being maintained in a certain state or condition. Hence, while sustainable can mean supporting a desired state of some kind, it can also mean maintaining undesirable conditions. Vernacular buildings are a fundamental characteristic of human civilizations that are sustained and transmitted by practices from one generation to others. Theoretical interpretations of how and why traditions are sustained across generations are still rare, despite the wide interest in sustainable development. This lack of interest is unfortunate, especially in relation to the concern about the conservation and preservation of traditional buildings in all regions of the world.

In 1980, the International Union for the Conservation of Nature (IUCN) published the 'World Conservation Strategy' (1980), which defined sustainable development as meeting basic human needs while maintaining essential ecological processes and life-support systems, preserving genetic diversity and ensuring sustainable utilization of species and ecosystems. In 1987, the World Commission on Environment and Development (WCED) (1987) provided what has become the most commonly referenced definition of sustainable development as 'development that meets the needs of the present without compromising the ability of future generations to meet their needs'.

Since 1987, sustainable development and sustainability have been widely used but their definition is elusive (Lawrence 1996). Steve Hatfield Dodds (2000) presented an overview of the main, sometimes conflicting interpretations of sustainable development and how these can be translated into policies and practices. According to Hatfield Dodds, the wide range of contributions on sustainable development can be classified into five categories that focus on:

1 Providing guarantees that economic activity does not over-exploit natural resources or exceed the capacity of the earth to adjust to the impacts of human activities on which sustenance is based.
2 Ensuring that ecological integrity and resilience to change is maintained by the amount and diversity of natural resources and other environmental assets.
3 Reducing inequalities between human societies and within specific human settlements by authorising institutions to be key actors in reconsidering the environmental and social consequences of the uses of natural resources by humans.
4 Maintaining human well-being and quality of life by promoting broader participation in decision-making, especially at the local community level.
5 Fostering ethical frameworks, moral values and attitudes that give more consideration to future generations and the non-human components of the world.

Given that there is no consensus about what sustainability means, sustainable development has not provided a framework for the co-ordination of research or policy across different scientific disciplines and professional sectors (Lawrence 1996).

In 1998, Peter Marcuse reviewed the concept of sustainability noting its application outside the domain of environmental policies (Marcuse 1998). He argues that this broader application has often led to confusion and misunderstandings, especially when sustainability has been interpreted as a goal for housing and urban development. He states that sustainability ought to be interpreted as a constraint that defines the effectiveness of policies and programmes that include equity and social justice as primary goals. Marcuse argues that programmes and projects that benefit few at the expense of many may be sustained, even though they hinder the well-being of disadvantaged individuals and groups.

Modern industrialized societies are not respecting the basic conditions of sustainable development defined by the 'World Conservation Strategy' over twenty years ago (McMichael 1993). In contrast, many traditional societies did abide by these principles for the 'common good'. The key question then is in what ways these principles could be reintroduced into contemporary societies. This chapter argues that vernacular buildings and rural hamlets ought to be considered as part of a large warehouse of natural and cultural heritage that shows how humans have adapted to extreme conditions of ecosystems over long periods of time. The mechanisms used for these adaptive processes can provide lessons for future generations. These mechanisms can be understood by applying a human ecology perspective. This perspective will be presented briefly in the next section of this chapter. Then a case study in Switzerland will be discussed to illustrate how a human ecology perspective can be applied in studies of vernacular buildings and sustainability.

A human ecology perspective

The term 'ecology' derives from the ancient Greek words *oikos* and *logos* and means 'science of the habitat'. It is generally agreed that this term was first used by Ernst Haeckel (1834–1919), a German zoologist, in 1866. The word ecology designates a science that deals with the interrelationships between organisms and their surroundings. Since the late nineteenth century the term 'ecology' has been interpreted in numerous ways. For example, in the natural sciences, botanists and zoologists use the term 'general ecology' to refer to the interrelations between animals, plants and their immediate surroundings.

In contrast to general ecology, 'human ecology' usually refers to the study of the dynamic interrelationships between human populations and the physical, biotic, cultural and social characteristics of their environment and the biosphere, as shown in Figure 6.1 (Lawrence 2001). However, this is not the original meaning of this term, which was first used in 1921 by Robert Park and Ernest Burgess in their contribution titled 'An Introduction to the Science of Sociology'. They defined human ecology as the study of the spatial and temporal organization and relations of human beings with respect to the 'selective, distributive and accommodative forces of the environment' (Park, Burgess and McKenzie 1925). This publication became a landmark for many other contributions that studied the spatial distribution of human populations, especially in urban areas.

A human ecology perspective can be applied to study vernacular buildings and human settlements. Human habitats define ecological and economic limits that circumscribe the livelihood of resident populations. In principle, the relationship between resources and human societies is mediated by information, knowledge and values, including religious doctrines and myths. These components of human culture include goals and ideals, technology, information and knowledge, as well as administrative, legal and political dimensions as shown in Figure 6.1. The ways that societies and groups develop and use technologies to fulfil their needs and sustain themselves is also a means for constituting and reaffirming societal goals, group and national identities, social norms and cultural values. From this perspective, it is possible to explain why the nourishment required by an Inuit differs significantly from that of an Australian Aborigine, or that of a Swiss farmer in the Alpine region. In principle, although the vital need for nourishment is common to all human beings, the amount of energy required for survival is relative and variable between and within human societies. In principle, nutrition is mediated by a range of biological, climatic, cultural and physiological mechanisms and rules that vary between ethnic groups, across cultures and within societies, as well as over the course of time.

The biosphere and local ecosystems define ecological limits on the resident populations. Whether and how these limits are interpreted in relation to energy supply and transformations, food production and water consumption, the generation of waste and recycling, or uses of renewable and non-renewable resources is variable over time. The relationship between available resources and human societies is mediated by information, knowledge and values that are used implicitly or explicitly to invent and use resources, create tools, harness energy and develop skills. Vernacular buildings, for example, reflect the conscious and/or unconscious know-how of local craftsmen and the inhabitants. It is important to underline that decisions are made based on choices, customs, conflicts, negotiations and compromises.

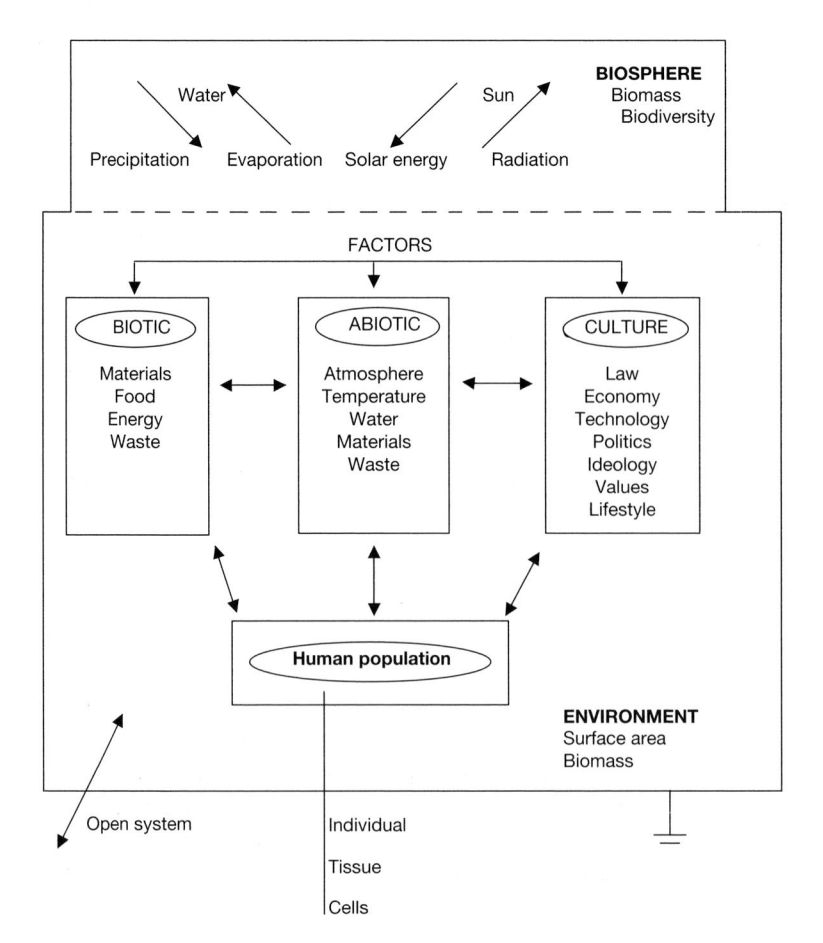

6.1
The holistic framework of a human ecology perspective showing the interrelations between biotic factors (genetic bio-space), a-biotic factors (eco-space) and cultural, social and individual human factors and artefacts.

A vernacular building is multi-dimensional and complex. In order to comprehend this complexity, it is necessary to apply an integrated approach. Nonetheless, even in a specific field such as vernacular architecture, there are many contributions that show the recurrent use of restrictive interpretations which their authors consider to be exclusive rather than complementary. For example, Lawrence (1990) discusses how many authors usually examine *either* the material constituents of vernacular buildings, *or* those non-material factors that are implicated in their layout, construction and use. Concurrently, there are too few contributions that analyse both these sets of constituents as well as the reciprocal relations between them. An ecological approach is one way of achieving this objective.

Concepts and principles of a human ecology perspective

One basic principle of biological life is that all living organisms (irrespective of their species) impact on their surroundings (Boyden 1987). The interrelations between organisms and their surroundings influence the volume and quality of the available local resources, the discharge of waste products and the creation of new resources. All living organisms change the conditions upon which they depend for subsistence by their existence. Humans are integral components of ecological systems and, therefore, they explicitly influence the living conditions of other species.

There are certain conditions and limits overriding the sustenance of human groups and societies that are defined by some fundamental principles that should be an integrated part of human ecology (Lawrence 2001). First, the biosphere and the earth are finite. Both natural and human ecosystems at all scales of the planet and its atmosphere are circumscribed by certain immutable limits, such as the surface of land, its biomass and biodiversity, the water cycle, biochemical cycles and thermodynamic principles about the production and transformation of energy, including the accumulation and radiation of heat from the earth.

Second, human ecosystems are *not* closed, finite systems because they are open to external influences of an ecological kind (e.g. solar energy, earthquakes), of a biological kind and also of an anthropological kind (e.g. disease and warfare). This means that human sustenance is the result of internal conditions and processes as well as external factors that have unpredictable impacts on human ecosystems.

Third, humans must create and transform energy by using materials, energy and acquired knowledge to ensure their livelihood (Boyden 1987). The increasing disparity between ecological and biological processes and products on the one hand, and the products and processes of urbanized societies on the other hand, is largely attributed to the rapid growth of sedentary populations, the creation of many synthetic products that cannot be recycled into natural processes, plus increases in energy consumption based on the use of non-renewable and renewable resources (such as wood from forests) at a greater rate than their replacement.

Fourth, human beings can be distinguished from other biological organisms by the kinds of regulators they commonly use to define, modify and control their living conditions (Laughlin and Brady 1978). Humans have several mechanisms that enable them to adjust to specific environmental conditions. Adaptation is a set of interrelated processes that sustain human ecosystems in the context of a continual change. Evolutionary adaptation refers to processes of natural selection and is only applicable to populations and it is inter-generational. Innate adaptation refers to physiological and behavioural changes that occur in individuals that are genetically determined and do not depend on learning. Cultural adaptation refers to adaptation by cultural processes that are not innate, such as legal measures or changes in lifestyle, and therefore it includes institutional adaptations (Lawrence 2001). The outcome of adaptation depends on a complex set of biological, ecological, cultural, societal and individual human mechanisms.

The organizing principles of human ecosystems are derived from people–environment–biosphere relations (Boyden 1987). Hence the substantive characteristics of the environment should be addressed in the same way as the cultural and social characteristics of human populations. For example, successful adaptations to ecological constraints include the way people adapt their culture for the 'common good'. An example of this is the transhumant economy of people living in the Alpine region of Switzerland.

Cultural and social regulatory mechanisms are transmitted by the tacit know-how of populations, including social rules and customs that are shared and respected in order to ensure sustenance. For example, famine is a recurrent event throughout human civilization in many regions of the world. The transhumant economy of many Alpine populations should be considered as a conscious means of adjustment to seasonal and geographical variations in the quantity, variety and distribution of edible matter and other local resources. Sedentary communities employ one or more cultural dispositions to buffer variations in the supply of environmental resources, such as drought, flood or plagues of insects that can have a severe impact on the supply of food and water. One means that helps to counteract these kinds of temporal variations is the storage of surplus produce. Owing to extreme seasonal variations, this custom is essential for the survival of the populations of the Alpine region of Switzerland.

Responses to disturbances of ecological systems are varied and unpredictable because they depend on the type and intensity of the external impact (e.g. a small, single incremental disturbance such as an avalanche, or a large, enduring impact such as a dam) and the internal properties of the ecosystems. These responses include short- and long-term change, with or without equilibrium states and internal transformations. In principle, ecological systems are not static but dynamic and change continually in terms of their composition, the interrelations between their components and their equilibrium conditions. The dynamic nature of ecological systems is partly related to their diversity and their variability. Some changes to ecological systems stem from external sources such as unpredictable climatic events (e.g. frosts, hurricanes or droughts). Ecological systems must adapt to these events in order to survive by self-regulation, other-

Learning from the vernacular

wise they will not be sustained. These internal responses do not only account for the magnitude of the disturbance but also the degree of variability that it has experienced historically.

The preceding examples show that adaptability and resilience are fundamental characteristics of human culture, which should be related to the characteristics of human ecosystems, including vernacular buildings. On the one hand, human groups may relocate or adapt their habitat in order to survive local environmental perturbations that may affect the supply of food. On the other hand, however, since the foundation of sedentary societies, human groups have primarily adapted to their environmental surroundings by modifying some constituents of their habitat and lifestyle rather than genetic adaptations.

Case study in Switzerland

The Alps form a geographical divide between northern and southern regions of Europe. This mountain range has a diversity of geological and climatic zones, especially in the main valleys oriented east-west. One of these valleys is the bed of the Rhône River, which flows from its source in central Switzerland into the Lake Léman and then into diverse regions of France. The Alps in Switzerland, including the Canton of Valais (Wallis), have been occupied by human settlements since at least the Paleolithic era. The Valais also comprises numerous mountain passes. This has meant that it has served as an important crossroads for trade routes and migration between northern and southern Europe. This historical role helps to explain the cultural and economic diversity of this small region of

6.2
A traditional house and barn (foreground) with modern buildings constructed in Törbel during the twentieth century (behind).

Roderick J. Lawrence

6.3
The know-how related to uses of traditional building construction methods has not been forgotten and is used to restore old buildings.

Switzerland, as well as other regions such as Ticino which is just south of the Alps. Mountain valleys and passes have served as protective areas for oppressed people who migrated from other regions of Europe. During the thirteenth and fourteenth centuries, for example, the Walisians migrated to the upper Rhône valley, then moved south to the Aosta valley and Ticino, as well as east to Graubünden in Switzerland, and the Tyrol in Austria.

The agricultural economy of the Alpine region, including the Valais in Switzerland, is the result of a detailed understanding of local geological and climatic conditions that vary considerably across relatively short horizontal distances, between altitudes and seasons. The most obvious example of cultural adaptations to these local conditions is the practise of a transhumant economy. This means that it is customary for farmers to move livestock (cattle, goats and sheep) from one grazing pasture to another in a seasonal cycle that begins with barns and fields in the lower valley in winter, and extends upwards to meadows at relatively high altitudes in summer, before returning down to the valley the following autumn (Netting 1981).

Until the late nineteenth century, and sometimes later, agriculture was the main economic activity of the Alpine region of Switzerland. The economy was self-sufficient and it ensured that all residents could survive with the local

6.4
The Alpine landscape surrounding Törbel is used for several purposes by the resident population including the cultivation of fruit and vegetable crops, herding dairy cattle and forestry.

resources available to them. This sustained and sustainable relation between the local population and their habitat can be used to identify key principles about vernacular buildings, human settlements and sustainability.

It has been common for the hereditary title to land and farm buildings to be divided equally among all heirs, and not just the oldest son of the property owner. This custom of property rights has led to the fragmentation of fields and buildings: not only was the land subdivided, but the use of buildings was shared by the inheritors. This tradition can be contrasted with that of pooling property rights to form common property (Stevenson 1991). This alternative could have alleviated recurrent problems of fragmented land and property titles that have made the economic viability of raising livestock and producing milk very difficult.

In the Alpine region, land uses extended from the river banks up to altitudes as high as 2,000 metres above sea level, which is about the limit for the growth of coniferous forests, of spruce, fir, pine and larch trees. In general, villages were constructed in the valleys adjacent to the rivers, whereas only single farm buildings or small hamlets were erected on sites at higher altitudes. It is noteworthy that the farm buildings and fields in the valley were generally privately owned, whereas the Alpine meadows and forests were often common property shared by the bourgeois in the village (Stevenson 1991). Each farmer

6.5
Natural disasters including avalanches and landslides pose continual threats to Alpine communities. The tacit knowledge of residents about the risks in using specific sites has been transmitted orally from generation to generation.

usually had the right to graze as many animals in the Alpine meadow in the summer as he cared for during the winter season in the lower valley. Each farmhouse had vegetable plots and fruit trees, and often fields nearby were used to produce crops and fodder twice each year. Given the specific conditions of the Valais, rye has been cultivated for centuries and vineyards have been common since they were introduced by the Romans.

The farm houses, barns and other buildings constructed in the Valais, and other regions of the Alps, relied almost exclusively on the availability of land and local materials for building construction. The two predominant materials are timber and stone. The long, straight trunks of fir trees are suitable for log buildings. Commonly, log buildings are constructed on a stone foundation which can be built as one storey, usually containing storage rooms and a cellar.

The most basic layout of residential buildings includes two rooms, a bed chamber and a kitchen, which can be found in separate buildings, placed adjacent to each other in a horizontal layout, or constructed vertically above each other. The vertical ordering of rooms offered few possibilities for extensions to the farmhouse. Nonetheless, this building typology remained common because it consumed a minimal amount of arable land, and it was appropriate for small building sites. In contrast, the horizontal layout of room offered greater potential for extensions, subdivisions and changes to the functions of rooms. Generally, the kitchen was the only room with a fireplace. It is not surprising that it was a multi-functional room. However, in farmhouses with a kitchen and a parlour, the kitchen hearth included a heating vent for the back-loaded parlour heating stove.

A specific example of an ecological approach is Netting's (1981) study of Törbel, a Swiss Alpine village. Törbel is situated in the valley of the River Visp, south of the River Rhone, at an altitude of 1,500 metres. It is an old settlement of log houses, barns and granaries which are roofed with slate slabs. The village is surrounded by hay meadows, grazing pastures and cultivated gardens, with sloping vineyards and forests of fir and larch beyond. Netting shows that this beautiful locality is not just a natural landscape but also a cultural one. The land has been appropriated and locally available resources have been used for the 'common good' throughout many generations. Netting analyses the village as a small ecosystem, based on a subsistence economy that reflects human culture including social conventions, rights and rules about the expansion, intensification and regulation of local resources and the village population. For example, he refers to the judicious use of timber for building construction, for heating domestic space and for cooking, in relation to the long-term conservation of the forest to ensure subsistence and minimize risks. Trees were not simply considered by the residents as building materials, nor as a source of energy, because they also anchored soil on sloping ground, protected the watershed, reduced the danger of avalanches and provided shelter for livestock.

Netting's fascinating study encompasses agricultural production, climate, demography, geography, local environmental conditions, resources, religion and social organization, rules and customs. Netting notes that two-thirds of all buildings were residential, while the remaining third included special purpose barns, granaries, chambers and cellars. These outbuildings reflected the importance of food storage, curing produce and preserving goods for the wellbeing of the inhabitants who relied on a subsistence economy.

The village is a small compact human settlement with buildings constructed from locally available stone and timber. The buildings have been constructed by local craftsmen helped by some inhabitants, using experience and know-how that has been handed down from generation to generation. The cultural resources involved include property rights (both private and common), rules and conventions about the amount of available land for building construction, and tacit knowledge about those sites at risk from landslides and avalanches. The location and orientation of vernacular buildings in this and other Alpine villages not only accounts for local climatic and geographical conditions, but also enables these buildings to be considered as human made places for the daily lives of the inhabitants. The conditions of the building site are given by nature, and result from long-term biological, climatic and geographical processes.

Netting's ethnography shows how a human ecology perspective can be used to interpret vernacular buildings in terms of the availability and uses of materials and human resources. The crucial principle that human ecology provides to explain specific cases is the co-evolution of people's lifestyles and values, which adapt to the limitations and possibilities offered by human habitats. A human ecology perspective explicitly considers the interrelations between the availability of natural material resources in or near a building site; the dynamic cultural resources of the inhabitants (including their socially accepted customs and rules about uses of land and other resources for their livelihood); and human energy and know-how applied to construct buildings (including the invention of building construction methods and techniques).

Basic principles for professional practice

The preceding sections of this chapter have presented theoretical principles and illustrated these by studies of vernacular buildings in the Alpine region of Switzerland. Many other regions and case studies could be added but this is not intended here. Instead, the reader can refer to numerous entries in the *Encyclopedia of Vernacular Architecture of the World* for examples in all regions of the world (Oliver 1997a). The purpose of this section is to extend the discussion in the previous sections by identifying and illustrating key principles for professional practice in the twenty-first century.

Compact human settlements

There are many types of human settlement layouts including linear, nodal, compact and dispersed (Oliver 1997a). The concentration of activities, the built environment and the resident population has many ecological and economic advantages compared with a more dispersed form of human settlement. This is not only applicable to the Alpine region of Switzerland but to other kinds of ecosystems such as deserts. Hence the first set of principles concerns 'compactness' and the ecological efficiency of buildings, services and infrastructure in human settlements. In essence, a compact form of human settlement uses less arable land, which is a precious non-renewable resource for the sustenance of all ecosystems (Wackernagel and Rees 1996). In addition, the compact human settlement has a lower unit-cost for most kinds of infrastructure and services such as roads, drainage, piped water and sanitation.

Today, information and statistics in European countries and other continents of the world show that the dispersion of human settlements has become commonplace during the twentieth century (United Nations Commission on Human Settlements 2001). This is unfortunate because arable land is a finite resource with extreme limitations, especially in a country like Switzerland. In contrast to this trend, architects and urban planners can promote ecological efficiency in existing urban neighbourhoods by not accepting to design new out of town shopping malls or housing estates. These kinds of peripheral developments on the outskirts of cities convert productive agricultural land and forests into new suburban sprawl that destroys both the ecological and the social fabric of human settlements. They also frequently create dependence on automobiles, thus isolating those who do not drive motor cars, notably children, the handicapped, the aged and the poor.

Building adaptability for reuse
The second set of principles deals with 'adaptability' of the existing building stock for the reuse of old buildings to serve the needs of contemporary daily life. Vernacular buildings in Switzerland and many other countries have been handed down from generation to generation. They have been adapted for a range of uses that have evolved over time. This custom has been particularly appropriate in the Alpine region of Switzerland, which has gradually diversified its local economy by developing tourism. Today, the principle of adaptability is too easily forgotten by architects, town planners and public officials who want to demolish, rather than renovate existing buildings. The mistakes that were made in many European cities in the 1950s, 1960s and 1970s should not be forgotten by architects, planners and policy decision-makers (Lawrence 1995). It should be noted that the vernacular buildings were rarely made redundant as quickly as many buildings designed by architects during the twentieth century.

It is extremely difficult to adapt existing buildings, neighbourhoods and transport systems that were constructed during a period of relatively low cost fossil fuels and steady economic growth during the last half of the twentieth century. There is a need to consider how to reduce uses of non-renewable resources, how to lower greenhouse gas emissions and lower solid waste disposal. This reduction will require creativity and imagination by professionals in order to renovate and reuse individual buildings as well as plan for sensitive in-fill projects on urban and suburban sites that help to link existing neighbourhoods and urban infrastructure.

Patterns and principles from history
It is possible and necessary to identify principles of good practice from historical examples of building construction and the layout of human settlements. Sustaining human settlements means ensuring the maintenance of those buildings and facilities that make a city or town a pleasant and safe place to live in. It involves guarantees to protect the natural and built landscape, which is a combination of unique cultural and ecological characteristics. It means a concern about ensuring that the cultural heritage of human settlements is handed down in good condition from one generation to the next.

There is much to be learnt from good examples of vernacular buildings and the layout of traditional towns. This knowledge can be used for diverse purposes such as the promotion of 'ecological technology' that can reintroduce natural energy flows and local materials back into building construction. At the beginning of the twenty-first century there is still insufficient professional knowledge that can help to explain why it is important to conserve and protect basic natural resources. Such knowledge is still rarely translated into professional practice. For example, building design and construction together with the layout of traditional towns, should explicitly account for:

- Water cycles that collect and reuse rain water and grey water in buildings and adjoining open spaces.

- Natural ventilation in contrast to mechanical systems of air-conditioning for all kinds of buildings.
- Reusable materials, such as wood clay and brick, should be used instead of non-biodegradable synthetic products in new building construction and renovation projects.

Innovative approaches of this kind not only help promote the local environment and protect the cultural heritage of human settlements. In addition they can be a catalyst for a new kind of ecology-oriented tourism and economic investments.

Interrelated scales form a web
The fourth set of principles concerns the interrelations between different geographical scales of all architectural and urban/rural projects from the scale of the room and building to the block and the neighbourhood, to the city or town, the regional, national and global levels. For example, energy consumption and dependence on fossil fuels is closely linked to the way we construct our buildings, layout our cities and towns, and service them by infrastructure and transportation systems. In turn these characteristics of human settlements impact on the quality of air in the local environment, ambient noise levels and the local climate (McMichael 1993). They also contribute to greenhouse gas emissions and the depletion of the ozone layer at the global level.

During recent decades, academics, policy decision-makers and city planners have ignored the interrelations between the characteristics of human settlements. They have often ignored complexity, especially the web of economic, ecological, health and other social characteristics of the built environment. This has meant that the 'one problem – one solution' approach was often applied in the twentieth century by architects and urban planners in order to resolve problems by identifying and applying 'the best solution'. For example, in the 1950s, urban planners and traffic engineers developed programmes and projects for public transportation that often gave a higher priority to private vehicles than to diverse kinds of public traffic (Lawrence 1995). Today it is increasingly recognized that more integrated policies and programmes are necessary.

Ecological and cultural diversity
The fifth set of principles stresses that professionals should not forget the ecological and cultural diversity of human settlements, even the small hamlets of the Alpine region of Switzerland. Purposive human behaviour has been studied by social anthropologists including Edward Hall, psychologists such as Irving Goffman and sociologists including Robert Merton. Their contributions consider the latent and manifest functions of human activities as well as intended and unintended consequences. They show that the members of human groups and communities co-operate successfully to sustain themselves because there is a social order founded on implicit or explicit conventions, customs and norms that

regulate human activities, including the sharing of information and tacit knowledge. These mechanisms have been illustrated by traditional methods of using resources in the Alpine regions of Switzerland.

At a more general level, there is an urgent need for studies of social values and lifestyles related to the components of human settlements that will enable policy-makers and professionals to predict and plan for social change. With this kind of understanding architects and urban designers could develop a new, innovative interpretation of the qualities of the built environment, including the ecological characteristics of the site and the cultural values of the resident population.

Today, policy decision-makers, social scientists and design professionals need to use a range of complementary methods for the collection of information and data in order to improve their understanding of the cultural and social determinants of housing projects and other kinds of development. The cultural values and the social and ethnic diversity of populations should be understood by using both qualitative and quantitative analytical methods, as well as participatory approaches (Barton and Tsourou 2000).

Participatory approaches
Citizen participation has been an integral component of the construction of many vernacular buildings and settlements around the world, including those in the Alpine region of Switzerland. These approaches were part of local customs and practices. Today, participatory approaches for decision-making about housing, urban planning, environmental conservation policies and public health have been increasingly advocated by international conferences and organizations. They have been applied at the local level by municipal governments and non-governmental organizations (NGOs) on the understanding that in democratic societies, complex issues should not be interpreted by one set of criteria or values. This trend was endorsed by Agenda 21, which advocates citizen participation in decision-making. Agenda 21 was formulated as a programme of action based on twenty-seven principles contained in the Declaration on Environment and Development that are meant to promote sustainable development in the twenty-first century. The declaration includes principles for sustainable development, including key themes of development, demography, human health, environmental quality, the economy and poverty. It stresses dimensions of social development such as education, equity, empowerment and human rights (United Nations 1993).

There is no consensus about the definition and methods of participatory processes (Barton and Tsourou 2000). Participation can be interpreted as a broad term that refers to dialogue between policy institutions and civic society, in order to formulate goals, projects and the allocation of resources so as to achieve desired outcomes. A wide range of techniques and methods can be used including civic forums, focus groups, citizen's juries, surveys, role-playing and gaming. These methods can be applied using aids or tools such as maps, plans, photographs, small- or large-scale simulation models and computer-aided design kits (Marans and Stokols 1993).

Communication, information and public awareness
Individual and community awareness, education and consciousness are prerequisites for a societal commitment to the redefinition of goals and values that ensure a more balanced use and a more equitable distribution of all kinds of resources. Without this commitment, based on a sound knowledge base and shared goals and values, recent requests for more public participation cannot redefine policy formulation and implementation in meaningful ways. Public participation and empowerment alone are not panaceas for current urban and broader environmental problems, but they can serve as vehicles for identifying what local residents consider as key issues concerning the renovation and upgrading of existing buildings and urban neighbourhoods. However, as Lawrence (1995) has argued, before individuals and community groups can effectively participate with scientists, professionals and politicians in policy formulation and implementation, there are long-standing institutional and social barriers that need to be dismantled.

Conclusions

Human settlements in the Alpine region of Switzerland have been sustained despite the extremely constraining conditions of local ecosystems. These severe conditions are critical factors that provide both constraints and opportunities that occur simultaneously. There are absolute limits to the availability of local resources for building construction. The ways in which these resources are used in a complementary way to provide opportunities is largely based on human ingenuity and customs handed down from generation to generation. Land use provides an example: the steep sloping sites are unsuitable for building construction until some plot is cleared and levelled. Simultaneously, the abundant rainfall during the warm months of the year provides ideal conditions for the growth of fodder and trees. Hence as much arable land as possible can be used for these purposes, whereas land use for building construction is minimized. Exactly where buildings are constructed depends on an in-depth understanding of land areas at high risk from avalanches and falling rocks. In this way, a combination of critical factors can be analysed in specific localities to help understand the siting, layout and construction of vernacular buildings.

An ecological approach applied to the study of vernacular buildings reminds us that the layout, construction and use of human habitats stems from the interaction between a wide range of factors. An ecological approach can illustrate how the layout, construction and use of buildings and settlements is related to lifestyle and values concerning the social organization of households and communities. In principle, the local human-made environment (e.g. the anthropos) is meant to reflect the orders of the universe (e.g. the cosmos) in order to guarantee its sustenance.

An ecological perspective also identifies the predominant constituents of building constructions since sedentary settlements were founded in the Alpine region of Switzerland centuries ago. Since then, buildings have commonly been constructed using timber and stone. With the advent and growth of industrialization, however, mass produced materials such as cement, steel and glass have gradually supplanted the basic constituents of traditional building construction. Concurrently, transportation made it possible to import and export building materials and transpose construction techniques. As the indigenous cultural know-how of traditional building methods declined, the impacts on the layout and construction of the built environment, plus the consumption of materials and energy increased significantly. At a more general level, today, there are choices between traditional materials and methods, synthetic materials and new technologies: the former usually enable the use and reuse of renewable resources, whereas the latter require more energy and more specialized expertise. They also produce more non-recyclable waste products. In addition, the use of imported building materials can have unforeseen negative impacts on human well-being. In sum, modern materials and methods may produce more unintended ecological costs which the biosphere and human populations must assume not only today but in the future.

This chapter suggests that inadequate responses to current ecological, economic and social problems are due to a number of reasons. These reasons include misconceptions about people–environment–biosphere relations, inappropriate institutional and professional practices, and the lack of a societal project for the 'common good' of current and future generations. It has also been suggested that some of these reasons could be corrected by a better understanding of the processes and practices used to sustain human settlements despite extreme ecological conditions. At the beginning of the twenty-first century, these kinds of basic principles could be reapplied to improve, then sustain human ecosystems, at the scale of specific buildings, hamlets, cities and regions.

Chapter 7

Lessons from the vernacular

Integrated approaches and new methods for housing research

Lindsay Asquith

Introduction

If the vernacular makes up 90 per cent of the world's buildings and consists of approximately 800 million dwellings (Oliver 2003), it arguably cannot be ignored within the context of future housing research. Despite this statistic, the vernacular *is* often ignored in both architectural education and from within the architectural profession. Furthermore, the study of housing also has a long way to go before it attracts the attention warranted, not only by practitioners, but also by educators. The design of homes is usually not only a minor teaching module in architectural education, but is often divorced from both the architectural profession and the future inhabitants of these houses. Housing design in much of North America and western Europe is at the very least, predominantly in the hands of developers, with design being based at worst on the financial bottom line and at best on outdated ideas of what the consumer wants a home to provide.

This chapter intends to marry these two marginal interests in the vernacular and housing and to illustrate how, when studied together with integrated approaches taken from architecture, anthropology, sociology and behavioural studies, innovative methods can be deployed and used as tools to aid good housing design for the future. If the housing needs of all inhabitants globally across cultures are to be met in the twenty-first century (in the UK alone, 200,000 new homes need to be built every year to reach demand) it is vital that methods and approaches from a range of disciplines are used as tools for analysis that cross not only the boundaries of each discipline, creating new theoretical responses, but are also adopted in education and practice alike.

The vernacular approach to housing studies

It is important to look firstly at why the study of the vernacular would have relevance to housing research and in turn future housing design. Paul Oliver (1997b: xxiii) states: 'All forms of vernacular architecture are built to meet specific needs, accommodating the values, economies and ways of living of the cultures that produce them.' In this respect the design of houses cannot become divorced from those that will eat, sleep, cook and play in them. In the vernacular, the builder is often from within the community, and may even be the inhabitant himself or herself, and is therefore aware of not only climatic and typological considerations, but also the values, rituals and beliefs that will shape the design of the dwelling. If the designer and the consumer both actively participate in the design of the dwelling, a unity of purpose is achieved and design is shaped by the community where traditions, rituals and norms are all applied in the design process. As Oliver (1979: 9) comments: 'There are very few unnecessary buildings in the vernacular and within the buildings themselves, very few unwanted spaces.'

Lessons from the vernacular are often used primarily to record and document building traditions and typological changes through history. This is especially true of vernacular architecture groups in the UK and the US. Surveys, plans and measured drawings are all used to record changes in plan type and houses are reduced to purely typological significance. The significance of the building to its occupants and how they feel about the interior of the dwelling, the spaces they use and the reasons why are rarely examined. Henry Glassie (2000: 67) concludes: 'If the intimate ordering of common life mattered in history as much as it does in reality, then the interior would matter, families would matter, communities would matter and women would be in the story.'

Research into housing should in future document how the house is used, by incorporating data that signifies actual space use, which can then be used to assess the changing needs of its occupants through time. Once the vernacular is seen not as a static building form, but as constantly evolving, reacting to changes in the communities that shaped its form, it will become higher on the agenda in architectural education and more considered in the world of the practitioner concerned with conservation and the sustainability of the built environment. Many studies have contributed a great deal to our knowledge of vernacular traditions across many countries and numerous cultures (Rapoport 1969; Low and Chambers 1989; Kent 1990; Turan 1990; Arias 1993; Oliver 1997a and 2003; Cierrad 1999; Amerlinck 2001), but the use and application of this knowledge is less discussed and has not been applied to housing research and theory, nor contributed to new methodological approaches until recently (Asquith 2003).

Lindsay Asquith

An integrated approach to housing studies: the vernacular response

Vernacular architecture is a subject without a discipline. Those that study it come from many disciplines, from anthropology, sociology and behavioural studies to human geography, history and architecture itself. The lessons that can be learnt from it are therefore all the more valuable as it can bring approaches and methods together and use them to develop methods and tools for analysis in future housing research. For the most part, methodological studies in housing research are scarce and those that use tools or methods for analysis are firmly entrenched in their own discipline. The need is for integrated approaches, not those exclusive to each discipline, and for future collaboration between disciplines that will result in innovative theory and new methods that will assist housing research today and throughout the twenty-first century.

Research into the domestic housing stock of most developed nations either concentrates on spatial type, the physical boundaries that frame the spaces we use (Alexander, Ishikawa and Silverstein 1977; Hillier and Hanson 1984; Hillier 1996; Hanson 1998; Dovey 1999) or, less commonly, the cultural and behavioural codes that determine the way space is used in the home (Hall 1969; Werner, Altman and Oxley 1985; Lawrence 1987; Arias 1993; Marcus 1995). Lawrence (1999) argues in his essay 'House, form and culture: what have we learnt in 30 years?' that domestic architecture and human behaviour should be examined on interrelated levels. The first level is that spatial and activity patterns in the home as universal to a culture should be examined, and this could be incorporated in what I would term the *anthropological approach*. Second, spatial and activity patterns as shared by a group or community or household in the form of daily routines and rituals should be studied, which will be termed the *sociological approach*. Third, individual spatial behaviour as determined by cultural or social traits, i.e. age and gender, should be investigated, which in this chapter will be referred to as the *behavioural approach*. The fourth level, not referred to in this instance by Lawrence, but one that is necessary and fundamental to any study of domestic architecture, is the influence of spatial type on space use, which shall be termed the *architectural approach*. These four approaches all have contributions to make in future housing research, not from within the boundaries of their individual disciplines but as integrated approaches from which new methods and solutions to housing problems can emerge. They are also approaches that have contributed to the study of vernacular architecture, a subject yet to be defined as a discipline (Oliver 1997a).

Anthropological concepts are often ignored in the study of western practices, especially those relating to the daily routines and rituals that shape the way we use our homes. Also, the practice in anthropology when considering the built form has been to concentrate on the house as a symbol of the culture that produced it. The building is seen as an artefact and is studied from within the boundaries of cultural knowledge. Future housing studies need to identify the common codes of practice, in the form of daily activities and routines that structure the spaces we use according to our cultural norms and practices that form the anthropology of space.

Sociological concepts, those issues that concern who we are, our roles in the community and within our families, are also important in defining methods for housing research. The sociological approach is more concerned with the *interpretation* of the idealized cultural concepts that form the basis of the anthropological approach. The ideas of what constitutes a 'proper family' have shaped the way individuals relate to each other in the domestic setting and these same ideas have influenced the design of housing (Munro and Madigan 1999). There has been little sociological research on age and hierarchy and the relationship between parents and children with respect to space use in a domestic setting. The need is to identify and comprehend space issues not just in terms of the role the family plays or the perceived roles of men, women and children, but in terms of what is important to each individual and how that is acted out in the home. The sociological interpretation of anthropological concepts in relation to domestic space use is a vital step on the way to understanding how the home and family function together, and what changes need to be made to future housing design. Young and Wilmott stated as far back as 1957:

> The problems are formidable, but if the purpose of re-housing is to meet human needs, not as they are judged by others but as people themselves assess their own, it is doubtful whether anything short of such a programme will suffice.
>
> (Young and Wilmott 1957: 165)

A behavioural approach to housing research is concerned with the perceptions, interactions, relationships and identities of the individual as he or she assumes their roles within the physical boundaries of the home. This approach moves forward from the anthropological and sociological concepts to an understanding of who does what, where and when, including or excluding whom, in order to assess the complex spatial pattern that illustrates life at home. Spatial behaviour needs to be regarded not as something static or culture-bound, but continuously variable, defined primarily by context. For example, the effects of age or gender, from a behavioural perspective can identify perceptions and cognitions that are context dependent, instead of being culturally specific. In modern society the house is no longer a text encoding the cultural rules of behaviour, or even the whole world view that can be passed on through time (Tuan 1977). The house is seen not to be representing group identity as much as the identity of the individual or individuals that reside within it.

Finally, an architectural approach to housing research looks at the physical spaces themselves. The house has often been overlooked as a built form worthy of attention by the architectural profession because it is so familiar. Domestic space has an intricate pattern, of which the users are not often conscious and are often only made aware when encountering a different spatial pattern from another culture (Hillier 1996). The architectural approach will also have more relevance once examined in unity with the insights gained from the anthropological, sociological and behavioural approaches. A configurational theory of architecture, how spaces are linked to each other, which forms the basis of studies using space syntax as a tool to illustrate space use is valid, but in linking it to sociological and psychological methods it can answer pertinent questions that relate to the very nature of domestic space. As Hanson remarks:

> The important thing about a house is not that it is a list of activities or rooms, but that it is a pattern of space, governed by intricate conventions about what spaces there are, how they are connected together and sequenced, which activities go together and which are separated out.
>
> (Hanson 1998: 2)

Research into housing needs, suitability and adaptability in particular with regard to space use, should not be the exclusive concern of developers or even designers or architects. Nor is it purely of value to anthropologists, sociologists and behavioural scientists. What the vernacular response to housing needs can and should do, because of its multi-disciplinary nature, is contribute to an integrated approach, based on shared knowledge rather than exclusivity. This approach should in turn be used to develop new methods or adapt existing ones, which will result in innovative approaches and theories in housing research.

From the vernacular to a conceptual framework

In *House, Form and Culture*, Rapoport (1969) examines and illustrates what he terms the *genre de vie* that affects built form. He lists some of the factors that he concludes have an important influence on the way a dwelling is constructed (1969: 61). They are socio-cultural factors as opposed to physical factors such as climate, technology, economy and materials. The five factors are:

1. some basic needs
2. family
3. position of women
4. privacy
5. social intercourse.

These factors are common to all cultures and should be applied in any assessment of housing need. A conceptual framework that includes these aspects should be developed in order to design new methods for future housing research. In this chapter, these key factors have been re-termed to form a set of variables that can best identify and illustrate how space is used in the home. Basic needs are the activities we perform and where. Family should be examined in relation to the concepts of age and gender within the roles and relationships inherent in family life. The position of women should be examined in relation to gender, not only in terms of the role a female partner or mother plays at home, but also with reference to her status outside of the home. Privacy should be examined in relation to age and gender but also as part of the study into spatial arrangement and configuration. Social intercourse needs to be assessed not just in terms of communication between family members (see Bernstein's theory on the sociology of language (Bernstein, 1970)), but through the structuring and re-structuring of time and the spatial type itself.

First, it is important to look at the role activities play in space use in domestic architecture. The type and combination of activities affects spatial patterns resulting from practical considerations as to where specific activities can take place: 'through considering differences among simple activities – cooking, eating, playing, sleeping – get to visions of lifestyle–values–world views – sub-cultures/cultures as they relate to the built environment' (Rapoport 1980: 17).

Room function needs to be examined in relation to domestic routine and ritual. The existence of many rooms in a house does not mean that they are all used, activities occurring simultaneously in one room when others are left, seemingly, without apparent use. Any examination or assessment of what activities are performed in the house needs to consider what the main activities are; the number of activities and where they take place; the rooms that accommodate the majority of activities; rooms for ritual use and rooms used not for their intended function.

The influence of gender on domestic architecture needs to form part of the conceptual framework for future housing research. This should not be predominantly from a feminist viewpoint, which leads to exaggerated differences between men and women, but should concentrate on gender relations, the relations between men and women in the home rather than their perceived roles. The spatial organization of a dwelling should not be seen to define the relationship between men and women, as this leads to gender stereotyping based on myths that associate men and women with certain spaces. Space is constantly structured and re-structured, defining gendered spaces only within a specific context at a given point, within a set of norms that are not immutable. Methods and models of analysis are needed that can adapt and record the complex and variable nature of space use according to gender. Activities need to be assessed in relation to gender, as does room use. The amount of time spent on activities as well as the amount of time men and women spend in the home is also an important consideration. Furthermore how men and women perceive space in the home in relation to space claim can inform as to the communication within the home and how status can affect spatial organization.

Age in terms of not only life-cycle, the generational impacts on house design, but also the ages of children and adults within the house is an important consideration in housing research. More emphasis needs to be given to the interaction between children and adults in the home and how their relationships may influence domestic spatial organization. As Shildkrout maintains:

> If one focuses on the interaction of people of different ages, all as dependent variables, without assuming that one group or the other 'makes the rules', one is forced to re-examine the society itself and study the significance of the participation of people of different ages.
> (Shildkrout 1978: 111)

Segregation of space according to age can occur almost unconsciously through changing needs, or it can occur through methods of control in the form of physical barriers as a result of the need for privacy. Research into housing requirements needs to look at what activities people of different ages perform within the home, the spaces they use and how much time they spend in the home according to daily routine. Also relationships between family members change through time and each stage cannot be looked at in isolation.

The way space is structured and re-structured according to time, influences spatial and social organization and needs to be examined. Social intercourse can be controlled by time as well as spatial configuration, so that groups with different rhythms and tempos can occupy the same spaces and yet hardly meet. Patterns of time and space use can be governed by external factors such as adult working patterns or the dictates of school life on children. Life-cycle changes do affect space use and studies that concentrate on generational changes in the way space is adapted and re-created do have value. However, the need is for more qualitative research into the effects of cyclical time, the daily, weekly, seasonal rhythmic patterns of time in relation to space. An analysis of time in relation to domestic space use needs to identify the rooms in houses that are used most of the time and those that are not. It also needs to identify periods of time when houses are busy or quiet and record when activities are occurring to establish patterns of space use in relation to routines and rituals.

The final concept that makes up the framework is the very space itself, the physical membrane that is the actual dwelling. By analysing spaces in terms of the way rooms relate to each other within the house (Hillier 1996) and establishing patterns of behaviour across spatial type, it is possible to evaluate how the configuration of spaces may transmit cultural and behavioural codes of social practice. Eating, sleeping, playing, encountering, avoiding, entertaining and interacting, and the way we perform these activities through cultural practice *and* individual choice, determines how space is used and claimed. Similarly, the physical form of the house, its boundaries and the way spaces are linked, influences where we perform these activities. Patterns of domestic space use occur almost unconsciously and often we are not aware of them until challenged. Room function needs to be analysed in relation to spatial type, but also space use according to gender and age in relation to spatial type should be examined for a deeper understanding as to the question how much spatial organization is determined by social factors and, conversely, how social relationships are affected by the form of space itself.

From conceptual theory to methodological designs

If we accept that activity, time, age, gender and the spatial type itself influence the way we inhabit our homes and all have a major part to play in current and future design considerations, the next step is to design an appropriate methodology. The methods chosen need to illustrate the multi-faceted nature of domestic space use and claim, not to identify any dominant factors, but to identify and establish all the factors that contribute to the way we use the space of our homes.

Lessons from the vernacular

The time diary

To record activity, time and space use in relation to each individual, which can then be examined with respect to age and gender, a time diary can be used. Diaries recording actual time have been used in housing studies (Ministry of Housing and Local Government Report 1969a and b; Wallman 1984) to effectively record household routines and space use in homes. The method is time-consuming for the participants and a level of understanding and commitment is required, but it can capture the essence of each individual's daily pattern of space use. Observation by a third party or through a camera can cause any participants in a research study to behave in a different way, not reflecting their normal daily pattern.

In the last few years my research has focused on the close study of seventeen families in the UK and the way they have used and claimed space in their homes across a variety of spatial types. The research used qualitative methods, interviews, questionnaires and time diaries, and I then evaluated the data using quantitative analysis to represent the findings. The method for recording activities is outlined in Figure 7.1. The data for all people living in one dwelling can be recorded over a time period and entered in this way. The data is then entered onto coded worksheets where activities can be grouped according to activity types as illustrated in Table 7.1.

A coding system was used for all the rooms in the house starting from the entry point. For example the hall was usually A, and the rooms in sequence from this point would be given letters accordingly. This is important as when representing the data in graph form, the rooms could be illustrated as they appeared, rather than randomly, and the depth of rooms from the entrance established.

7.1
Time sheet for an adult female and her ten-year-old son.

135

Lindsay Asquith

Once all the data from the case study families is entered from the worksheet to the template, graphs can be drawn to illustrate a number of key concepts that will be illustrated in a later section.

Spatial mapping
The time diary records activities and room-use as they actually happen. However, what spaces each individual associates with or claims is an important aspect in the structuring of space within the home, which can be hierarchical depending on age, gender or status outside of the home. To record space claim from the perspective of individual household members, a simple plan of each home with all

Table 7.1 Worksheet

Day	Person	Activity	Activity type	Room	From	To
7-May-02	Female adult	Shower	Personal hygiene	Ensuite	06:55	07:10
7-May-02	Male child 2	Washing	Personal hygiene	Bathroom	07:00	07:15
7-May-02	Male adult	Shower	Personal hygiene	Ensuite	07:00	07:20
7-May-02	Male child 1	Dressing	Dressing	Bedroom M/S1	07:00	07:20
7-May-02	Female adult	Dressing	Dressing	Bedroom A/P	07:10	07:15
7-May-02	Male child 3	Dressing	Dressing	Bedroom M/P	07:10	07:20
7-May-02	Female child	Dressing	Dressing	Bedroom F/P	07:15	07:30
7-May-02	Male child 2	Breakfast	Mealtime	Dining area	07:15	07:30
7-May-02	Male child 1	Breakfast	Mealtime	Dining area	07:20	07:35
7-May-02	Male child 3	Breakfast	Mealtime	Dining area	07:20	07:30
7-May-02	Male adult	Dressing	Dressing	Bedroom A/P	07:20	07:30
7-May-02	Male child 2	Brushing teeth	Personal hygiene	Bathroom	07:30	07:35
7-May-02	Female adult	Breakfast	Mealtime	Dining area	07:30	07:45
7-May-02	Male adult	Breakfast	Mealtime	Dining area	07:30	07:40
7-May-02	Female child	Breakfast	Mealtime	Dining area	07:30	07:40
7-May-02	Male child 3	Talking	Talking	Dining area	07:30	07:40
7-May-02	Male child 1	Brushing teeth	Personal hygiene	Bathroom	07:35	07:40
7-May-02	Female child	Brushing teeth	Personal hygiene	Bathroom	07:40	07:45
7-May-02	Male child 3	Brushing teeth	Personal hygiene	Bathroom	07:40	07:45
7-May-02	Female adult	Brushing teeth	Personal hygiene	Ensuite	07:45	07:50
7-May-02	Male child 3	Talking to Mum	Talking	Bedroom A/P	07:45	07:55
7-May-02	Female adult	Combing hair	Personal hygiene	Bedroom A/P	07:50	07:55
7-May-02	Female adult	Making tea	Preparing food	Kitchen	17:00	17:20
7-May-02	Male child 3	TV	TV	Playroom/old dining room	17:00	18:00
7-May-02	Female adult	Making tea	Preparing food	Kitchen	17:35	18:05
7-May-02	Female child	Eating tea	Mealtime	Dining area	18:00	18:40
7-May-02	Male child 3	Eating tea	Mealtime	Dining area	18:00	18:40
7-May-02	Female adult	Eating tea	Mealtime	Dining area	18:05	18:40
7-May-02	Male adult	Eating tea	Mealtime	Dining area	18:10	18:40
7-May-02	Male child 2	Eating tea	Mealtime	Dining area	18:10	18:30

Note: The data for Table 7.1 is taken from the same family as exemplified in Figure 7.1.

Lessons from the vernacular

rooms drawn could be shaded in to identify the perception of the space claim of each individual. This could be done to illustrate both individual space claim and also how individuals saw the space claim of others, which could highlight any associations with gender, age or status. Figure 7.2 is a spatial perception map filled in by an adult male to record his individual space claim.

This data is also entered into a work sheet but is given a numerical value where claim is evaluated as a percentage of the total room. A sample of the worksheet in relation to this particular adult male is included as Table 7.2. This qualitative data can be used in a number of ways to illustrate space claim within the home.

Spatial configuration diagrams

To relate the findings that formed the basis of the anthropological, sociological and behavioural conceptual framework to an architectural approach, a tool that can identify how spaces within the home relate to each other spatially, needs to be adopted. This tool is commonly known as the j-graph and is a method derived from mathematics and geometry, and adapted by Bill Hillier for space syntax research. The spatial configuration diagrams or j-graph can map the potential connections and movements between spaces more readily than the frequently used architectural plan (Hillier and Hanson 1984). By incorporating the syntactic structure of each home into one simple diagram, a set of otherwise complex spatial relationships can be more easily understood. For example, terraces houses have

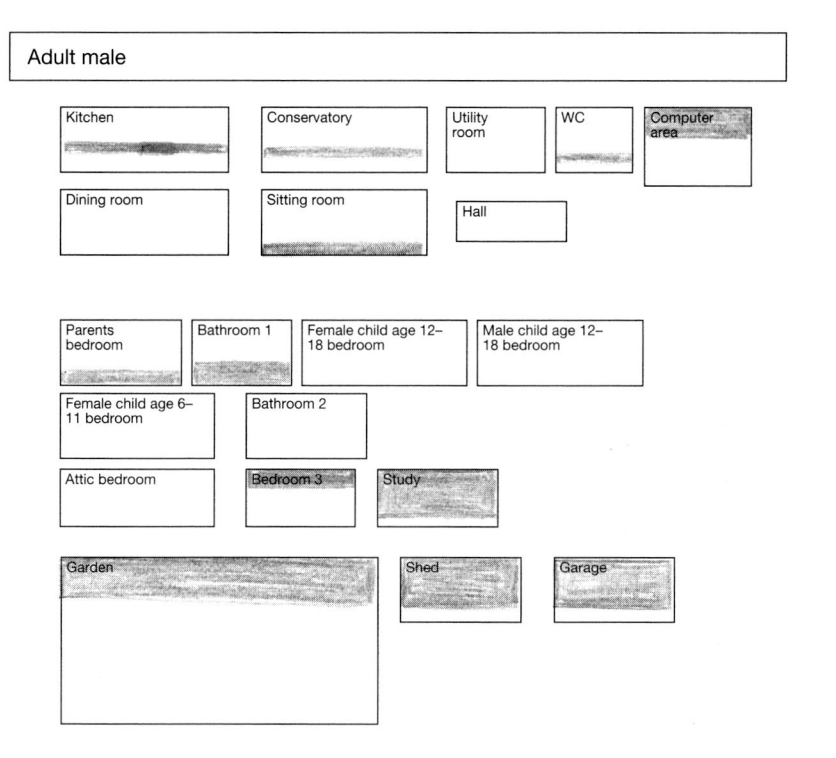

7.2
Perception map of an adult male in relation to his own space claim.

Table 7.2 Space claim worksheet of adult male in relation to downstairs areas and outside only

Perception by	Perception of	Type	Hall	Sitting room	Dining room	Kitchen	Computer area	Utility	WC	Conservatory	Garden	Shed	Garage
Adult male	Adult male	S	10	20	0	20	35	0	20	20	20	90	85
Adult male	Adult male	F	10	20	0	20	35	0	20	20	20	90	85
Adult male	Adult female	F	30	20	80	50	0	100	15	20	40	0	15
Adult male	Child female 1	F	30	20	20	20	10	0	10	10	15	0	0
Adult male	Child male	F	10	20	0	5	45	0	20	0	10	5	0
Adult male	Child female 2	F	20	20	0	5	10	0	35	50	15	5	0
Adult male	Nobody	F	0	0	0	0	0	0	0	0	0	0	0

Note: Type S is the perception of the adult male for himself only and type F is for when he considers the whole family.

primarily a linear syntactic structure (rooms in sequence with no other pathway between spaces) and are often much deeper than other house types, and this may determine different patterns of space use and claim. The house type that accords with a looped or ringy structure (rooms that connect with many choices of pathway) may also determine how space is used because it enables many more social encounters to occur.

However, to fully link the spatial type to the methodological framework it should be integrated with the other methods. In my PhD research I developed the drawing of the spatial configuration diagram by integrating it with the data obtained from the time diary. Figure 7.3 is an example of a spatial configuration diagram, which adopts this method. Each room is represented by a bubble, but the size of the bubble is indicative of the amount of time actually spent in each room by all household members. From this adaptation of the conventional j-graph it is easier to see at a glance which rooms are used the most, and relate this information to the position of the room in relation to others and the point of entry. This example illustrates that the kitchen, which is well-connected to other rooms, is the most used.

From qualitative beginnings to quantitative measures

The data collected by the methods explained above can be examined and evaluated in a number of different ways, depending on the aims and questions of the research. For example, if the aim was to discover the main activities across all the households, all the data could be combined and the information evaluated to produce Figure 7.4. This shows that the top six activities in these homes, in descending order, are watching television, mealtime, housework, computer, reading and preparing food.

However, research into individual homes illustrated that in many homes there was more than one television, often as many as four, and many were in bedrooms. Therefore the provision of just one room for the sole purpose of watching television is no longer a user requirement, with many family members choosing to watch television in privacy. Likewise, computers were found in many areas of the home, from bedrooms to dining rooms, and with

Lessons from the vernacular

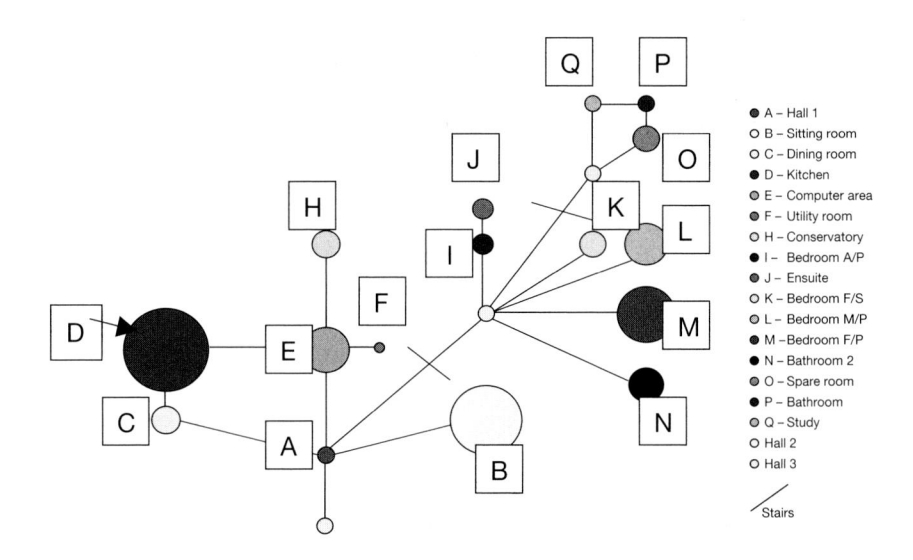

7.3
Spatial configuration diagram illustrating the tool used for spatial measurement adapted and developed in the research.

laptop computers being more common, a separate enclosed space known as a study is not always required. Similarly mealtimes occur in several areas, ranging from the kitchen to sitting rooms and even bedrooms. However, all families did eat together on at least one occasion in the time period analysed and most preferred to eat in the kitchen if it was large enough to accommodate a table, or in a dining room that was connected to the kitchen. Dining rooms on a linear syntax or on a branch were often only used for ritual occasions (Christmas, Easter) or for functions other than mealtimes. Any analysis of the major activities in the home should therefore not look solely at the activity being performed, but also at the rooms used and where those rooms are in relation to each other.

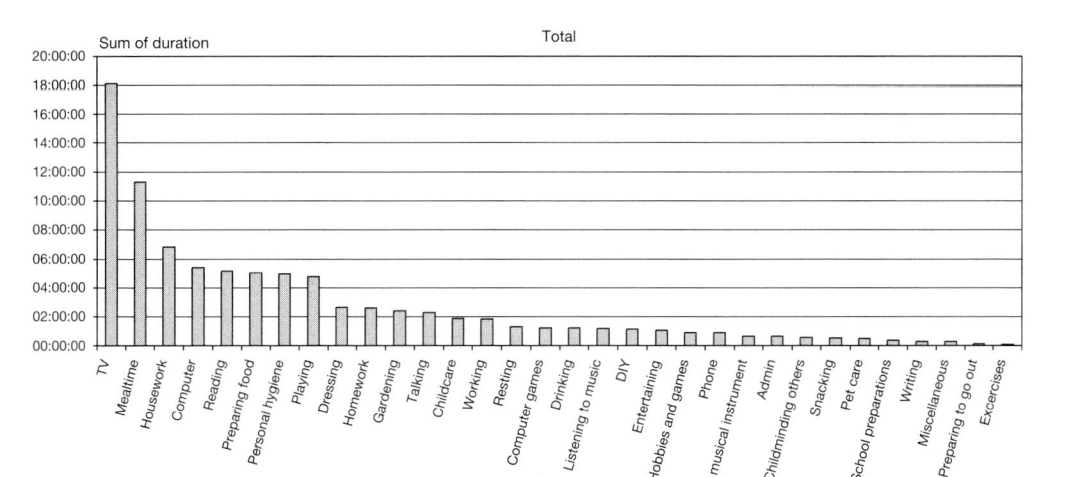

- A – Hall 1
- B – Sitting room
- C – Dining room
- D – Kitchen
- E – Computer area
- F – Utility room
- H – Conservatory
- I – Bedroom A/P
- J – Ensuite
- K – Bedroom F/S
- L – Bedroom M/P
- M – Bedroom F/P
- N – Bathroom 2
- O – Spare room
- P – Bathroom
- Q – Study
- Hall 2
- Hall 3
- Stairs

7.4
Graph illustrating the main activity types in the case study homes.

139

Furthermore, an assessment of time in relation to activity can highlight the temporal nature of space use. The essence of domestic life is expressed through its temporal characteristics, the daily, weekly, seasonal patterns of time (Lawrence 1985). The time diary can also be used to show bubble charts, which relate activity to time and these, in turn, can also relate time to room use to show periods of occupancy. Activity and time for one family showing activity patterns of weekday and weekend are illustrated in Figure 7.5 where the size of the bubble again represents time spent on a specific activity.

Gender can be examined from the recorded data in a number of ways. Individual rooms commonly associated with women, i.e. kitchens, bedrooms and sitting rooms, can be assessed to ascertain whether men or women record more time here than their male counterparts. Likewise spaces commonly associated with men, such as sheds, garages and studies, could be examined. Working patterns can also be analysed in relation to not only space use but the activities performed, in that men and women are now in many instances spending similar amounts of time in the home (or away from it) as more women work full or part-time than in previous generations. The majority of the case study families in the research indicated that women are still responsible for the larger share of housework, childcare and food preparation, irrespective of her status outside of the home. Figure 7.6 is an analysis of these tasks in relation to an adult female working full time.

An analysis of how gender may influence space use also needs to take into consideration how much space men or women actually claim for themselves (see Figure 7.2, on p. 137), which may be as a result of status or assumed spatial hierarchy as appropriated through identification with cultural rules, norms and values. Status outside of the home may also affect how space is appropriated. Results from the study showed that equality outside of the home did have some influence on the way space was appropriated. Women at work in full-time jobs or part-time jobs with equivalent or higher status than their male partners claimed less space in the home than those predominantly at home, and often claimed less space than their partners. Working patterns should therefore be important in any future assessment of domestic space use in relation to house design, as they could be deemed more important than assumed gender roles.

Age as a determining factor in space use is also important and can be examined in a number of ways. The different space needs of adults and children occupying the same dwelling is rarely assessed and those of children of different ages even less so. Childhood is a developing state, not a static concept and more emphasis should be given to the interaction between children and adults in the home. The data gathered can be examined to assess room use or the activities performed by each age range, and where these are performed. The case study data illustrated that children's bedrooms had to accommodate twice as many activities as those of adult bedrooms and in all cases the child's bedroom was far smaller than his or her parents'. The perceived status of the 'master bedroom' needs to be examined in relation to its actual use in future design considerations. When children are young, from babies to the age of about eleven, they tend to

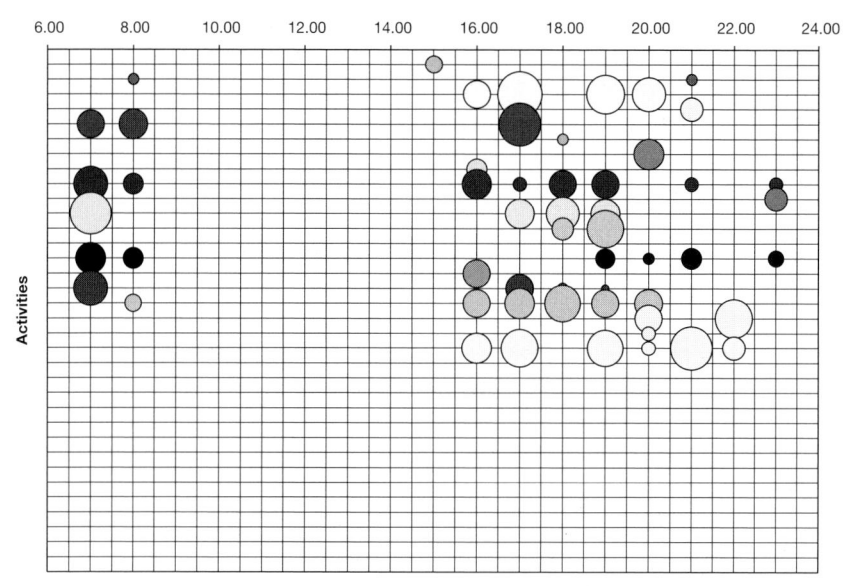

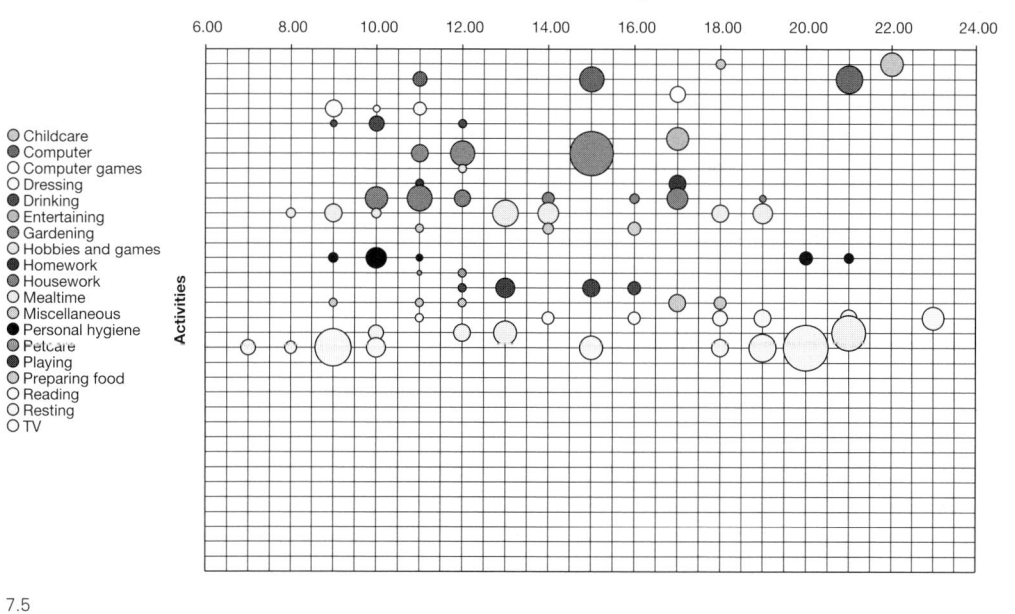

7.5
Time and activity patterns.

Lindsay Asquith

Figure 7.6 Household task allocation.

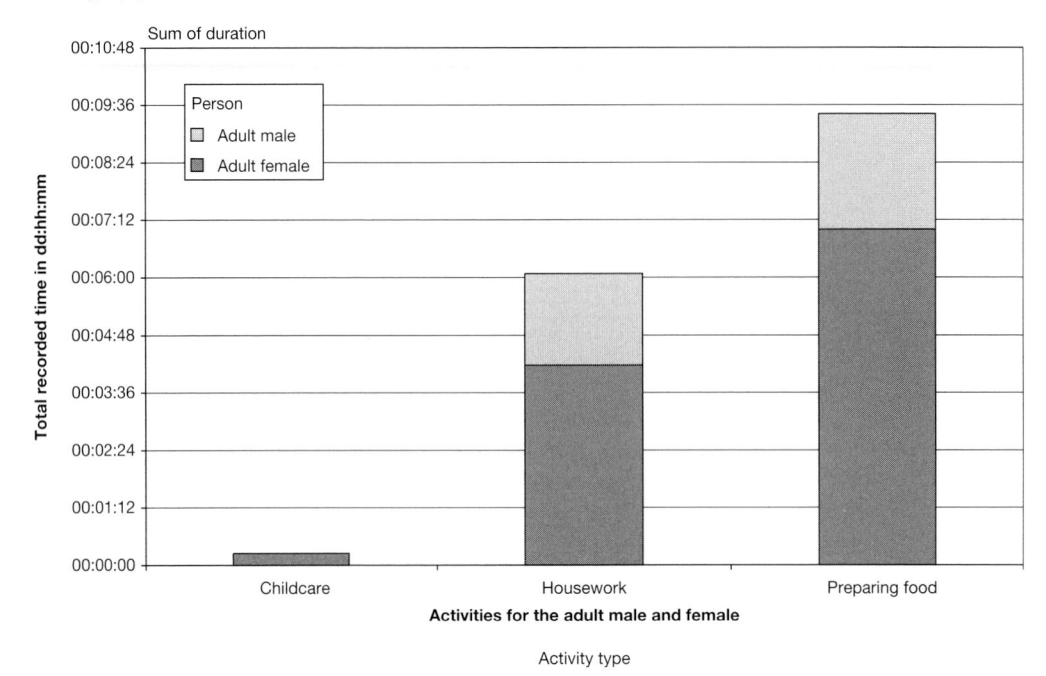

play or socialize in the communal areas of the home, but as they grow older their own bedroom becomes the central focus for many of their activities. This is illustrated in Figure 7.7 by the data gathered on the three activities common to most children; playing, reading and homework, in relation to where these activities occurred in a family with a child aged eight. Analysis on homes with older children showed the same activities occurred in only one room, their own bedroom.

The spatial configuration of the eight-year-old girl's family home is illustrated in Figure 7.8. The communal areas of this home are utilized more by this age group of children than are the private spaces of their bedrooms. Also the looped syntax of the sitting room, dining room and kitchen enhances the communication between family members, ensuring that young children do not feel isolated and parents can be close at hand while engaged in other activities.

New methods from traditional approaches

The combination of methods outlined in this chapter that resulted from an integrated approach to housing research has produced a new approach for qualitative and quantitative research into house design. There are many reasons why mass-produced housing has failed in the past and will continue to do so in the future. One of the major reasons has to be that when design becomes divorced from the user, people are required to live in houses with fragmented spaces, for which they have little understanding and are forced to adopt other cultural norms and values (Oliver 2003). It is now vital that research into housing that

Lessons from the vernacular

Person Child female 1

The spaces used for playing, reading and homework

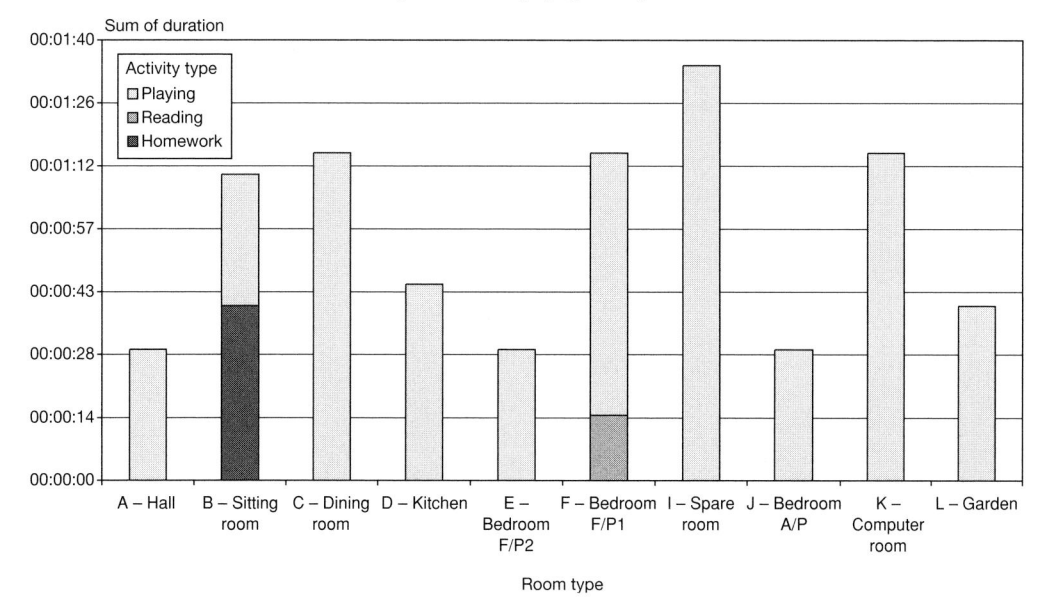

7.7
Space use of an eight-year-old girl.

communicates the needs of the user to the architect and developer is carried out to ensure a future of good housing design. Like vernacular builders, architects and developers need to be involved in all aspects of the production process, to develop design models that reflect the needs of the communities that will ultimately dwell in these houses. As Oliver (1992: 12) states: 'In any design process for the dwelling of the future, the issue is not of a rapport between design and residence but at a profound level between designer and resident.'

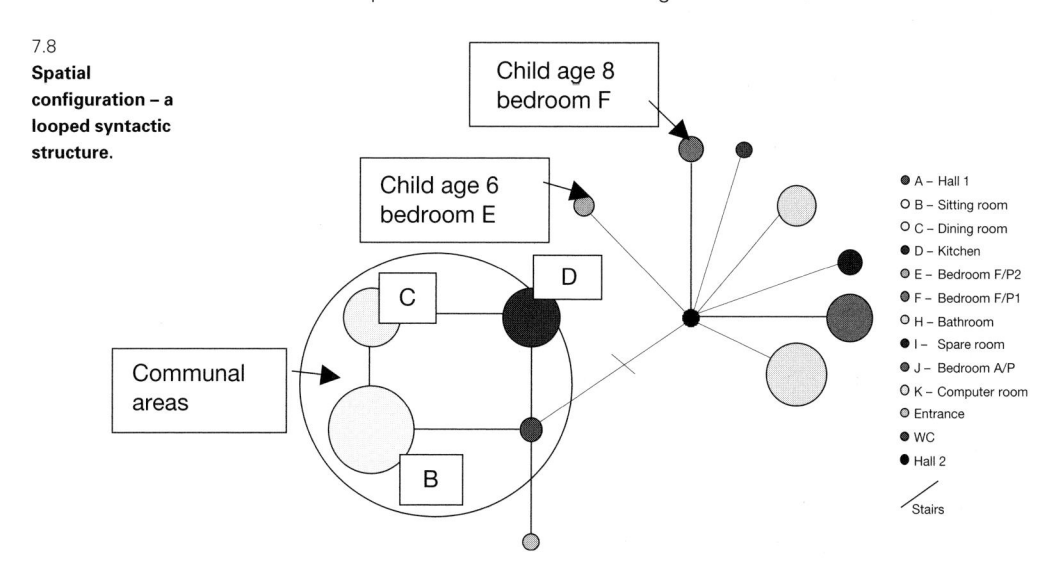

7.8
Spatial configuration – a looped syntactic structure.

143

Lindsay Asquith

The research that contributed to the development of these methods for understanding behaviour in a domestic spatial setting, while also taking into account spatial configuration and family type, is only a beginning. It has highlighted the need for future development of not only these methods, but also others to establish a methodological protocol in housing research. Only then will there be the essential understanding of the housing needs of diverse societies in a multitude of regions and circumstances.

Chapter 8

Sheltering from extreme hazards

Ian Davis

> *A town is made of buildings, but a community is made of people; a house is a structure but a home is much more. The distinctions are not trivial, nor are they sentimental or romantic: they are fundamental to the understanding of the difference between the provision of shelter which serves to protect and the creation of domestic environments that express the deep structures of society.*
>
> (Oliver 1981: 41)

Instant shelters

Shelter following disasters is a theme that provides endless fascination for legions of intrepid inventors, relief officials, architectural and industrial design undergraduates, and product manufacturers. Typically, these designers observe poignant media images of the plight of disaster survivors wrapped up in blankets, and are driven to their drawing boards. Sometimes the motivation is humanitarian, while on other occasions there is a clear expectation of rich commercial pickings (Davis 1978 and 1981).

In 1972 I embarked on the first PhD research on the subject of post-disaster shelter (Davis 1985). From this time onwards I have received numerous letters, brochures, phone calls and visits from product inventors and designers who tend to share a deep frustration that they are unable to get their cardboard, plastic or metal shelters into full-scale production as a global product, so that they can be transported in the hold of those cargo planes that the media delight in screening as they fly off to some remote disaster site in the dead of night.

On one occasion an enterprising retired postman from Somerset in England turned up in my office to ask me to accompany him to the college car park. He then took me to the trailer of his bicycle and proceeded to pull out various items in aluminium, rope and canvas sheet in order to erect an instant disaster shelter, as a crowd of curious students gathered round to watch the most unusual spectacle. He was not a happy man. He told me that he had spent a full two years riding his bike and trailer and its shelter contents all over Europe. His quest was to try and convince UN Agencies, the Red Cross and countless other agencies that single handed he had solved the global post-disaster shelter problem. Alas, nobody had shown even a remote interest in his invention packed away within his bicycle trailer, and rather sadly few officials had even accepted his invitation to join him in their car parks for demonstrations. I admired his energy and enterprise, but he failed to get the letter of endorsement he wanted to strengthen his case.

Some designers have had even more ambitious intentions, which are more in the league of *Thunderbirds* and International Rescue missions:

> Moss the tentmaker will not be satisfied until someone buys his favourite idea – an already tested shelter that can be rushed to an earthquake, or other disaster stricken areas. Carried over the site by helicopter and released in mid air, it opens like a parachute and drops safely to earth, ready for immediate occupancy.
>
> (*Time* 1976: 60)

Almost thirty years on, aspirations for the ultimate shelter product continue unabated. Two examples found on the internet demonstrate the process. An Australian Rotary Club is currently seeking donations for 'Shelter Box' (a British invention), for use in the 2004 Asian *tsunami* relief operation:

> Shelter Boxes provide shelter and warmth for ten people for a prolonged period. Each box will be uniform in size, weight and content, and will contain a high quality ten person tunnel tent manufactured from lightweight but resilient and robust materials plus ten high-quality sleeping bags . . . Once on site Shelter Boxes can be used in many useful and varied ways.
>
> (www.shelterbox 2005)

The *tsunami* also captured the attention of an American Professor in a School of Architecture and his student class:

ARCHITECTURE CLASS TO CREATE SHELTERS FOR *TSUNAMI* VICTIMS

A plane bound for Thailand on Sunday, January 9, carried a University of Houston Gerald D. Hines College of Architecture professor and his dream of building shelters for *tsunami* ravaged villagers. Bill Price will survey the damage and identify a village for his students to adopt.

'The idea is for our students to design three prototype shelters for the *tsunami* survivors and have them shipped over', Price said. 'I return from Thailand in a week and have the plan firmed up for the class to consider'. Price teaches fifth year studio, an advanced design class. His students will create the shelter design as part of their class.

'The college believes in hands-on design, and this will most certainly be an example of that', Price said. 'We want our students to be aware of how their designs impact the community. We have a great opportunity to turn a learning experience into a humanitarian contribution'.

(Ramirez 2005)

Inevitably such instant shelters appear to have more to do with the needs of those who generate the concepts and precious little with the harsh, pressing shelter needs of survivors of disasters who need far more than physical protection, especially when it arrives in a novel shape, just one metre high that you have to crawl into in a similar manner to an arctic tent. A common theme of these instant shelters, aside from their lack of interest in cost factors, local cultural constraints or building traditions, is their exclusive focus on providing protection from the elements. All the designers of these units that I have met have failed to recognize the complexity of shelter as 'house' and 'home', as well as satisfying all the other vital functions that are discussed briefly in this chapter.

Shelter functions

In the late 1970s a team of researchers, under my direction, prepared the first United Nations Guidelines entitled *Shelter after Disaster*. This required an international team to gather data on shelter needs and provision from a global range of natural disasters. The guidelines were published in 1982 (United Nations Disaster Relief Organization 1982), and in the subsequent years there has been no significant follow-up research on shelter needs following natural disasters, although this has taken place for displaced persons in refugee and displaced person contexts (Corsellis and Vitale 2004). Currently, in 2005, the UN Office of Conflict and Humanitarian Affairs (UNOCHA) is preparing to undertake a project to update the 1982 guidelines.

When the results of the data collection for the 1982 study were analysed, it was possible to identify nine functions for shelter:

Emergency shelter serves several vital functions:
- Protection against cold, heat, wind and rain.
- Storage of belongings and protection of property.
- The establishment of territorial claims (ownership and property rights).
- The establishment of a staging point for future action (including salvage and reconstruction as well as social organization).
- Emotional security and the need for privacy.
- An address for the receipt of services (medical aid, food distribution, etc.).
- Shelter within commuting distance of employment.

- Accommodation for families who have temporarily evacuated their homes for fear of subsequent damage.

(United Nations Disaster Relief Organization 1982: 8)

A number of critical sheltering needs have occupied the efforts of a wide range of people and agencies following the Indian Ocean *tsunami* disaster of 26 December 2004. The issues illustrate and underline the theme of this chapter, as it explores some of the multiple roles of shelter and functions as noted above.

Factors affecting shelter in Aceh, Indonesia, following the *tsunami* of 2004

Many issues face the survivors of this disaster: where are they to rebuild their homes, and who will assist them? Should they remain close to the place where they lost their homes as well as family members? What if they are fishermen dependent on living in close proximity to the sea? How can they balance everyday survival needs with the remote risk of another *tsunami*? These demanding questions concern the location of their homes and their relationship to the protection and re-establishment of secure livelihoods. A letter from an aid official working in Indonesia six weeks after the disaster, who requests anonymity, states:

> the issue of reconstruction planning in Sumatra is a hot topic and the Government needs to make up its mind damned quick, because the rumours as to what might happen are part of the driving force which makes people leave the security of their temporary camps/hosted families in order to return to their smashed homes, or their house foundations. The Government is rumoured to be considering the enforcement of a two hundred metres to two thousand metres 'no build' zone beside the coast. The national law is, and was to create a band just seventy-five metres, but this was flouted anyway . . . The Government must have an equitable means of recompensating those who have lost their land plots. This is further complicated by the fact that land ownership here is not at all clear, and I have been told that the land registration titles that did exist have been destroyed in the *tsunami*.

Then, to add to the complex human factors of governmental priorities and survivors rights, the survivors are now more vulnerable with an additional hazard threat to contend with:

> Additionally, I am told that the tectonic plate has shifted by tilting down more steeply to the north west, so that the coast has dropped from about one to one-and-a-half meters. This means that the areas subject to flooding have dramatically increased since the *tsunami*.

(Personal communication with the author, 9 February 2005)

In these short eyewitness observations there are a bewildering array of key issues that surround the issue of sheltering from extreme hazards. The comments highlight an interlocking collection of complex concerns that all relate to the creation of sustainable shelter and settlements. These issues include the demanding time dimension of recovery, livelihood security, human rights and occupancy titles, governmental inertia to enforce past land-use controls, and concerns to improve future public safety; and all these issues are compounded by major environmental changes induced by the *tsunami*. Many of these issues are familiar concerns that regularly surface in reconstruction scenarios, and they assert the human rights issue in the provision of shelter and reconstruction.

In this instance it is the conflict of families wanting to take immediate actions to secure their homes, probably running into sharp conflict with the authorities who are sensibly concerned with public safety issues. Both demands are fully understandable, but they can run in opposition. The governmental delay in decision-making on the 'no-build' coastal strip may freeze essential community recovery actions until detailed risk assessment exercises are complete. The dilemma facing the Government of Indonesia is that in deciding to create, and enforce, a 'no-build' coastal strip, this will have adverse local economic consequences as it will force residents from their original house sites, with inevitable consequences to them not being able to continue to pursue their traditional livelihoods. The issue is: how can long-term safety be balanced with the re-establishment of livelihoods?

Underlying the dilemma is the recognition that the location of homes may be a direct function of the livelihoods of fishermen, who will not be able to operate if they do not live in close proximity to their boats and the sea. Then the livelihood of the fishermen is closely linked to all the support industries, such as making and repairing nets, boat building, and processing and selling fish. In 1986, I faced a similar issue in the Bicol region of the Philippines where, after a devastating cyclone, the government sought to relocate fishing communities away from the beaches where they lived in highly vulnerable conditions to a safe site on a hillside 2 kilometres inland. In this instance, the naive plan was to turn these people from fishermen and net makers into rural handicraft workers. Later it became clear that it was not a humanitarian concern that was driving the project, but rather the actions of local hotel owners who had wanted for many years to get access to the beaches for their tourists.

Several thousand miles away from Sumatra an identical issue is under discussion in Sri Lanka, where the government is also considering the enforcement of a safe 'no build' zone beside the beach, as protection from future *tsunamis* and the flood surges that can accompany cyclones. But here, once again, concerns are being voiced by local advocacy groups that the pressure to clear fishermen off the beach sites is not motivated purely by safety concerns, but rather derives from the powerful forces of hotel companies who for years have been wanting to clear unsightly fishing communities and all their paraphernalia from the beaches that they want to see occupied by their sunbathing tourists reclining beneath wafting palm trees.

Ian Davis

Shelter priorities of survivors in Sri Lanka following the *tsunami* of 2004

In Sri Lanka, in February 2005, discussions concerning recovery options are in progress between the Government, the United Nations and international funding bodies such as the Asian Development Bank. A number of possibilities have been identified for the transitional shelter that is being proposed as an interim measure prior to full reconstruction. It has been suggested that each family should have a number of location alternatives, and these can be set within an order of preference that was determined from consultation with the survivors. Alternative A was the clearly stated preference of the surviving community, with Alternative B being less favoured, and so on. The order is significant in terms of the functions of shelter:

> A: return to their own land.
> B: relocation on a new plot identified by and supplied by the Government.
> C: living with a relative, a host family or in rented accommodation, prior to moving to a permanent site (alternative A or B) and reconstructing their homes.
> D: living temporarily in a transitional settlement, also called camps, prior to moving to a permanent site (alternative A or B) and reconstructing their homes. Transitional settlements should be considered a last resort, in part due to the difficulties in maintaining protection and livelihoods.
>
> (Personal communication with the author, 10 February 2005)

It is significant that the survivors priority scale for post-*tsunami* shelter in Sri Lanka in 2005 is similar to the findings from the research data in *Shelter after Disaster* (United Nations Disaster Relief Organisation 1982). Within these guidelines we recorded the averaged results from the fieldwork conducted on post-disaster shelter needs and provision:

> Survivors show certain distinct preferences for their shelter in the aftermath of disaster. The evidence suggests that their priorities are:
> 1 To remain as close as possible to their damaged or ruined homes and their means of livelihood.
> 2 To move temporarily into the homes of families and friends.
> 3 To improvise temporary shelters as close as possible to the site of their ruined homes (these shelters frequently evolve into rebuilt houses).
> 4 To occupy buildings which have been temporarily requisitioned.
> 5 To occupy tents erected in, or next to their ruined homes.
> 6 To occupy emergency shelters provided by external agencies.
> 7 To occupy tents on campsites.
> 8 To be evacuated to distant locations (compulsory evacuation).
>
> (United Nations Disaster Relief Organization 1982: 6)

The priorities of the Sri Lankan survivors emphasize some contradictory priorities of families. Some had a priority concern to return to their own locations despite the *tsunami* devastation, while others displayed an understandable openness to voluntary relocation following the devastation of their dwellings. Living with families and friends was a moderately popular option, while living in tents or temporary structures was considered most undesirable. Thus the first priority and the last conform precisely to the 1982 UN shelter findings cited above.

In terms of the functions of shelter, these priorities suggest that families have a strong identification with the location of their homes, and that they prefer to live in overcrowded conditions, sharing with relatives or friends rather than accepting the offer of a temporary shelter or tent. It is not easy to detect any strong concern for safety in these survivor's priorities, other than the acceptance of some to be relocated by the government, possibly motivated by fear of another disaster.

Stereotypical assumptions concerning shelter functions

Up to the present time the popular perception of the public, as well as relief officials concerning shelter following disasters, has been the need to provide their occupants with physical protection from adverse climatic elements, such as extreme cold, heat or wet conditions. This need is of obvious importance but the examples cited above following the 2004 *tsunami* disaster go some way to show a far wider range of concerns. For countless relief officials, there has often been incomprehension that the physical provision of temporary shelters or tents has often been rejected by survivors as being the least desirable shelter option. In the stereotypical view, physical protection from the elements has been perceived as shelter's sole function, and attempts to talk with the recipients of shelters to ascertain whether there happen to be any other unmet needs are an extreme rarity. Furthermore, there is evidence that officials continue to make the facile assumption that people are prepared to dispense with their normal patterns of living, or their traditional forms of dwellings, under the extreme pressure of emergency conditions. However, this frequently repeated notion that people will accept anything after a disaster event, again and again has been demonstrated to be a myth, given the reality that a rich diversity of functions, in response to well-articulated needs, remain the normal pattern for a typical shelter and settlement environment in emergency or transitional situations.

However, just twenty years after the first international research on the subject was published in 1982, belated recognition is now being given to shelter possessing a far wider role than a physical covering, and not surprisingly these functions are found to be similar to those of more permanent dwellings. For example, in the 2004 edition of the Sphere Project's *Humanitarian Charter and Minimum Standards in Disaster Response*, the diverse functions of shelter are described:

> It should be acknowledged that shelter, in addition to providing protection from the climate, security and privacy for individual households, etc., also serves other purposes. These include the establishing of

territorial claims or rights, serving as a location at which to receive relief assistance, and the provision of post-disaster psycho-social support through the reconstruction process. It can also represent a major household asset.

(Sphere Project 2004: 221)

Limited recent attention given to the subject of disaster shelter has indicated that shelter has a range of closely interconnected functions. These can conveniently be classified within five broad contexts: physical, environmental, social, economic and political functions. Some are familiar, well-documented roles such as offering environmental protection from the onslaught of an anticipated hazard (such as an impending flood, wind-storm or volcanic eruption). Other social roles are less well-recognized, yet remain essential in effective disaster management, such as the need for shelters in order to provide an address for the receipt of vital services (food, cash grants or loans and medical assistance).

Most of the functions described are essentially 'positive roles'. However, certain political constraints regularly apply to influence the design of settlements, and these pressures are certainly not in the interests of the shelter occupants. For example, politicians, as well as local community leaders, have frequently dictated a requirement for shelters to fulfil a symbolic political function to make certain they 'appear temporary' (with all the possible deprivations that such status may confer). This design requirement may have no rationale, other than to reassure anxious local communities who live in the vicinity and fear that the 'temporary' settlement will become permanent, thus posing a real or imagined threat to their own status and livelihood security.

Livelihoods and shelter linkages

The examples cited from recovery actions following the *tsunami* stress the importance of the link between shelter and livelihoods that is associated with the location of settlements. But the links continue into issues of public safety and vested interests. Both cases indicate that the process of sheltering in the wake of a disaster has to be recognized as a political as well as a humanitarian, economic and physical process.

In Cambridge a group of dedicated engineers, planners and architects under the leadership of Tom Corsellis have created an organization called Shelterproject. As noted above, they have developed a definitive set of guidelines concerning the needs of transitional settlement for displaced populations (Corsellis and Vitale 2004). One of the rich qualities of this set of practical guidelines is the broad view the authors take on shelter. This provides an excellent context concerning wider functions and linkages to associated sectors. One of these dimensions concerns the link between a livelihood approach and shelter provision:

Beyond security, survival and health, transitional settlement has important effects upon the ways in which individuals and families take decisions and action, and seek to provide for their own needs. The support of transitional settlement can stimulate the recovery of their livelihoods, particularly as shelter and the land connected to shelter are major capital assets.

(Corsellis and Vitale 2004: 21)

In reconstruction planning a frequent conflict can be observed. This is on the one hand the desire of many political leaders to achieve a rapid recovery, especially if there are forthcoming elections. In this scenario, governments often sign contracts with large firms of building or engineering contractors to rebuild destroyed settlements. The perceived strength of this approach can be found in just two directions: speed and efficiency. However, the inherent weaknesses are the virtually total lack of involvement of the survivors in their recovery. This approach can be called the 'commercial option'.

In contrast, the 'livelihood option' leads towards a far more holistic approach to reconstruction. Three aspects of recovery can be integrated: the involvement of survivors in various roles, economic recovery and physical rebuilding. Survivors' involvement can include various useful possibilities that include their participation in decision-making concerning the settlement and dwelling design, and their active roles as building workers, where they can be trained in safe construction (Aysan *et al.* 1995). One of the major values of this approach concerns psycho-social recovery. If a widow or widower has lost their partner as well as children, then work is essential as a part of their recovery therapy. However, after disasters there is often an acute shortage of work due to the collapse, or interruption of local industry or commerce.

So the opportunity to use survivors as part of the local workforce can bring rich dividends. It can put money in their pockets that can fuel the recovery of the local economy as well as their own family rehabilitation. It can also enable them to learn new marketable skills in safe building. Thus the strengths of this approach are extensive: buildings are built, skills are developed, local economies are strengthened, bereavement can be softened and participants will 'own' their new settlement in a manner that is not possible when they simply move into the commercial option.

The only major weakness of the livelihood approach is that it inevitably takes a lot longer to accomplish than the commercial approach. However, there are rich gains to be secured from an understanding of the functions of shelter in this approach, such as educational opportunity, income source, economic generation, therapeutic value and the builder of communities as well as physical structures.

The underlying principle behind the livelihood option is to propose that officials responsible for reconstruction strategies and policies need to recognize that local skills and resources, drawn from the affected area, should be used before looking outside the area for support. If this policy can be adopted, the local economy can significantly benefit from reconstruction and recovery activities.

Two shelter contexts

To conclude it may be useful to examine the differing functions of shelter in a pair of very different disaster situations. The first example concerns shelter in Kobe, Japan, following the earthquake of 1995. In Kobe it was possible to observe the distinction between the typical 'sheltering' needs (social and psychological), as opposed to 'shelter' needs (physical provision) following a major disaster. The 350,000 people who were displaced by the earthquake found initial shelter through three approaches: First, by moving into the homes of friends and relatives (between 20,000 and 40,000 persons); second, by occupying schools and community buildings (about 182,000 persons); third, by improvising structures in open spaces such as parks and sports grounds; and finally, by repairing their dwellings and remaining within them (Davis 2001).

The second example was observed in 2002. Joe Ashmore, working with his colleagues in the shelterproject in the Martin Centre of Cambridge University, undertook a detailed study of shelter needs and response to internally displaced persons (IDPs) in Afghanistan (Ashmore 2003). Their paper analysed the diversity and adaptation of shelters in transitional settlements. A key conclusion from this project supports the concept outlined by Paul Oliver in quotation at the beginning of this chapter, that shelter is a far broader issue than mere physical protection and provision:

> in Maslack, many IDPs noted that food security and the desire to return were their main concerns, not their shelter structure. When asked how their transitional shelter could be improved, a common response was nothing, because return home was the preferred – perhaps the only desired – option. 'Home' refers not only to the location where they normally live, but also to the services, availability of work and social infrastructure which were present before forced displacement. This result indicates the wider scope of any emergency situation, that the IDPs comfort, security, and wishes go beyond the physical shelter objects which can be provided.
>
> (Ashmore 2003: 285)

Conclusion

The discussion in this chapter has been wide-ranging. The essence in observing the rich series of functions of shelter is that they grow from taking a 'process' rather than a 'product' view. This can be effectively emphasized by the use of the word 'sheltering', as opposed to the use of the word 'shelter' with its implication of a finite physical solution. Thus a broader and deeper understanding is needed if progress is to be made in developing recovery strategies. But the persistence, even growth in the quest to adopting the 'commercial option' for rapid reconstruction, or the design of instant shelters as described at the outset of this chapter, are significant. They indicate very clearly that we are still a very long way from any recognition of sheltering as moving beyond a tangible 'shelter' or 'house' to become a home, financial asset, income generator, place to receive services, bereavement therapy, skill generator and workplace.

Chapter 9

A journey through space
Cultural diversity in urban planning

Geoffrey Payne

Introduction: ethno-centrism in space

This chapter explores the ways in which people from different cultural and political backgrounds have evolved rational and ingenious solutions to meet their shelter needs and the lessons this may offer for professionals working in the field. It is also a personal journey through these issues which reflects on a range of international research projects and practical experiences over a period of more than thirty years.

The concepts with which we perceive and use space are like a language; we use them intuitively and almost without thinking of the structures involved. Yet there are hundreds of mutually unintelligible languages, all with their own distinct dialects, so why should there not be diverse ways of perceiving and using space? At a time when western attitudes and practices are extending their reach through globalization, it may be that non-western ways of perceiving and using space provide people with an important sense of their own identity. They may even offer lessons for application or adaptation in other contexts and should form an integral component of development programmes for upgrading existing informal settlements or planning new developments funded, or supported by, international donor agencies. To achieve this, a more holistic and multi-disciplinary approach to settlement planning will be essential for success.

Geoffrey Payne

Much of the literature on this subject in English has tended to assume or imply that European and American concepts of space are universally relevant. In his seminal book on urban form, Morris reviewed the origins of urban settlements but focused on European and American examples, adding a chapter to the third edition (Morris 1994) on the Islamic city, together with appendices on China, Japan, Indian *mandalas* and Indonesia (Morris 1972, 1979 and 1994). While reflecting a welcome awareness of different traditions, the inclusion of these alternatives did not provide an analysis of why these societies had evolved such distinct urban traditions. Similarly, Lynch identifies characteristics of good urban form (vitality, sense, fit, access, control, efficiency and justice), which, while internationally relevant, are illustrated primarily through western examples (Lynch 1981). The launch of the journal *Urban Design International* in 1997 reflected an increasing interest among western urban designers in other traditions by which non-western societies have organized the built environment. However, a rich vein of material exists which exerted a substantial influence on earlier generations and has not perhaps received the recognition it deserves.

Space and place

My personal initiation into these alternative ways of perceiving and organizing space was initially through the work of Rudofsky which catalogued the vast repertoire of ingenious, yet rational settlement and building forms developed by communities around the world without any professional involvement (Rudofsky 1964). A more analytical approach was adopted by Gunter Nitschke in a series of seminal articles published in *Architectural Design* (Nitschke 1964a, 1964b and 1966). These demonstrated that early Japanese urban planning was based on Chinese concepts, which were in turn derived from religious considerations intended to reflect a cosmic order at the urban scale. In this sense, they were similar to the *mandalas* of India, which provided a map of the cosmos and determined the spatial ordering of key symbolic features. Within this system, residential areas were organized on three scales within a rigid grid into the '*bo*', '*ho*' and '*cho*', in which a block of about 120 metres ('*cho*') formed a quarter of a '*ho*' and that in turn was a sixteenth of a '*bo*'. This was the basis for the organization of the early capitals of Nara and Kyoto. However, this somewhat rigid geometry did not suit Japanese conditions, in which large areas of level ground were rarely available. As a result, the plans as built reflected a different ethos based on the Japanese concept of '*ma*' or 'place', which Nitschke defined as the 'simultaneous awareness of the intellectual concepts form + non-form, object + space, coupled with subjective experience' (1966: 117) (see Figure 9.1).[1] He identified three different ordering principles, namely: 'apparent disorder', 'geometric order' and 'sophisticated order', and likened the first category to 'fortuitous order, in which 'man's efforts to impose his own order on nature are unsophisticated' (see Figure 9.2). Nature is the basis of all systems of order, man accepts it as the controlling element – contours, natural falls, river beds and ravines determine boundaries and divisions, roads, the form of villages and buildings. Man acts intuitively, unconsciously as it were, as an extension of nature. Nitschke defined geometric order as a form in which 'man

9.1
Plan of Imai, Japan, a sixteenth-century city.

seeks to impose an intellectual concept or order on nature. Number and geometry are used as the means of control in this conscious stage'. Using early Shinto shrine precincts as examples (see Figure 9.3), he demonstrated how the effect of geometry 'leads to "perfection of type", one aspect of beauty in Japanese terms and how this is restricted by geometry and doomed to lead to rigidity, to a dead order' (Nitschke 1966: 120). However, even in these cases, the designers modified the purity of the designs by moving the entrance gates to the shrines off-centre, as an acknowledgement that perfection was the preserve of the gods. According to Nitschke, sophisticated order:

> emerges only when man has fully absorbed and worked through the principles of geometric order – which pertain to a static, immutable world – and discovers the order of an organic, constantly changing universe. This stage is not altogether unlike the first, but the intuitive grasp of nature has been replaced by perception and a conscious application of her principles. This stage is super-conscious.
> (Nitschke 1966: 118)

He explained the transition from one stage to the next as appearing to be 'simply a progression from unconscious asymmetry through symmetry to conscious asymmetry', but considers it 'in fact far more complex. In each phase there is a different consciousness of space, or rather *place*, which is the determinant in the shaping and placing of all forms' (Nitschke 1966: 118). As examples of 'sophisticated order' Nitschke cites Katsura Palace (see Figure 9.4) and Nijo Castle, in Kyoto, both of which 'permit new elements, of the same, or different quality and size, to be added or taken away as required; in other words, it permits "change" in its three aspects of growth, fulfilment and decay, to take place'. One is reminded of the Chinese/Japanese character for 'change' *eki*, which also stands for 'ease' (1966: 133). He continues:

> Each phase of growth is complete and beautiful in itself; nothing seems to be missing. The eternal architectural paradox [that] of giving an [sic] impression of completeness within incompleteness, is solved. Life itself silently solves this paradox all the time, a human, an animal and tree does not at any stage suggest visual incompleteness. The structure visible in the plans is not only adaptable to change, but is even secretly stimulating growth and life. Though so formally perfect, it turns out to be the opposite of formal perfection which would mean death.
>
> (Nitschke 1966: 133)

Nitschke also demonstrated that form had its counterpoint in space or the void *(ku)*, which made the context of a form an integral component of its development. For a student of architecture in middle England, Nitschke therefore offered another cultural perspective on concepts of space, or rather place and therefore place-making, and demonstrated that these concepts evolve with our relationship to the world around us and each other. In other articles (Nitschke 1964a and 1964b), he reviewed the work of the 'Metabolists', an emerging Japanese group of architects similar in some ways to the Archigram group in the UK. Many of their ideas revolved around the need to incorporate change, growth and decay in the built environment. They shared a common desire to create forms which were both complete at any one point in time, yet capable of change, as found in nature and embodied in the concept of '*Ma*'.

The impact of these articles determined me to visit Japan and in 1967 I was able to raise funding to visit for three months. Despite being unable to meet Nitschke, his approach established the foundation for a lifelong curiosity about how different cultures and social groups perceive and organize space. Among other significant examples was the fact that many Japanese classical gardens, such as the Katsura Palace garden, were often extremely small, yet created a sense of infinity by ensuring that it was not possible to see the whole garden from any single vantage point. Similarly, the layout of temples was designed to create a changed state of mind by literally leading you from the everyday, profane environment across a (spiritually cleansing) bridge and along a stream, at strategic points of which would be located places for rest and reflection, before

9.2 **Plan of Igaia, Japan.**

A journey through space

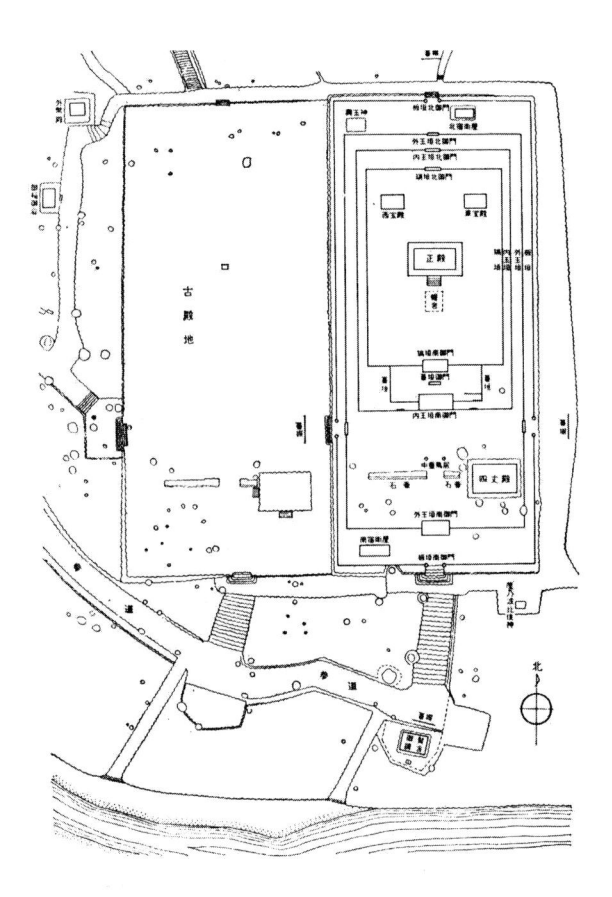

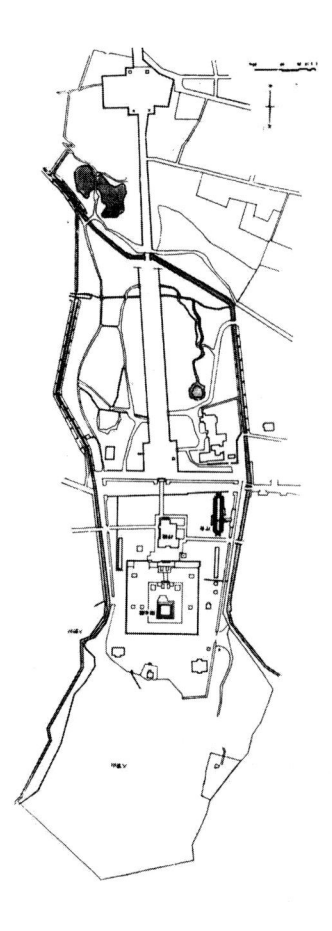

9.3
Inner sanctuary of the Ise Shinto shrine in Japan showing an adjacent alternate site. On the right, the Grand Shrine at Izumo.

climbing steps to the temple itself. The spatial organization of the temple also reflected the social order, since only the Emperor and the High Priest were permitted to enter the inner court, while the second prayer gate, defined with a curtain of pure white silk, was the furthest a normal worshipper was ever allowed. The Japanese concept of place also distinguished between physical objects and their social meaning. For example, key elements of many famous temples, including Ise, were rebuilt on adjacent sites every twenty years, but the new buildings were still regarded as ancient because they still housed the gods.

The visit to Japan exerted a powerful influence over my subsequent approaches to design and planning.[2] The search for rules which permit flexibility and growth led to an interest in games, particularly chess, since these provided a restricted set of pieces and movements, yet could be played at different conceptual levels to produce an almost limitless range of outcomes. The issue of structures which stimulate options for local variations suggesting an organic process of growth and change was addressed nearly a hundred years ago by Patrick Geddes who also used the analogy of chess as a basis of finding ways of improving urban environments and reflecting local cultural norms (Tyrwhitt 1947). In a series of highly innovative studies in India, he advocated the need for

Geoffrey Payne

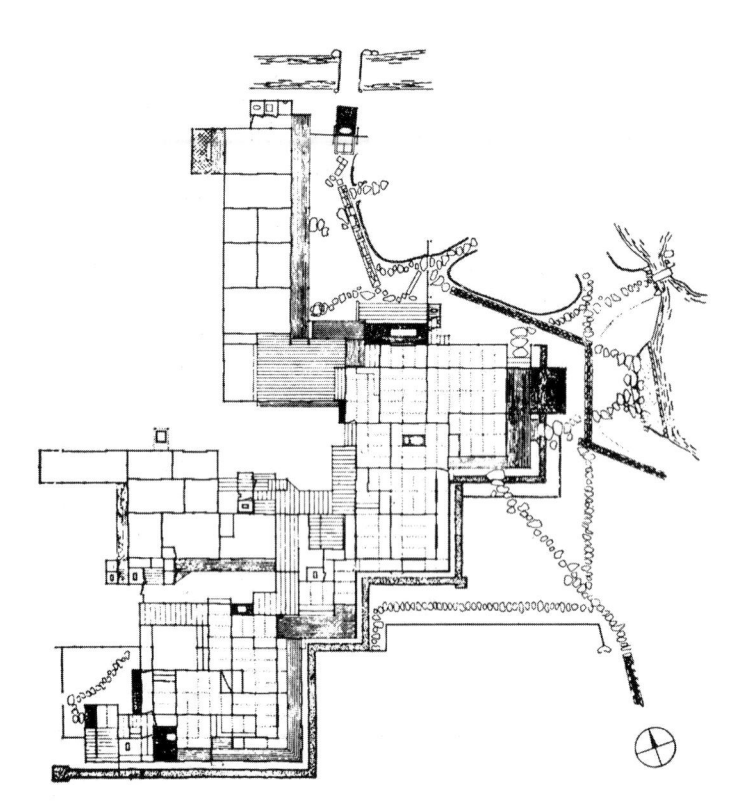

9.4
Plan of Katsura Palace, Kyoto, Japan.

diagnostic surveys and pioneered the concept of 'conservative surgery' in which the role of the professional was to work with the grain of local traditions rather than remove all the pieces and start again. This interest in space as perceived and used, not just seen, was reflected in my final year student project.[3] This explored options for creating a flexible planning structure within which a declining coal-mining community near Nottingham could meet its changing needs, within a social and economic environment that was also changing rapidly (Payne 1969).

Planners versus people

I applied for a Commonwealth Universities Research Scholarship and after travelling through Iran and Afghanistan, I finally reached Delhi and registered at the School of Planning and Architecture in 1970. In the late 1960s, Delhi was already growing at a rapid rate as migrants from rural areas arrived in search of a better life. The demands on housing agencies exceeded the ability to provide social housing for the poorer households and led to the emergence of numerous squatter settlements scattered throughout the capital. The professionals involved in housing and urban development, together with politicians, administrators and many middle-class voters regarded these as a scar on the city's grand colonial environment and supported measures to remove them. This culminated years later in a crisis which forcibly removed up to half a million slum dwellers and contributed to the collapse of Indira Gandhi's government.

A journey through space

Sadly, few people at that time were concerned at the plight of these newcomers to the city or willing to find ways of absorbing them into its economy. In the event, however, the poor did not wait for the authorities to help them and created their own settlements on land which nobody else had claimed or developed and set about finding work in the service sector. Located just behind the capital's commercial centre of Connaught Place, one of these unauthorized settlements accommodated 2,000 people in huts on less than 1 hectare of land (see Figure 9.5).

How did they organize such limited space in ways which enabled them to survive? This question combined the concern for spatial organization aroused in Japan with the social concern aroused in the Nottingham thesis study and led to an intensive period of physical and social surveys together with observation of spatial use patterns at different times of the day and different seasons. What became evident was that the residents had evolved a symbiotic relationship between private and public space which enabled the latter to be used for different purposes at different times.[4] Small '*chowks*' or open spaces under trees became communal meeting, working and commercial spaces due to the shade they offered, while the main thoroughfare was used for cooking in the evenings and grazing animals, as well as circulation. Other huts were grouped around small communal courts and all huts were close to the main thoroughfare, a *chowk* or a court. It was this symbiotic arrangement which enabled people to live at such high densities without apparent tensions and created an environment which, while lacking even basic amenities, was convivial in the sense that Geddes had recognized so many years earlier (see Tyrwhitt 1947).

To architects and planners indoctrinated by the professional views that only environments that they had created were ordered, such solutions were understandably anarchic. Yet the environments created by such professionals imposed a superficial order based on concepts of planning inherited from the British during the colonial period or influenced by foreign architects such as Le Corbusier. These schemes lacked conviviality and segregated housing from work and recreation, unlike the more dynamic and multi-functional informal

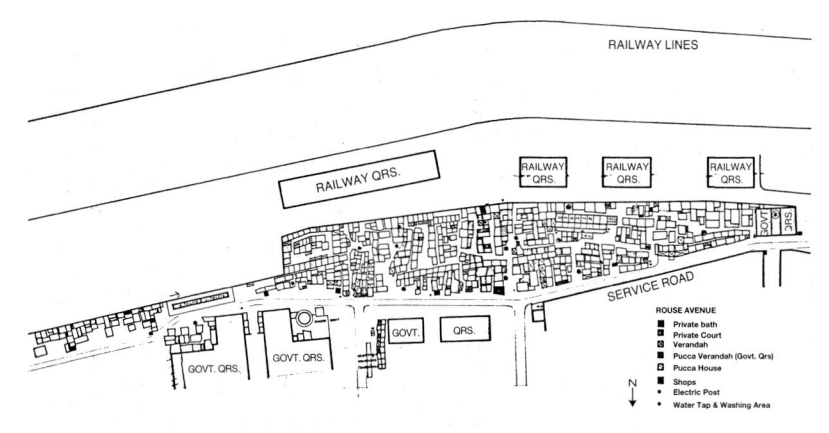

9.5
The Rouse Avenue squatter settlement in central New Delhi, before its demolition in 1976.

settlements. They created sterile environments which often depended on substantial subsidies and therefore (fortunately) could not be built in sufficient numbers to meet ever increasing needs. Le Corbusier's proposals for Chandigarh, for example, reversed the traditional arrangement of mixed land use and medium to high density with narrow streets that suited the climate, and replaced them with segregated land uses, low densities and wide roads. Housing accommodating the low-income groups was on the cheapest land furthest from the main employment centres, thereby imposing heavy travel costs within a grid layout suited for European levels of car ownership, not the public transport system needed by the majority of local people. The plan for Chandigarh was largely a solution to Europe's urban planning problems rather than those of India. However, the real cost was far greater than the inconvenience it imposed on its residents, since it encouraged local professionals and succeeding generations to deny the rich indigenous traditions for organizing space that had evolved over many centuries and replaced them with a half digested set of alien and largely inappropriate values. This permeated into popular consciousness and encouraged many people to regard traditional environments and building designs as backward and therefore to be discouraged by the expanding middle classes of many developing countries. In this way, local aspirations for progress became synonymous with denying local achievements and adopting foreign values and mores. This evoked uncomfortable memories of visiting Hiroshima to find a city rebuilt largely in the image of the west that had destroyed it decades earlier.

The extent to which ethno-centric western professionals and their local counterparts intentionally or unintentionally undermined local confidence in non-western ways of perceiving and using space went largely unchallenged in academic courses in either the west or developing countries themselves. Even urban design, which originated out of a recognition that architectural preoccupations with their individual buildings, and planners with their two dimensional plans, had left a conceptual void in the three-dimensional urban environment which we actually experience and use, did little for many years to recognize the rich variety of spatial traditions still remaining but under threat. This prompted me into awareness that housing was not the preserve of any one conceptual framework or profession and could only be tackled successfully by multi-sectoral, culturally diverse teams working together and willing to listen to the people they sought to help. If my training had been based in the UK, it was therefore in India that I received my real education.

Cultural concerns go mainstream
This interest could easily have been a temporary personal experience had it not been for meeting Paul Oliver on my return from India. Following an invitation to give a lecture at the Architectural Association School of Architecture on my Indian studies, Paul offered me a post teaching in the Graduate School. This exposed me to his own writings (Oliver 1969 and 1971) and those of John Turner, Amos Rapoport, Bill Hillier, William Mangin, Anthony King and others engaged in the field of housing, spatial organization and the role of communities. John Turner had

already published seminal papers on informal settlements in Peru and created models explaining the logic behind their development (e.g. Turner 1965 and 1967), which was later articulated into an influential book (Turner 1976). Meanwhile, Rapoport had followed up his book on *House, Form and Culture* (Rapoport 1969) with publications identifying the concept of a 'cultural core', which involved communities retaining certain modes of acting and spatial organization considered central to their sense of identity in order to more easily accommodate change in areas considered less critical (Rapoport 1979). Meanwhile, King was extending his work on colonial urban ways of perceiving and using space as a means of social control (King 1976) into the domestic environment by demonstrating that the design of the bungalow was itself based on traditional Bengali designs, adapted to impose separation of the local servants from the colonial masters of the house (King 1984). For one academic year, I was able to combine a full-time teaching post at the Architectural Association with being a full-time masters student up the road at University College London, where Bill Hillier was pioneering an environmental studies course, the basis of which was to become space syntax theory (see Hillier and Hanson 1984).

Teaching at the Architectural Association included the opportunity to undertake international study visits and in 1974 we visited Ankara, Turkey, where almost half of the city's population were living in unauthorized settlements. Ankara was different in several ways to Delhi and other cities in urbanizing countries. First, and perhaps most importantly, the government had sought to superimpose liberal, secular structures onto Turkish society and to use these to emulate the process of economic development achieved in the west. High priority was given to modernization, industrialization and urbanization, an approach fundamentally different from the anti-urban bias found in many other countries.

While the new capital was planned according to international standards current at the time and incorporated provision for low-income housing areas, these were rapidly swamped when migration accelerated and the city regularly doubled in population every decade. Under the Ottoman Land Act of 1858, villagers in the under populated Anatolian plateau were permitted to occupy unused state land, providing they developed or cultivated it. Naturally, rural urban migrants were quick to exploit traditionally legal ways of occupying and developing public land in the city, even though different laws applied there. The outcome was a rapid expansion of informal *gecekondu* settlements, mostly on public land.[5] Initial settlements were planned with lots of narrow, twisting lanes and dead ends in order to discourage strangers, especially officials or the police, from entering and to disorientate those who persisted. By exploiting legal loopholes which prevented completed dwellings from being demolished without a court order, settlers became more confident and planning layouts gradually changed to a more open, regular pattern (see Figure 9.6).

This became even more regular in later re-blocking as *gecekondu* settlers sold their informal rights to developers skilled in obtaining official permissions for apartment buildings or *apartment-kondu*. Such legal plurality or ambiguity, together with resource constraints and limited local powers, led to a

Geoffrey Payne

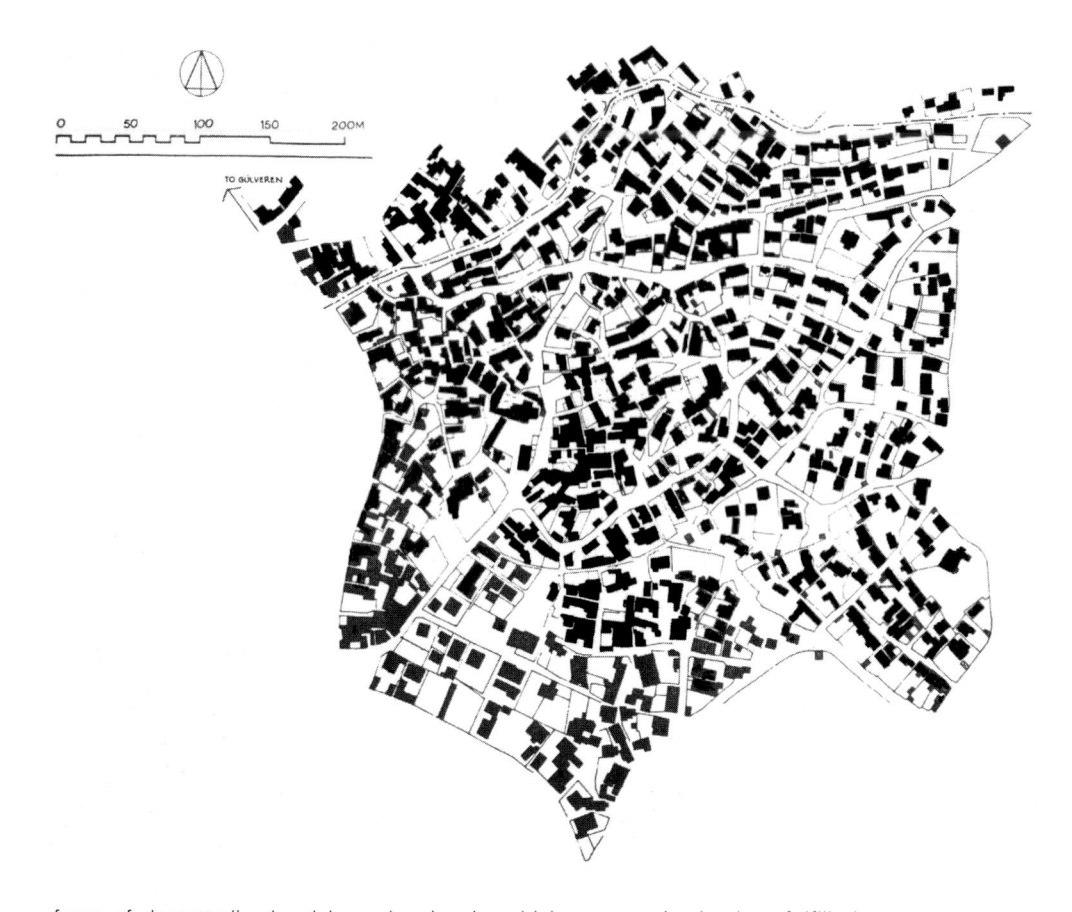

form of decentralized, ad hoc planning in which community leaders fulfilled similar roles as they had done in isolated rural communities. The result was that many communities generated their own development proposals for transport links, services provision, layout regularization and so on, and had a major influence over the implementation of such developments (see Figures 9.7a and 9.7b). Although the process was strongly influenced by political considerations, communities nonetheless had a major role in determining the form and nature of local development and suggested that, in this respect at least, Turkey had much to offer other countries, including the UK.

From theory to practice

In the late 1970s, an opportunity arose to put some of the research findings and recommendations into practice as part of a consultancy team developing urban land and housing projects in Ismailia, Egypt. Because of two wars with Israel, the Suez Canal region had been evacuated between 1967 and 1974, but when the Israeli army withdrew from Sinai, the Egyptian government was determined to redevelop the three main cities of Port Said, Ismailia and Suez City. As in Turkey, here was another case of a government actively encouraging urban growth and seeking to manage it effectively, rather than inhibiting it. The UK consultants

9.6 Part of the Bahçelerűstű squatter settlement, Ankara, Turkey. Note the irregular inward-looking layout in the older area to the west, developed when the threat of official demolition was strong and the more regular, outward-looking layout in the later area to the east after the threat had receded.

9.7a and b
The main road in Şehit Mevlut Meriç mahalle, Ankara in 1976. The same view in 1981 after a community led upgrading programme.

prepared master plans and were then invited back to prepare 'demonstration projects' to show how the objectives of the plans could be implemented.

The team commissioned to prepare proposals for settlement upgrading and new developments in Ismailia had an advantage over the other teams in that the city was surrounded by an unlimited expanse of flat, government owned desert land.[6] Another key advantage was that the Governor was amenable to new ideas. A series of diagnostic surveys of the housing market were carried out and revealed that a majority of all households, and most poor families, lived in informal settlements, of which the largest was named El Hekr after the land tax or ground rent paid to register a claim to unused state land. In fact, this practice

Geoffrey Payne

was based on the same Ottoman Land Act that applied in Turkey, whereby people were entitled to occupy and develop unclaimed state land on payment of a modest ground rent.[7] The official acceptance of this process resulted in layouts which were an extension of the formal grid layout used by the city's French planners who developed the city initially as the headquarters for administering the Suez Canal which they were constructing.[8] However, while the main north–south roads followed the formal plan, east–west links took on a much more informal character, providing a range of semi-public, semi-private spaces for communal use (see Figure 9.8). These evoked similarities with traditional urban layouts throughout the Middle East, which consist of compact layouts with mixed land use, medium rise and medium density, the qualities that most good urban environments seem to embody.

In preparing proposals to upgrade the existing settlement and expand it in ways which were efficient, affordable and appropriate to the needs of local people, the consultancy team undertook detailed physical, social and economic surveys of the existing settlements and communities. A shortage of office space forced the architects, planners, engineers, economists, sociologists and management experts involved in the project to interact continuously. This contrasted sharply with the ways in which housing projects were formulated and implemented routinely in Egypt and other countries, by which planners would identify potential project sites, architects would prepare detailed plans which would then be passed to engineers to ensure conformity with standards and regulations, and then passed to the accounts department which would invariably lack adequate funding, necessitating either delay or a repeat of the complete process.

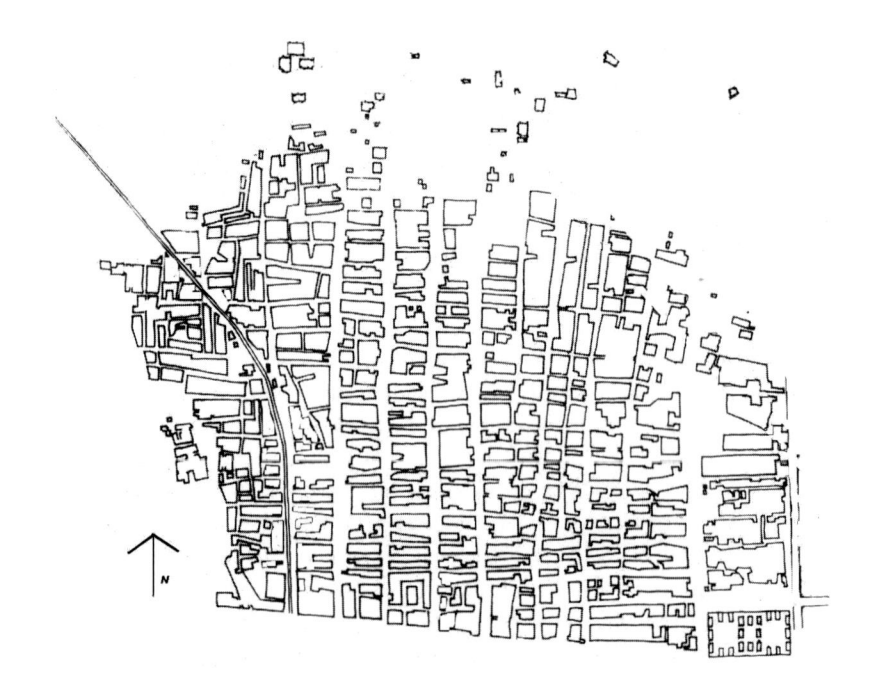

9.8
The El Hekr informal settlement in Ismailia, Egypt, in 1977 before upgrading and expansion.

Multi-disciplinary teams were able to avoid such wasted efforts and provided a more stimulating professional environment to work in, since each team member was able to see issues from a different perspective than their own and broaden their understanding and effectiveness as a result.

Years later, when directing a masters degree course on Building and Urban Design in Development (BUDD) at the Development Planning Unit, University College London, this approach was adopted with considerable success, despite difficulties from some colleagues who felt that recruiting anthropologists and economists to a design course was inappropriate. Vindication, for me at least, was achieved on a study visit to Morocco to study and make proposals for improving a small town near Marrakesh. All the architecture and planning students studying the existing informal settlements presented graphically sophisticated analyses which focused on the spatial attributes of proportion, texture, scale and appearance. The anthropology student, however, presented a plan of the existing layout onto which were marked a series of red dots. His explanation was that these indicated spaces which were an extension of the domestic environment where he had seen traditional, and usually poor, women working, socializing and supervising children. His conclusion and recommendation was that if planners sought to liberate such people, they should make such spaces an integral component of all new residential areas. None of the architects had been trained to interpret space in this way. Similarly, the economics student, working with an architect student, impressed her colleague with the need to consider costs and affordability to a level which would otherwise never have occurred. It is a sad commentary on the way in which professional institutes validate academic courses dealing with housing and urban development that they reinforce professional boundaries rather than seek to loosen them. This results in many universities throughout the world training students to adopt narrow perspectives on what is essentially a multi-faceted subject.

The Ismailia projects made a great effort to create proposals which were broadly self-financing and provided a range of options for plot size, levels of initial infrastructure provision and construction based on observing how local people developed land and housing and financed such development themselves. As such, we were students learning from those whose expertise was managing to build their lives, and their houses, with the minimum of resources. A critical feature of this process was the need to minimize initial expenditure and maximize future options. Accordingly, the team's proposals sought to permit plot sizes, levels of initial infrastructure and building construction to lower standards than officially permitted, but at least as high as those found within the existing informal settlement. Furthermore, detailed plans included a range of informal communal spaces for the new development and allowed residents in the areas to be upgraded to have control over such spaces (see Figure 9.9). The emphasis was on initiating and reinforcing a process of incremental development which was locally sustainable without dependence on external subsidies.

In the event, the projects were implemented in a very different way than was envisaged. The government rejected mud-brick options and insisted on

Geoffrey Payne

Initial development

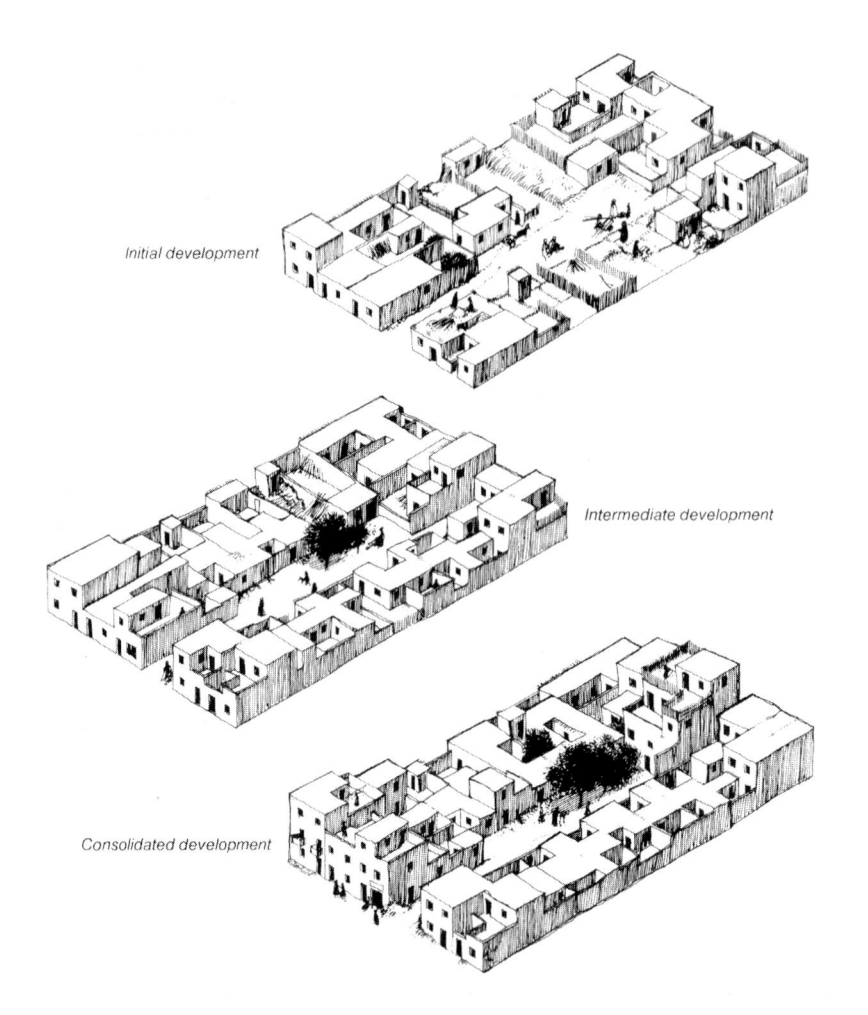

Intermediate development

Consolidated development

9.9
Proposals to adapt the existing layout principles of El Hekr in new development areas in order to incorporate communal open spaces and incremental plot development.

initial conformity to full building standards. It also stipulated that loans be repaid within seven years rather than the fifteen to twenty-five years proposed. This raised both the initial costs and monthly repayments to levels which poor households had no chance of meeting, and thereby excluded them from the project. In addition, another donor agency offered to provide full infrastructure, which made the area potentially the best serviced part of the redeveloping city. Finally, the limited number of clerical staff appointed to process applications was inadequate to undertake checks on eligibility and resulted in many higher income households obtaining plots and building apartment blocks on a speculative basis. In this sense, the project was less successful than had been intended, though there was an ironic twist to the story in that President Sadat, who had a holiday home in Ismailia, was so impressed by what he was advised was largely self-financing housing for the urban poor, that the agency implementing the project was expanded to become the Ismailia Land and Housing Development Agency with a brief to apply the principles throughout Ismailia. While we had therefore failed to implement the projects as envisaged, we had inadvertently succeeded beyond any reasonable expectation in demonstrating the benefits of the approach.

Lessons learned

The Ismailia projects demonstrated clearly the intimate interactions between social, cultural, economic, technical and political aspects of settlement planning. Nowhere were these aspects more closely interwoven than in Jordan, where many thousands of Palestinian refugees have been living in informal settlements for decades. Since the attitudes of both the government and the refugees themselves was often ambiguous in terms of wanting full citizen rights in Jordan, in case these prejudiced their eventual return home, whole communities were living in limbo. In one case known as East Wahdat, the residents lived on a steeply sloping site in tin huts. However, they were not necessarily poor. Their inadequate housing was due to the owner of the land permitting them to occupy it on the condition that they did not construct permanent houses, leaving him the prospect of developing it to its full commercial value at some future point. As part of a World Bank loan, he and other landowners agreed to receive payment for the land and enable the community to develop it in ways which would stop sewage running in open tracks down the main pathway, and enable them to realign their plots and rebuild their houses when they were ready and able.

Preparing plans to upgrade the settlement was not made easier by the difficult topography and the informal way in which the settlement had evolved. These were compounded by an inaccurate survey and local suspicions as to the intentions of foreign planners and surveyors visiting the site. On a personal level, travel within Amman involved time-consuming journeys into the city centre and out on another radial road, limiting productive time on site. To overcome these constraints, it was decided to develop the proposals locally and a friendly resident with a smooth concrete floor to her hut agreed to allow us to spread the map of the area out and use her house as our office. Needless to say, neighbours were soon coming to the door to see what was going on and how they would be affected by the project. However, they were mostly unable to make sense of the map and its complex contour lines, plot and building lines, etc. This inhibited their ability to offer suggestions and was in danger of reinforcing suspicions. The solution to this problem proved simple. Pieces of expanded polystyrene, bits of wood or matchboxes were found and placed on the map to indicate buildings, and people were immediately able to identify their own house, suggesting that three dimensions are more easily understood by lay people than two. With increasing interest from local residents, it proved possible to complete a provisional plan in less than ten days thanks to local participation. Many residents suggested ways of routing paths that would not require them to rebuild, while others offered ideas based on intimate familiarity with the site which no amount of outside analysis would have yielded. After years of exposure to the literature advocating participatory, bottom-up approaches to local development, here was tangible evidence that it was soundly based. The result was a project which, after considerable refinement by the project team, resulted in a highly attractive modern version of

Geoffrey Payne

a traditional Arabic urban environment, with narrow pedestrian lanes to provide shade, but also access for cars (see Figure 9.10a and b).[9]

The Middle East was urbanized long before Europe and has a long tradition of sophisticated urban forms which reflect cultural, climatic, economic and other factors. Unfortunately, contemporary planners, reflecting perhaps the desire to demonstrate modernity in western terms, have done too little to retain,

9.10a
The East Wahdat informal settlement, Amman, Jordan in 1981.

9.10b
Proposed layout after upgrading, retaining the traditional winding, narrow pedestrian dominated circulation pattern.

adapt or re-interpret such traditions and have tended to either replace them completely or provide a pastiche of traditional elements which ignore the principles upon which they were based. Organizations such as the Aga Khan Foundation, Princes Trust and the Building and Social Housing Foundation (BSFH) have helped instil a more self-confident respect for past achievements and encourage local professionals to build on what remains.[10] Consultants and writers such as the late Jim Antoniou have also made a major contribution in this respect.

Of all the examples in which local professionals have developed a more egalitarian and productive relationship with local communities, two stand out as exceptional. In Karachi, Pakistan, the Orangi Pilot Project (see Khan 1996; Hasan 1999) pioneered a supportive role in which young professionals offered technical advice on disposing of sewage from individual houses to the end of lanes and eventually to main outlets for connection to the city sewerage network. Although the authorities were slow to respond and install the connection, the community of almost a million people were able to demonstrate that they could act as urban managers and planners as well as house builders. The professionals identified and channelled local initiative. In another example, the Kampung Improvement Programme in Surabaya, Indonesia (see Silas 1984) harnessed community initiative to install narrow paths, services and community facilities to high density settlements (see Figure 9.11). Modest standards and regulations were adopted rather than imposing conventional ones that

9.11
View of upgraded *kampung* settlement in Surabaya, Indonesia.

would have required massive relocation. As a result, the community improved their houses and managed a ten-point community development programme that included literacy, health, livelihood generation, security and environmental development.

Customary concepts

Different, but equally intense, attachments to land exist in all parts of the world. In Papua New Guinea, for example, 97 per cent of all land is held in customary tenure by local ethnic groups. Local chiefs enjoy considerable power in terms of allocating land to *wantoks* or tribal kith and kin who in turn receive the right to occupy, use and sometimes transfer their assets. However, the concept of ownership is completely alien and the tribe considers itself the custodian of its land, which is held in trust for future generations.

Although this system of managing land has worked effectively for many generations, the introduction through colonial rule of western concepts of property ownership, land titles and individual rights has undermined not only the traditional land-management system, but other aspects of tribal identity and social relationships. When a small plot on the edge of an expanding town is suddenly capable of generating a substantial rental income or capital value, this stretches bonds of family loyalty and offers of free accommodation for *wantoks* may not be so readily forthcoming. This erodes traditional mutual support systems and throws families onto the formal market in ways which they are not necessarily well-equipped to cope with.

One visible expression of this fragmentation of Papuan society has been an alarming increase in alcohol abuse and violence, which is in turn restricting progress in economic and social development. This is not to suggest that the introduction of individual property rights (or individual rights in other aspects of life) is wrong per se. However, it has proved impossible for the new statutory systems to develop appropriate responses at the scale required to meet the needs of a predominantly poor and increasing urban population. This has revealed the extent of the disparity between systems of land management which enjoy social legitimacy and those that seek to replace it. This came sharply into focus in the island of Bougainville, where one of the world's largest copper mines was discovered. Possibly realizing the value of the land they had lost, some members claimed that they had not agreed to the loss of their land in the first place, rendering the agreement null and void in their eyes. Other tribal members then claimed that it was against tribal custom to sell land, since they did not regard themselves owners, but only custodians, and were therefore not entitled to deprive subsequent generations of their birthright. The result of this dispute and the unresolved relationship between customary and statutory systems of governance is that the gap between those able to succeed within the new system and those forced to the edge is increasing. Papua New Guinea is perhaps simply an extreme example of a pattern common throughout the developing world and, to some extent even within 'developed' countries, where formal governance systems are failing to meet the needs of the poor.

Customary land-management systems are not, of course, unique to Papua New Guinea. They are widespread through the Pacific and sub-Saharan Africa, where they are coming under threat from western systems based on individual rights and market forces. In Lesotho, for example, all land is held by the King in trust for the nation and has traditionally been managed on the King's behalf by local chiefs. This system has worked well, despite occasional cases of mismanagement, and has enabled people to obtain easy access to land for housing and sources of livelihood. However, the government's decision to render the role of chiefs in land allocation illegal within urban areas has exposed failures of the new statutory system. Conventional methods of public-sector land acquisition have attracted numerous charges of inadequate or delayed compensation, which has encouraged customary groups to continue allocating land for housing unofficially. At the same time, the failure of public-sector land development and housing projects to meet even a modest proportion of housing needs in expanding urban areas has forced those needing homes to seek help from the chiefs or those allocated land by them. Even middle-income households no longer follow official procedures which require the Commissioner of Lands in the Ministry of Local Government to personally sign every lease in the country and include other requirements which collectively take up to seven years on average to complete. The result is that virtually everybody follows a procedure which is illegal, yet enjoys widespread social acceptance. This is hardly conducive to good governance.

Land-use patterns in the Lesotho capital Maseru reflect these processes in the form of low-density settlements spreading many kilometres from the city centre. The traditionally close attachment to the land felt by local communities means that land is not regarded primarily as a commodity to be traded, and this has restricted land prices to a minute proportion of total housing costs. Accordingly, people regard the official minimum plot size of 600 square metres as inadequate and plots of more than 900 square metres are common within the urban boundary. For many low-income households, this is justified in that it enables them to keep animals and grow some of their own food. However, it also makes densities so low that the provision of basic services, such as piped water supply, sewerage and surfaced access roads become uneconomic, stretching distances to the main employment areas and making travel an expensive and time-consuming process. With Maseru growing at an estimated 7 per cent a year, it is likely that its existing population of 250,000 will double within a decade. Unless densities increase, at least in some areas, the health and accessibility problems may become unmanageable. In this instance, market forces may be welcome as a means of encouraging more efficient use of available land on the urban periphery.

In most other parts of the world, market forces are already well entrenched. In Peru, for example, collective efforts have been a means to realizing individual benefits for decades. The squatter invasions reviewed so fully by Turner (1965, 1967) and Mangin (1967) in the 1960s have become established parts of the urban land and housing market, with many households eventually obtaining property titles from the authorities. The organizers of such invasions took care to plan their settlements on state-owned desert land using a simple grid pattern, often with the help of sympathetic local students. This made the allocation of titles and the provision of services relatively easy and provided everyone with similar size plots. The first thing that most families did was to build a brick wall across the whole frontage of their plot, both to protect it and also to ensure privacy for whatever form of development they undertook within the plot. In these areas, there is no intermediary space between the fully private and the fully public space as found in Arabic settlements, simply the articulation of the two categories. Unless empirical research can demonstrate the opposite, this suggests that the spatial language used by these settlers does not have a rich vocabulary.

Conclusions

The global trend for market forces to penetrate all sectors of economic activity has undoubtedly threatened the ability of many indigenous systems of governance to meet increasing and changing needs. However, governments have so far rarely risen to the challenge of replacing traditional forms of governance or land management in ways, or at the scale, required. The result is that a large proportion of the urban populations of the developing world find themselves in a legal limbo. In a sense, we should not be surprised at this problem. After all, it took Europe nearly two centuries to transform itself from a predominantly feudal, rural society to a democratic, urban one. European countries had the added benefit of small populations, powerful economies and influence throughout the world which enabled them to export their problems. For example, troublesome religious minorities could be encouraged to leave and establish their own communities in North America, criminals could be deported to Australia and the second sons of the aristocracy could be sent abroad to govern the colonies. Countries undergoing this transformation today have larger populations and do not control even their own economies, let alone the world's, and cannot export their problems. In fact, they are finding it difficult to cope with those left by the previous colonial powers. Furthermore, they have only had five decades to develop a legal, institutional and technical framework appropriate to their diverse situations. While globalization is tending towards conformity to a predominantly western world view, there is an urgent need for countries to examine objectively those elements of indigenous systems of governance and land management that have worked in the past, and ways in which they can be adapted to meet contemporary and future needs.

Interestingly, there are signs that two ancient Asian concepts of space are being taken up in the west by people seeking a harmonious environment. These are the ancient Chinese art of *feng shui* (wind and water) and the equally ancient Indian science of *vastu*. These approaches date back at least 3,000 years and have exerted a major influence on planning and design from individual rooms to entire cities. According to Kingston (1996: 18), 90 per cent of all properties in Hong Kong are built according to *feng shui* principles, and its effectiveness has even been cited as explaining Hong Kong's prosperity.

A key element in *feng shui* is that vital energy known as *ch'i* needs to be managed carefully in order to create a harmonious environment. *Feng shui* does not encourage cul-de-sacs as they 'expose houses to an onslaught of *ch'i* rolling in from the street' (Collins 1996: 31). The most active *ch'i* is in the front of a house, which is therefore not a good place to locate a garage or a bedroom. Conversely, the back of a house is not the best location for a workroom (ibid: 45). A key feature of the approach is the need to create a balance of the *yin* (feminine aspects such as dark, cool, soft, wet, earth, moon) and *yang* (masculine aspects such as light, hot, hard, dry, sky and sun). The ideal location for a house is midway between mountains and the flood plain. Since this cannot often be realized, symbols of these can be created by planting hedges or creating ponds as in the example of a Japanese house where a vertical rock is placed in a one metre square corner of the plot representing mountains, a rockery with several small rocks in another small area at the side and a gravel area raked to resemble the sea laid in the front of the house, all within a small Tokyo house plot. *Feng shui* also provides a framework (*bagua*) for relating personal aspects of life, such as love, health and work in different locations.

Similarly, *vastu* 'aims to align living spaces with unseen spiritual and natural laws' and 'align both the home and garden with the cosmos' (Pegram 2002: 9). The *Vastu Perusha Mandala* 'maps the complex forces of the universe and provides an architectural blueprint for the home'. As with *feng shui*, a key objective of *vastu* is to strengthen or reinforce positive forces and appease or minimize negative ones so that the '*gunas*' or three types of energy are in balance. Both these concepts of perceiving and using space are considered by some contemporary advocates to be globally applicable as 'the energies of the cosmos exert their influence around the whole globe'. However, the impact of such energy may not be the same throughout the world, since in the southern hemisphere the sun travels in a northern arc. It may also not be so readily applicable in homes equipped with air conditioning and other modern conveniences.

The examples reviewed in this chapter demonstrate that there are many ways in which people perceive and use space and that simplistic attempts to change these can be counter-productive. It also provides reassurance that although western models of planning and design based on commercial land markets are penetrating most parts of the world, alternative concepts such as '*ma*', *feng shui* or *vastu* show that a purely materialistic or utilitarian approach is not sufficient to meet peoples' need for something extra. We should, however, beware of claiming universal applicability for *any* single spatial language, given the diverse needs of widely different communities. Instead, diversity of built form and settlement layout should be encouraged as a physical manifestation of social and cultural diversity. Even in a globalized world, there are still many different languages and ways of living.

The literature generated by Nitschke, Turner, Oliver, Rapoport and others amply demonstrates not only the rich diversity of spatial languages and forms which people have developed, but their potential contribution in organizing the upgrading of existing settlements and planning new ones. The key is therefore to find ways of harnessing this creative energy and assimilating it into official development programmes. It is these which need to change, not the populations of our urbanizing societies. It is therefore advisable for change to be introduced at a rate that society can accept and that is seen to build confidence, rather than erode it. The best way this can be achieved is by encouraging the active participation of local communities in the development process at all levels and in all sectors.

On a more practical level, it is important to break the blinkered hold which many professional institutions exert on academic institutions educating future generations of planners, architects and other disciplines. Until we encourage a more holistic and locally sensitive approach to addressing issues of urban development and housing, we cannot complain if the next generation of professionals repeats the failures of the past.

Notes

1. According to Nitschke (1966: 152), the Japanese word for design, *ma-dori*, literally means, 'grasp of the *Ma*'.
2. A number of my papers on planning and design in Japan were published in a special issue of *Arena*, the Architectural Association journal, 1968, volume 83 (921).
3. Undertaken with a friend, Peter Cookson Smith.
4. For more details of the settlements studied see Payne (1977).
5. *Gecekondu* literally means to 'land by night', as in the case of mushrooms.
6. Clifford Culpin and Partners, later reformed as Culpin Planning.
7. In the nineteenth century, Egypt was still part of the Ottoman Empire.
8. The plan followed the traditional colonial grid pattern criss-crossed with diagonal roads and public squares.
9. Traditionally, streets in Arab towns were based on the need for a fully laden camel to pass unhindered, with wider streets (dual carriageways!) wide enough for laden camels to pass in opposite directions. Halcrow Fox were awarded the 'Aga Khan Award for Architecture' in 1992 for this project.
10. BSHF is based in Coalville, Leicestershire, UK.

Part III

Understanding the vernacular

Chapter 10

Vernacular design as a model system

Amos Rapoport

Introduction

The publication of the monumental *Encyclopedia of Vernacular Architecture of the World* (*EVAW*) (Oliver 1997a) both helps define vernacular studies as an accepted field and marks the end of what might be considered the first stage of vernacular studies. Although conceptual and theoretical issues have occasionally been addressed and are discussed in *EVAW* (volume 1), it can fairly be said that, so far, vernacular studies have been at the 'natural history' stage, describing and documenting buildings, identifying their variety, classifying them and so on. In fact, many fields have two ways of working, often two stages, a natural history stage and a problem-centred stage leading to explanatory theory. This is not only necessary but essential, and the lack of a natural history stage in Environment–Behaviour Studies (EBS) has inhibited the development of the field, by premature borrowing of specific approaches from other disciplines. A certain minimum amount of data is necessary in order to begin the development of concepts, principles, generalizations, mechanisms and so on. Too much such data, however, can become counterproductive, making integrative work and conceptual and theoretical development difficult (e.g. Rapoport 1997, 1998, 2000a, 2000b and 2004; cf. Bunge 1998: 438).

With the publication of *EVAW*, therefore, the time has come to move to the next 'problem-oriented', comparative, integrative and more conceptual/theoretical stage. *EVAW* helps not only by addressing theoretical issues, but by organizing data. I have recently emphasized the small 'scale' of 'culture', i.e. groups (Rapoport 2000a, 2001, 2002a, 2004: Ch. 3), leading to the great variety of vernacular environments; this variety can be derived from *EVAW*. I counted 1,278 areas/groups, each with its own distinct built environment, and there are probably more.[1] This not only leads to essential comparative studies, but raises the important question of why this should be the case. My answer is that it is due to the role of latent as opposed to instrumental aspects of activities, i.e. of meaning. Other important questions and hypotheses follow that can be studied by using the full range of vernacular environments.

One way to move to a more 'problem-oriented' stage is to consider vernacular design from the perspective of EBS, understood as a scientific discipline that studies Environment–Behaviour Relations (EBR). The question then becomes the relationship between EBS and vernacular studies, in which vernacular design provides an essential body of evidence for EBS (Rapoport 1990a), whereas EBS relates the vernacular to environments more generally and can provide a set of concepts, a context, new types of data, and a framework for studying vernacular design.

In this chapter I address these issues by revisiting a question posed elsewhere: Why study vernacular design (Rapoport 1999a)? The brief answer was: to learn from it. Here I expand, elaborate and develop it further, arguing that this is best done by looking at vernacular design as a *model system*.

Some preliminaries

Before addressing that question, however, it is necessary to restate briefly some definitions, assumptions, etc., which concern the nature of vernacular design, and how it is to be conceptualized.[2] Also important is the physical relationship between vernacular and other types of environments.[3] Neither can be fully analysed or understood in isolation; one must consider the ways in which they contrast and reinforce each other (Rapoport 1990a: 14–17, 1993a: 42–3 and Fig. 5, 1999–2000: Fig. 10.9). I do this by briefly synthesizing some aspects of my work.

Defining the domain (its ontology) is an essential first step in research (Rapoport 1990a: Ch. 1, especially 9–11).[4] It is not sufficient to study just buildings; one needs to study systems of settings within which systems of activities take place. Together these form cultural landscapes which comprise fixed and semi-fixed features, which make the whole of material culture the domain. Since non-fixed features (people, animals, vehicles and so on) are always present, behaviour automatically becomes part of the domain. At the most abstract level vernacular design produces particular types of environments, which can be conceptualized as organizations of space, time, meaning and communication.

In moving to conceptual and theoretical development, clear definitions are essential. 'Vernacular' has been used too loosely to be useful (Rapoport 1990c: 67–8). In general there is also a need to dismantle such complex, overly broad concepts. In redefining the vernacular I discussed the role of taxonomy and suggested replacing monothetic, ideal type definitions by polythetic ones; these involve multiple attributes, not all of which need to be present in every case.[5] Initially I used seventeen process characteristics (how vernacular environments come to be) and twenty product characteristics, i.e. the attributes of the resulting environments (Rapoport 1990c). This has since been raised to twenty-one – the addition dealing with levels of meaning (Rapoport 1990d). This makes the important point that these attributes represent a preliminary, hypothetical open-ended set. Furthermore, the attributes are not ranked nor is there discussion of how they are to be 'measured'.[6]

Although this dismantling and redefinition concerned the vernacular, it has proved applicable to many (all?) environments and is, therefore, inherently comparative, the differences among environments being identified, and their characters clarified, through the specifics of the various attributes. This makes possible conceptual linkages among various environments. In fact, the initial list defined vernacular by contrast to high-style. It has since been applied to 'tradition' and 'traditional environments' (Rapoport 1989), and spontaneous settlements (Rapoport 1988a). These latter, in both their process and product characteristics, seem the closest to traditional vernacular that exist currently and may also prove extremely useful as a model system.

Why study vernacular design

Why study vernacular design in 2005? One reason is simply because one likes it. I, and, I suspect others, certainly began that way. In the case of the use of model systems in biology one also finds a fascination with, and 'love' of the specific organism chosen. But in neither case is it enough. In biology, 'naturalists' love, and need to be in touch with, nature as opposed to theoreticians, interested in concepts, ideas, generalizations and explanation. The two often fail to understand each other, and both are needed (Bateson, cited in Gould 2002: 408–9). This is also the case in vernacular studies, where some need to be in touch with real buildings (e.g. Hubka 1984, 2003) as opposed to, say, my interests. A 'bridge' between the two is needed and the approach of this chapter can provide that bridge.[7]

The primary reason for the study of any environments is to learn from them (Rapoport 1982a, 1999a). This general statement, however, needs to be developed to show how lessons might best be learned.[8] The first, and most obvious, reason for studying vernacular environments is that they comprise most of what has ever been built. In order to develop explanatory theory in EBS it is essential to generalize, and that requires the largest and most diverse body of evidence.[9] That includes all types of environments, all of history (including prehistory and possibly animal origins), all cultures and the whole environment (not just buildings). Inescapably, most of that evidence will be what is broadly called 'vernacular'. Even today only about 2 per cent of buildings are designed by architects – the remaining 98 per cent cannot be ignored (Rapoport 1990a: Figs 1.1 and 1.2; 13–14).[10] Thus vernacular environments provide an unequalled, and only possible, 'laboratory' with a vast range of human responses to an equally vast range of problems; cultural, technological, of resources (including materials), site, climate, ways of making a living and so on. This increases the 'repertoire' of both problems and successful and unsuccessful solutions, of processes and products, of ambience, and at scales from semi-fixed elements to cultural landscapes (Rapoport 1990d, 1993a, 1993b, 2002c).

Amos Rapoport

Moreover, relationships among professionally designed elements are rarely designed in the traditional sense of the term, so that the environment as cultural landscape, including semi-fixed elements, is always vernacular (Rapoport 1993a, 1999–2000).

Studying the full range of environments also changes our understanding of basic concepts – with important implications for theory development and valid cross-cultural comparisons. For example, a dwelling needs to be defined as a system of settings within which systems of activities take place (Rapoport 1990e, 2004: Ch. 2, especially Figs 6–8). Differences among these systems change how urban fabric is conceptualized and our understanding of how buildings fit into it (Rapoport 1982b, 2004: Ch. 2, especially Fig. 8). Similarly 'city' needs to be redefined, as that which organizes a region, when more diverse evidence is used (Rapoport 1977, 1993b), as is the case for 'privacy' (Altman 1975; Rapoport 1977) and many other concepts.

Once it is realized that vernacular design cannot be ignored, still the most common attitude, one can still deny that it can provide any useful lessons. Accepting that something can be learned, the most common approach is to copy certain formal qualities (shapes, massing, details, etc.), often based on a romanticized version of the vernacular. In general this approach has not, does not and is increasingly unlikely to work.[11] The only valid approach is to derive more or less general lessons and principles by analysing vernacular environments using EBS concepts, models and the like, and applying these lessons to design. The difference between these two can be diagrammed as illustrated in Figure 10.1.[12]

10.1
Learning from the vernacular.

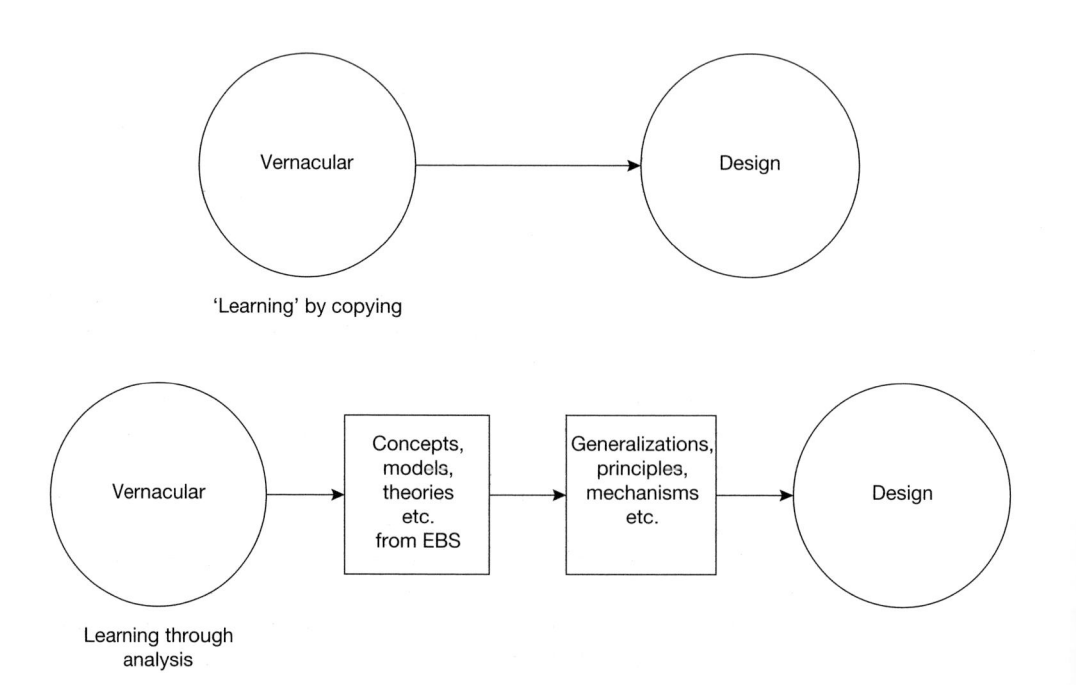

'Learning' by copying

Learning through analysis

Vernacular design as a model system

I have argued elsewhere that the ability to derive useful lessons requires a certain level of abstraction, and requires moving away from the 'natural history' stage to more problem-oriented, conceptual ways of addressing the topic. Moreover, the potential applicability of such lessons has a major theoretical implication; it leads to a particular view of 'culture', one that is not completely relative but constrained by evolution as suggested by sociobiology, behavioural ecology and evolutionary psychology (Rapoport 2000a, 2000b, 2001, 2004: Figs 37–9; cf. Brown 1991; Goldsmith 1991; Barkow, Cosmides and Tooby 1992; Mithen 1996; Pinker 1997; Wilson 1998; Lopreato and Crippen 1999; Segerstråle 2000 and Alcock 2001, among many others). This is, often reluctantly, being accepted by at least some anthropologists and other social scientists. With these questions in mind the study of vernacular built environments might play a role in resolving issues such as the extent of constancy/variability, the possibility of different cultural expressions of constants, etc. (Rapoport 1977: 337–9, 2001: Fig. 2.1; 32, 2004: Fig. 39; Clottes and Lewis-Williams 1998; Lewis-Williams 2002a and 2002b).

Any lessons can be both specific and general. The most specific might address the design for specific groups or in specific locations (Rapoport 1983a). General lessons might include expanded repertoires of possible product characteristics, how environments create and/or reinforce various forms of identity, i.e. the mechanisms linking components of identity and elements of cultural landscapes (e.g. Rowntree and Conkey 1980; Arreola 1984 and 1988; Macsai 1985). Learning can also be about process since the dynamics of both product and process can be studied using the polythetic definition of vernacular design. Although it has been suggested that lessons regarding process are unlikely, given the very different contemporary social and technological conditions (Sordinas 1976), contrary views can also be found (Hakim 1986 and 1994; Akbar 1988). This then becomes an interesting researchable question, leading to potentially important insights into the nature of design and the role of rules which I discuss later. Still more general lessons include how environments communicate meaning and changes in the levels and types of meaning communicated (Rapoport 1988b, 1990b, 1990d). Insights can also be gained into the differential importance in that process of fixed, semi-fixed and non-fixed features in different types of environments, different contexts, at different times, for different groups, etc.

There can also be lessons about responses to climate and energy use (Rapoport 1969, 1987, 1995a), sustainability (Rapoport 1994), the variability of standards (Rapoport and Watson 1972) and notions of environmental quality (Rapoport 1983a, 1995b, 2004: Fig. 27), the nature and attributes of distinctive ambience (Rapoport 1992), preferences for various product characteristics and many other topics.

Because vernacular environments are those most clearly linked to 'culture', they are essential in clarifying the ways in which 'culture' and environments are related.[13] More generally, vernacular design, historically and in developing countries, as well as spontaneous settlements, offer a most useful entry point to the study of EBR. This is because in many such environments many aspects of EBR are more clear-cut, more 'extreme', because of higher criticality, and consequently 'black and white', rather than the more ambiguous cases of many present day environments which can be thought of as being in various shades of gray. Having understood certain EBR and the mechanisms involved in the former, these can be sought, identified, characterized and studied in, say, popular design in contemporary, large-scale, complex societies. If necessary they can then be modified in the light of culture change, for types of environments that do not exist in the vernacular (Rapoport in press b). This brings me to my principal point; that vernacular environments can usefully be considered as *model systems*. That requires a preliminary discussion of the concept of model systems.

The idea of model systems in general

Because of lack of space I can only summarize briefly the use of model systems. Modelling in general involves simulation using static or dynamic, physical or conceptual means; mathematical, numerical, computer, agent-based and so on (e.g. the behaviour of beach sand (Hoefel and Elgar 2003)). The use of model systems involves the use of one system to study phenomena in another, apparently very different system; for example, colloids for condensed matter physics (e.g. Yethiraj and Van Blaaderen 2003), or lower animals to identify neural correlates of human consciousness (*Nature* 2003) or to help establish what makes modern humans different (Carroll 2003: especially 852–3), for which primate model systems have also proved very useful.

In selecting such systems it is essential to make explicit the ways in which they both resemble and differ from the object of study. The development and use of model systems involves several cycles, integrating natural history (descriptive) data with conceptual, theoretical and empirical work. In this, previously ignored data can often be used, providing the large and varied body of evidence required, including extreme cases – one potential role of vernacular design (e.g. Rapoport 1990a).

Model systems are highly versatile. They can be used to test existing theory, to develop unified theories, to generate hypotheses, and to synthesize and integrate material from various sources and disciplines (Dugatkin 2001: xii, xx; Lyon 2003; cf. Rapoport 1990f, 2001, 2004) to reveal unexpected connections, clarifying highly general processes, such as evolution (e.g. Carroll, Grenier and Weatherbee 2001; Gould 2002).

The use and role of model systems in biology

The use of model systems has been most developed in biology and biomedical research where they play a central role, using specific organisms to study general processes (e.g. Kandel and Squire 2001; Gao and Shubin 2003) or other organisms, especially humans. Parts of organisms can also be used, e.g. the retina as a model system for neuroscience (He *et al.* 2003) or the use of *Drosophila* neuroblasts (Pearson and Doe 2003).

There are a number of almost 'standard' model systems in biology, used for a great variety of purposes, including human diseases (e.g. glaucoma (Alward 2003)). These include the fruit fly *Drosophila* (especially *D. melanogaster*), used since the 1920s and the basis of much of research into genetics and, more recently, brain disease in humans and ageing (e.g. 'Science on line', *Science* 2002b: 917).[14] Ageing research has also relied on *C. elegans*, mice and yeast (Hasty *et al.* 2003: 1355; Warner 2003); the filamentons fungus *Neurospora crassa*, used in genetics for over sixty years; yeasts (e.g. *Saccharomyces cerevisiae*); the zebra-fish; the frog *Xenopus*; the sea-snail *Applysia,* used in behavioural neurobiology (e.g. the work of Eric Kandel; Sokolowski 2002: 893); the nematode (planarian worm) *Caenoharhbditis elegans*, used for many purposes, including understanding Alzheimer's and Parkinson's diseases and treatments for them (e.g. Syntichaki *et al.* 2002); and, in botany, *Arabidopsis thaliana*, which has also shown that plants share much with the animal kingdom.[15]

New systems are constantly sought and introduced, e.g. guppy-like fish (in the genus *Poeciliopsis* (Morell 2002)), cichlid fish, farm animals, even well-studied populations of animals (e.g. macaques in Gibraltar and chimpanzees in Africa). Recently, bumblebees have become important model systems, their choice specifying both their useful characteristics and the topics they can illuminate (Heinrich 2003). More generally, to enable the study of simple organisms to be relevant to humans, the reasons and criteria for the choice of the system, and its differences from humans, must be made explicit.[16]

In a special section on neuroscience (*Science* 2002a), a number of model systems (*Drosophila*, the Zebra fish, *Xenopus* and mice) are shown to have been used to identify and study fundamental mechanisms of perception and cognition, many highly relevant to EBS: how the visual system works; how different senses interact in perception; how cognitive maps are formed and in which parts of the brain; and how topographic mental maps are developed and refined (cf. Silva 2001).

Clearly, many different model systems are used to study similar questions. One can, therefore, integrate and synthesize materials from various sources into new frameworks (e.g. Dugatkin 2001: xii; Lyon 2003; cf. Rapoport 2004). The use of a variety of model systems also throws light on the constancy of mechanisms; even organisms as primitive as archaea, bacteria, fungi and yeasts, as well as sea-snails, flies and worms, have been able to identify mechanisms relevant to humans.[17] Also, by comparing the results from such apparently divergent model systems one can derive convergent principles.[18]

There is an immense literature on the use of specific model systems, and also on their use in general. Thus, Dugatkin (2001) not only reviews twenty-five model systems used in behavioural ecology, ranging from insects and spiders to primates, but provides a most useful general analysis of their conceptual, theoretical and empirical characteristics and how they are built, used and tested. This means that one does not need to start from scratch, or to reinvent the wheel, in attempting to use vernacular design as a model system for EBS (cf. Rapoport 1990f, 2000b).

Vernacular design as a model system
In considering the possibility of using vernacular design as a model system for EBS, I start with the three basic questions of EBS (e.g. Rapoport 1983b, 1997,

2000a, 2000b, 2004), and suggest that the vernacular offers the most useful starting point for their development, especially in informing empirical research on current environments.

These questions define the domain of EBS and each can be dismantled to any degree of specificity to deal with the extensive materials it subsumes. This typically produces, as its first product, what one might call a lexicon of potential components which is open-ended and which can be studied and ranked, and can often be dismantled further. There are thus levels and scales of dismantling and one can determine empirically which components play a role in different situations, groups or environments, or are useful in addressing specific questions or problems:

1 The biosocial, psychological and cultural characteristics of humans that influence the characteristics of built environments

A central issue is to identify the relative role and interaction of constant and variable aspects in human characteristics and behaviour. This requires both the longest possible time line and maximum variety of groups and their environments, and only the comparative and integrative study of the full range of vernacular environments, including archaeological data, can provide this (Rapoport 1990a, 1990b). It then becomes possible to suggest alternative hypotheses for any constancies and differences found, an essential step in the development of theory in EBS (cf. Dillehay 2002; Nuñez, Grosjean and Cartajena 2002). The important recent developments in research on possible evolutionary constraints on culture need to be applied to the possible interplay of variable and constant aspects of built environments. A potentially useful analogue is recent research on rock art. By comparing the full range of such paintings among the San in Southern Africa where ethnographic data can be used, and in Palaeolithic Europe, a most interesting hypothesis is presented, which posits an universal mental process among humans based on neuroscience research, i.e. species-specific characteristics, which may be expressed in culturally specific ways (Clottes and Lewis-Williams 1998; Lewis-Williams 2002a, 2002b).[19]

This first question also raises the topic of human group characteristics. I have already mentioned the small size of relevant groups (Rapoport 2000a, 2002a, 2004) and their potential importance in explaining the great variety of vernacular environments, e.g. in *EVAW* (Oliver 1997a).[20] Vernacular environments are highly group-specific and, since culture helps define groups, it makes it possible to identify the relative role of cultural vs. other variables in EBS. The large variety of environments, situations and so on, makes possible the study of many and diverse groups, their characteristics and how these relate to characteristics of built environments, ways in which groups and group boundaries are defined, and the role of built environments and material culture in these processes. It might then be possible to identify (possibly constant) mechanisms of EBR and the range of their culture specific expressions. This is also helped by the fact that groups in such cases are more homogeneous and more tightly defined (Rapoport 1990e: Fig. 2.2; 12, 2004: Fig. 47).

10.2
Constant and variable aspects of 'culture', with the possibility of variable expressions of constants (hypothetical).

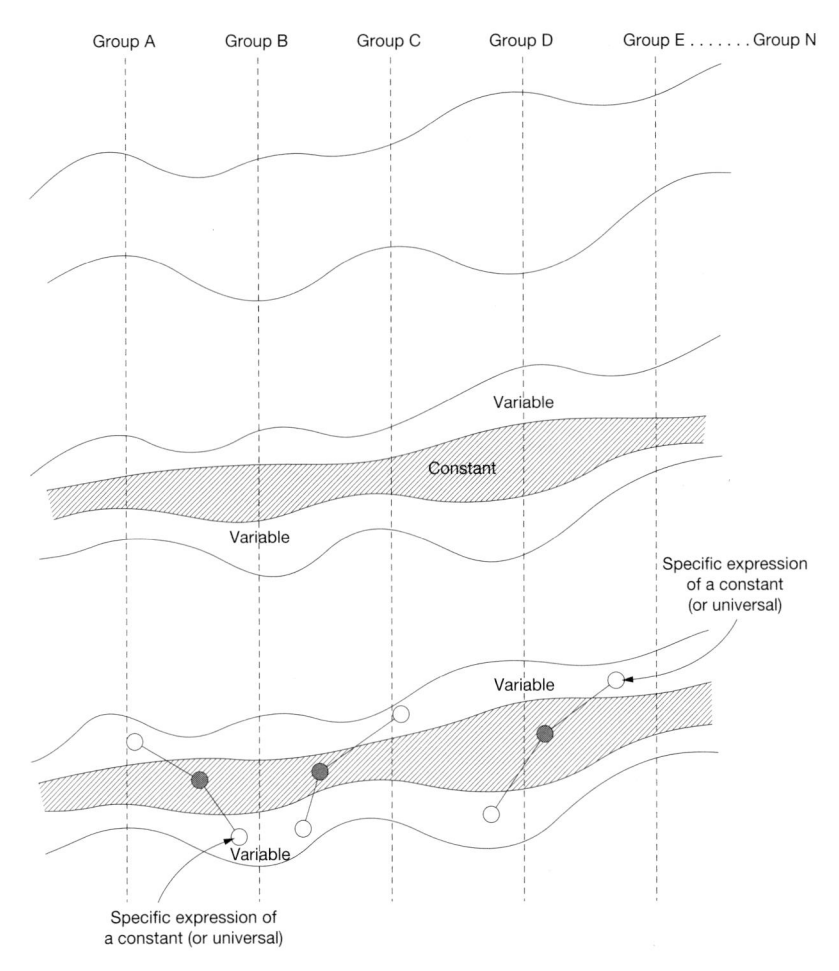

2 The effects of environments on humans

Given the limited resources and technologies of vernacular environments, their impact on people becomes greater. Since criticality is higher in these cases, any effects can be studied more easily. The great number of groups, i.e. their variety historically and cross-culturally, provides a range of levels of 'competence' of such groups. The range of 'competences' is further increased by the variety of culture-contact and culture-change situations including modernization and urbanization, conquests, migrations and so on. This would not only clarify the effect of environments on people, but might even make possible some predictions about both the magnitude and type of such effects (Rapoport 1983a, 1995c).

Since choice is the major effect of environment on behaviour (Rapoport 1983b), the choices made are highly enlightening. This is particularly the case in vernacular environments, where the choices are more limited (Rapoport 1969, 1990c). As a result the choice process and the choice criteria used can be identified much more easily than in the case of contemporary complex affluent societies where choice is virtually unlimited. Since design can

10.3 Differences within and among groups.

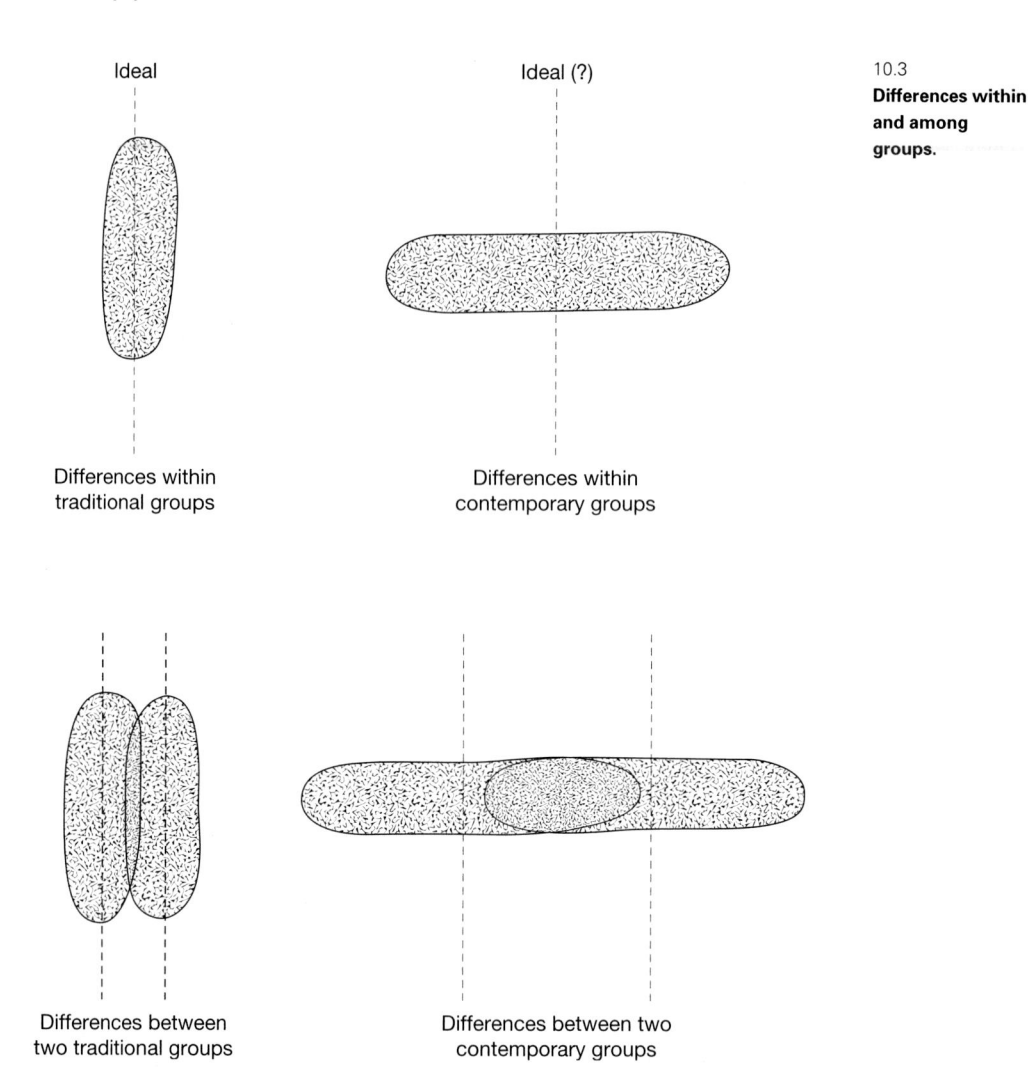

Ideal — Differences within traditional groups

Ideal (?) — Differences within contemporary groups

Differences between two traditional groups

Differences between two contemporary groups

be understood as a choice process, this provides an opportunity to understand better the design process, since both the initial possibilities considered and, as a hypothesis, the range and number of choice criteria in any given case are greatly reduced. The availability of the lexicon of process characteristics should also be helpful (Rapoport 1988a, 1989, 1990c and 1990d).

3 The mechanisms that link people and environments
We have already seen that mechanisms are likely to be important in studies of constancy. Knowing and understanding mechanisms is essential for explanation in any domain. As pointed out elsewhere (Rapoport 2000b: 120, 2004), a first attempt to list possible mechanisms in EBR suggests that their number is limited, although, being open-ended that list could be expanded if new mechanisms are identified. Mechanisms can also be dismantled, e.g. meaning into different levels

and types (Rapoport 1988b, 1990b, 1990d). Some of these mechanisms may also interact and work together.[21]

This first list (or lexicon) includes:

Physiology (comfort, adaptation, hence standards, etc.)
Anatomy (ergonomics, different ways of doing things, different postures)
Perception (including 'aesthetics', understood as the perceptual aspects of environmental quality)
Cognition (structuring and ordering the world, mental maps, orientational systems and their importance (i.e. making movement easy or difficult))
Affect evaluation (hence preference, and choice and the significance of little choice in vernacular environments and of greatly increased choice now)
Meaning supportiveness (physiological, anatomical, psychological, behavioural, social, cultural)
Some of the components and expressions of culture (derived through dismantling; cf. below)

For the reasons discussed earlier, vernacular design can be a useful model system to study all these mechanisms, although I only discuss a few aspects of some of those.

For example, the impact of *physiology* is more clearly linked to built environments in the vernacular in the absence of sophisticated technologies. Since air-conditioning leads to maladaptation to heat (Liao and Cech 1977; Rapoport 1987, 1995a), vernacular environments can then show the limits of possible human adaptation to heat, cold, etc., and clarify the extent to which standards are actually based on physiology and their variability, notions and importance of comfort, and hence ideals (Rapoport and Watson 1972). This, in turn, can help understand both effective climatic designs and the presence of apparently anti-climatic solutions (Rapoport 1969), due to tradeoffs made between comfort and privacy, status and so on (Rapoport 1987, 1995a).

More generally, one can then test the hypothesis that standards are closely related to meaning, with major implications for modernization and sustainable development, e.g. by identifying reasons for the 'obsolescence' of built environments (Rapoport 1994). Changing notions of environmental quality and changes in these can be studied, especially through the development of spontaneous settlements over time (Rapoport 1987, 1988a, 1994, 1995a, 1995b; Sastrosasmita and Nural Amin 1990).

In the case of *anatomy*, the great variety of statures and, more important, postures clarifies the range of possible ways of carrying out activities (Rapoport and Watson 1972). It also makes clear the need, first, to identify ways in which particular groups carry out tasks and where (Rapoport 1990e) and, second, possible changes due to culture-change/modernization, with implications for the need for open-ended design (Rapoport 1995d). Vernacular design itself, both at the urban and building scale, and with its large and varied body of evidence, and specific process and product characteristics, is a most useful model regarding open-ended design (Rapoport 1977: 355–68 and Figs 6.7 and 6.8; 1988a, 1990c: 92, Fig. 4.15; Banning and Byrd 1987; Salama 1995; Tipple 2000).

The specific product characteristic most relevant to open-endedness, what I call 'stable equilibrium', is related to *perception*. Although this tends to be the least variable mechanism (Rapoport 1977: 30–8 and Ch. 4), vernacular design can clarify differences in perceptual acuity and the need for redundancy, and the differential importance of different senses among various groups. In general, however, vernacular environments as such do not seem especially useful. On the other hand, recent research on perception using, as discussed earlier, a variety of biological model systems, throws light on the nature of processes of visual perception, the role of other senses and the understudied topic of how they interact as they do in 'ambience' (Rapoport 1992). Once these mechanisms in general are well understood, the great range and variety of sensory qualities of vernacular environments will enable studies impossible when using only contemporary high-style environments. This also enables the study of 'aesthetics', understood as the perceptual aspects of environmental quality.

Two examples of mechanisms linking people and environments

Consider, in somewhat more detail, two topics as examples of how vernacular design might be used as a model system.

Meaning
In its study, the use of vernacular environments is essential for several reasons. For example, I have hypothesized that high-level meanings (canonical meanings) (Blanton 1994: 8–13) are likely to be present and very important in vernacular environments, especially among non-literate groups or those with low literacy rates, but relatively, or entirely, absent in most contemporary environments (Rapoport 1988b, 1990b: Epilogue, 1990d, 2000b: 25–126 and Fig. 1). I have further suggested that middle-level meanings (indexical meanings) (Blanton 1994: 8–13) gain in importance in contemporary environments. Low-level, instrumental meanings are always present, but need to be more clearly marked in contemporary environments; i.e. they need higher redundancy, because of the heterogeneous nature of groups, and the presence of many diverse groups (see Figure 10.3). The only way to test these important hypotheses is through the full range of evidence discussed earlier, most of which is vernacular.[22]

Spontaneous settlements, the closest to traditional vernacular (Rapoport 1988a), enable the study of the role of meaning in ways impossible in other cases, under conditions of culture change and the resulting rapid and dramatic changes in meaning – of forms, materials, colours, semi-fixed features, spatial organization, etc. This makes it possible not only to test these hypotheses, but to learn under which circumstances and, therefore, in which environments they are important (Rapoport 1990d). There are also design implications, e.g. regarding 'sustainability' (Rapoport 1994). In discussing this topic I have previously used many examples. Two new examples strengthen the argument.

The first is what, in the archaeology of Israel, is called the 'four-room house'. When first discovered in 1927 it was though to be a temple, and a church service was held in its ruins (Bunimovitz and Faust 2000: 33)! Hundreds have now been found and they remained unchanged for 600 years (between 1200–586 BC), while the plan was used in a great variety of geographical and climatic regions and, apparently, for many very different functions; rural and urban dwellings, tombs, administrative complexes and monumental buildings (Bunimovitz and Faust 2000).[23] After rejecting a 'functional' (i.e. instrumental) explanation, Bunimovitz and Faust argue that the 'four-room house was a symbolic expression of the Israelite mind', i.e. of their ethos and world-view; at the same time it also helped to structure that mind (2000: 36). Various specific aspects of meaning are discussed, including ritual purity and egalitarianism, and also its role as an ethnic marker. Although numerous parallels could be found in vernacular design, this important role of high-level meanings would not be found in contemporary, especially high-style architecture.

Some examples have important implications. First, they help explain the possible symbolic relation between ethnic and other group identities (culture) and built environments. Second, they help explain the great variety of vernacular environments, because latent functions (meaning) vary more than instrumental functions. Third, because they apply not only to buildings but to cultural landscapes and material culture generally, including costumes, hairstyles and so on, these also can communicate identity; those that survive (tools, weapons and, above all, pottery) enable archaeologists to study the distribution of groups, conquests, migrations, etc. At a theoretical level, the use of material culture to communicate identity has important implications for understanding processes and mechanisms (e.g. Wobst 1977). These may still be operating, as in personalization, spontaneous settlements, colour use, landscaping and so on (Rapoport 1988a, 1990b, 2000a and references therein). All aspects of material culture are used to establish group boundaries (as are religion, language, music, diet, etc.). Built environments and material culture can also play an important role in 'defensive structuring', an important mechanism in group responses to stress (Siegel, cited in Rapoport 1977, 1990b).

The second, contrasting new example shows the loss of high-level meaning in those contemporary buildings where they would be expected: churches (Brown 2002).[24] The photographs shown are indistinguishable from shopping malls (which are the model used), gymnasia, convention centres, etc. In these 'mega churches' one can eat, shop, go to school, bank, exercise, children can play . . . they have become part resort, part mall, part town square and so on (Brown 2002). Even the religious core for worship resembles an auditorium, theater or convention hall, with no religious symbols visible in the photographs. This last point is also made implicitly in a recent criticism of new Catholic churches by a traditional Catholic, Michael S. Rose (Bernstein 2002). This is based on the absence of any symbolic elements, i.e. of high-level meanings. Their absence seems clear; whether they can still be present, or are needed by the public, i.e. whether the criticism is misplaced (as I suggest) is an interesting question which is raised, in the first instance by studying meaning in vernacular and traditional high style environments.

Culture
This is part of all three questions and vernacular design is essential in the study of the relation between culture and built environments (Rapoport 1969), because it is my hypothesis that such relations are stronger and clearer in the vernacular (Rapoport 2000a, 2002c, 2004b). Since 'culture' as such cannot be used in EBS research, because it is both too general and too abstract, it needs to be dismantled, and over some time I have developed a particular way of doing so (e.g. Rapoport 1998, 2000a, 2000b, 2002c, 2004, in press b).

Among the mechanisms discussed earlier, one refers to 'components and expressions of culture', and the vernacular is an ideal model system for each of those, although, as always, as part of the full range of environments. Among the many such aspects of culture I will briefly refer to a few, and discuss one in more detail as a final example.

Considering social expressions of culture (the 'too abstract' axis), vernacular environments clearly respond to the largest range of kinship systems, family structures, roles, statuses, social networks and so on. Moreover, these have been much studied by anthropologists, so that data are available and can be related to the corresponding built environments, as cultural landscapes, of hunter-gatherers, nomads, agriculturalists and in urban contexts. Hypotheses generated can then be tested in contemporary large scale, complex societies (e.g. Rapoport 2000a), developing countries (Rapoport 1983a) and among the many immigrant groups in numerous cultural contexts around the world, with important planning and design implications for developing countries, immigrant groups and cases of rapid culture change generally (Rapoport 1983a, 1995e, 2004).

The great variety of vernacular cases also makes possible the study of the components of culture (the 'too general' axis), such as values, norms, etc.; these become clearer by contrast with one another and are also related to the nature of groups. For example, seen cross-culturally all contemporary groups tend to conform, although with slight differences (Sunstein 2003), but the groups encountered in vernacular situations can be expected to conform much more (see Figure 10.3). This clarifies the role of social norms (which lead to conforming), and can identify this constant process and its implications for the shaping and use of environments. Norms lead to rules and *rule systems,* the topic I now discuss in somewhat more detail, albeit still relatively briefly, starting with process.

I start with the 'choice model of design' (Rapoport 1977: Figs. 1.6–1.8, 2004: Figs. 28, 32, 33). The application of choice criteria follows rules, and the consistent application of rules leads to style in buildings and to identifiable cultural landscapes, implying the presence of shared models or schemata (Rapoport 1969, 1989, 1992, 1993a, 2004).[25] These can be much more easily studied in vernacular situations where groups are more homogeneous, rule systems stricter, schemata more widely shared and changes in them slower. This has, as one example, been

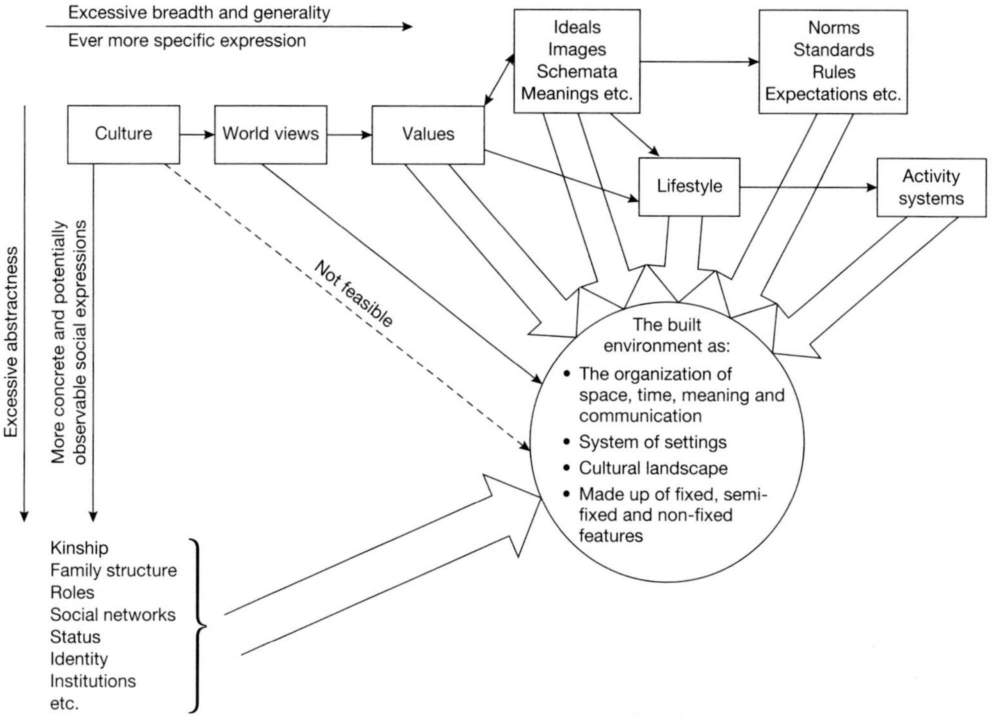
Excessive breadth and generality
Ever more specific expression
Excessive abstractness
More concrete and potentially observable social expressions
Culture
World views
Values
Ideals
Images
Schemata
Meanings etc.
Norms
Standards
Rules
Expectations etc.
Lifestyle
Activity systems
Not feasible
The built environment as:
• The organization of space, time, meaning and communication
• System of settings
• Cultural landscape
• Made up of fixed, semi-fixed and non-fixed features
Kinship
Family structure
Roles
Social networks
Status
Identity
Institutions
etc.

10.4
Dismantling culture, relating its expressions to the built environment (the width of the arrows correspond approximately to the feasibility and ease of relating the various elements).

done for Korea, using the best preserved case (Cheju Island), where a single schema is applied to buildings at all scales and of all types and to the whole cultural landscape (Nemeth 1987). This schema is based on geomancy (cf. Yoon cited in Rapoport 2000b), also used in other traditional environments including Chinese; through the use of *feng shui*, these are re-emerging among immigrant communities in the US and the UK.

Rule systems have not been much studied, even in vernacular situations, nor has the transition from informal, unwritten rules to more formal, written, legalistic rules. But some studies do exist; one recent study identifies the rule system developed by a sixth-century AD architect, Julian of Ascalon (Hakim 2001). Parts of that system relate to my example of Sarajevo (Rapoport 1969: 5), significantly also during the Moslem period. Possibly more relevant are studies of Islamic cities, especially their medinas, where the rule systems used have been identified. These consist of a system of formal and legal rules based on Muslim religious law (*Sharia*) and modifying informal rules (the *Urf*) based on local customs (Hakim 1986, 1994; Akbar 1988). Yet *medinas* are often seen as prototypical examples of vernacular 'organic' environments. This work offers the possibility of using simulations of the rule system identified (possibly using agent-based modelling), to see whether they reproduce the actual pattern. It is also significant that other existing studies of rule systems have used vernacular environments, both generally (e.g. Oliver 1989) and in specific cases: rural Virginia houses (Glassie 1975), Caribbean houses (Edwards 1971), San Francisco houses (Vernez Moudon 1986), Pennsylvania houses (Low and Ryan 1985), Buffalo bungalows (Downing and Fleming 1981), Vermont small towns (Williams, Kellogg and Lavigne 1987). By being used as model systems to study that topic, specific cases can easily begin to be integrated, also with cases from other cultures. This leads to hypotheses and conceptual frameworks, as well as the idea of using rule systems consciously to achieve certain product characteristics: i.e. to design rule systems rather than environments (Essex County Council 1973; Vernez Moudon et al. 1980; cf. Rapoport 1993a), with immediate practical implications for planning and design. Also, having thus identified the important role of rule systems, other questions can be asked, showing the heuristic value and fertility of using the vernacular as a model system. For example, one can begin to understand the development and role of more formal rule systems based on safety and health regulations, the impact of labour laws, bank loan criteria, insurance and legal rules, etc., all of which have a major impact on contemporary built environments.

Rule systems may not always be easy to identify, but once the basic principles are understood, controlled analogy can be used with better studied rule systems, e.g. manners, kinship systems, avoidance relations, and hence privacy rules, laws of purity and other rules regarding social relations, interactions and behaviours (who does what, where, when including/excluding whom, and why) (Rapoport 1990a: 96–9, 1999b). These could also be related to the environments then existing, and the hypothesized rule systems could be tested through simulations and, if necessary, replaced or amended; similarly for contemporary situations.

Since the use of the full range of environments as model systems provides the maximum opportunities for longitudinal studies, changes in all the well studied social rule systems, and in sex roles and notions of identity can be studied, and hence also their impact on built environments through their two-way relation with the control and regulation of social relations (i.e. organization of communication).

Rule systems play a role not only in process characteristics, but also in product characteristics. Rules are encoded in settings and, if and when understood, act as mnemonics for appropriate behaviour; in that sense they are also closely linked to behavioural rules. Both of these sets of rules are much clearer, because they are 'tighter', in vernacular design, making it easier to identify the mechanisms involved (Rapoport 1990b). Since these latter are likely to be constant, possibly with specific expressions, hypotheses derived can then be tested in contemporary situations (Rapoport 2000a, 2004).

The consistent application of rules leads to style; different styles are a result of different rule systems, and so are different orders. I argue that chaotic environments are impossible. Since they are products of human culture and follow rules they reflect some order organization of space, time, meaning and communication. But orders can be very different; those described as 'chaotic' are either not understood, inappropriate or not liked by the observer, so that Muslim cities have been described as chaotic, and US cities are lacking order (Rapoport

1977). I have given examples of different orders, geometric vs. social orders, different geometric orders, orders interpreted as 'slums' (e.g. Spalding 1992), orders used to define counter-culture groups (Barnett 1977) and so on (Rapoport 1977, 1984, 1990b, 1993c, 1999–2000, 2004 and references therein). Using the full range of evidence provided by vernacular design one can begin to identify the types of orders and their range (enlarging our repertoire), and also to trace their implications, e.g. for what I have called the 'stable equilibrium' of vernacular design and its implications for open-ended design (Rapoport 1977: Fig. 6.7; 358, 1990c: 91–7). This can also prevent the misinterpretation of environments that can also occur in other fields. For example, the interpretation of rock painting of the San of Southern Africa and of the Upper Palaeolithic in Europe, was greatly hindered by looking for the order found in Western art (Lewis-Williams 2000b: 172; cf. Lewis-Williams 2002a). This was also the case with Australian Aboriginal rock painting and non-western art generally.

Conclusion

The use of model systems implies that there are general principles of EBR, and that these, and the mechanisms involved, may be constant, although possibly expressed differently. It also implies that findings of EBR research are generally applicable or transferable, raising an important question of great practical importance; the extent to which research done mainly in the US (and western Europe) on specific groups and environments, applies to other societies, groups, environments and contexts (Rapoport 2002b). This is a question that EBS has hardly begun to address, and the use of the vernacular as a model system should be helpful.

By using the full range of 'vernacular' defined polythetically, comparative approaches follow almost 'automatically'. The ubiquity of vernacular environments, the fact that they are most of the environment and their variety in all ecological, geographic, cultural and other conditions, not only makes comparative studies easier, it inevitably leads to them by raising questions. If constancies are detected, as seems likely, these become very important, particularly since one can also study how they apply in specific contexts so that variations become different expressions of such constants. As in the case of other model systems, integrative studies also follow (Dugatkin 2001), not just among various vernacular environments, but between the vernacular and EBS, between still existing, used environments and archaeology (Rapoport 1990a, 1993b, 1999b; cf. Kehoe 2002). For all these, spontaneous settlements, changes to government housing and other processes in developing countries are essential and need urgent study. Similarly, in the study of rock painting, ethnographic data from Africa proved essential for understanding Upper Paleolithic paintings in Europe (Clottes and Lewis-Williams 1998; Lewis-Williams 2000a, 2000b). In that case, moreover, current research in a totally different domain (neuropsychology) proved critical, introducing a universal mechanism (altered states of consciousness) with specific cultural expressions. This is also the case in EBS, where the last twenty years of research in cognitive science, neuroscience, etc., in humans and various model organisms, greatly help research in environmental cognition (e.g. Silva 2001).

This way of using the evidence clarifies the nature of the lessons that can be learned, especially since that learning involves the use of explicit hypotheses and models (in the other sense!), concepts and conceptual frameworks from EBS. Such lessons can range from very broad and theoretical, e.g. the constancy of mechanisms involved in EBR, to highly specific, e.g. how to build new (either 'popular' or high-style) environments within a vernacular context, or how to preserve cultural landscapes without turning them into museums.

Environmental design research has had few longitudinal diachronic studies, and those usually project into the future. The use of vernacular environments enables the study of dynamics and change in both process and products characteristics, as does the study of the sequence of changes occurring in developing countries (Rapoport 1983a) and contemporary popular 'suburban' environments in different cultures. Hypotheses generated can then be tested using other research methods – experiments, interviews, simulations, etc.

In the same way that different model organisms, or even parts of them, are used for different purposes in biology and biomedical research, the different environments within the full range can be so used. Thus, for culture change one might use developing countries, spontaneous settlements or past culture contact situations: conquests, migrations and so on. For the importance of high-level meanings one might compare environments of non-literate groups, or those with low literacy, with others. To study hypotheses about changes in levels of meaning and in the importance of semi-fixed elements, a sample of the full range of environments might be used; similarly for uniformitarianism and the constancy of various EBR mechanisms and responses. The analogy for the latter is that the degree of conservation in genetics only became clear from the full range of model organisms; archaea, bacteria, planarians, humans, i.e. through comparisons. In all these cases one needs to study the same systems in different environments, selected from the range available.

I have already emphasized the need to be explicit in the criteria used in the selection of model systems, making clear both their similarities and differences to the system under study. The analysis of the rationale and criteria explicitly used in the many available biological and biomedical studies would offer useful guidelines for judging the appropriateness of particular environments as models. Whether the choice was, in fact, valid would, of course, be an empirical question. The very successful use of model organisms makes another very important point. The usefulness of archaea, bacteria, nematodes, yeast, etc., for studies related to humans suggests that one cannot assume a priori that, say tribal environments are irrelevant for contemporary urban situations. Explicit arguments pro and con are needed.

One of the important criteria used in evaluating theories, but also relevant when evaluating approaches, is their fertility; whether and to what extent they lead to new questions, new hypotheses, new generalizations and so on. I think that a case has been made that treating vernacular environments as model systems would rank high. For example, there are even implications for sustainability and climatically forced ecological change. Historically, numerous groups have had to cope with such events, e.g. the classic Maya, the Akkadian empire (for which evidence is available), the Anasazi and Hohokam, and many others. Their different responses, the successes and failures, hold important lessons, both generally and regarding built environments. There are even cases where of two groups, in one location and subject to the same climate change, one failed whereas the other succeeded (e.g. Vikings vs. Inuit in Greenland during the 'Little Ice Age'). Clearly, many questions follow once vernacular environments are considered as model systems.

Notes

1 Especially if one includes archaeological data, as one must (Rapoport 1990a, 1998, 2000a, 2004).
2 I will not deal explicitly with the third question discussed in Rapoport (1999a) (how the vernacular is to be studied), although, of course, indirectly I do address that topic.
3 The more important conceptual relationship will be discussed later in this chapter.
4 The other aspect of domain definitions (the questions regarding it), is the principal topic of this chapter.
5 Like all ideal type definitions, Rudofsky's very early definition of the vernacular as 'architecture without architects' (1964) does not really work. However, it helps to deal with vernacular design in the broad sense, including tribal, archaeological, vernacular, spontaneous and popular environments. Thus, reference to 'vernacular' without qualification in this chapter always refers to all of these environments.
6 For a recent attempt to use this approach, which regroups the original list into eight process and eight product characteristics, and tries to rank them as high, medium or low, see Kausarul Islam (2003).
7 This relates to the natural history and problem oriented stages discussed earlier.
8 This is different from my discussion of general approaches to the study of vernacular design (Rapoport 1999a: 58–60).
9 The same point is made that in developing evolutionary theory one must consider the largest possible body of evidence – the 'predominant pattern of life's history' (Gould 2002: 939).
10 Personal communication, Simis Zgoutas, President of the International Union of Architects (UIA), Berlin, 25 October 2001.
11 This is not unlike the use of Classical, Gothic, Romanesque or Islamic styles in certain forms of high-style design.
12 This has important implications for the preservation of vernacular environments. The preservation of actual environments, although important, is not sufficient. Nor is the documentation of such environments (also important, as such environments increasingly, and ever more rapidly disappear). It is essential to preserve the lessons and principles that such environments embody, a form of preservation that has not yet been considered or investigated (cf. Rapoport 1990a: especially part III).
13 Although this cannot be developed here in detail, it is essential to note that both 'culture' and 'environment' must be dismantled before they can be used (Rapoport 1998, 2000a, 2000b, 2004, in press b).
14 Increasingly, model organisms (especially the mouse) are being genetically engineered for specific purposes. Their value is reflected in advertisements in scientific journals (as just one example: Charles River Laboratories (*Science* 2002b: 1045)). This development of the 'design and engineering of mice for specific purposes' (Marx 2003), especially for cancer research, is not possible in the case of the vernacular as a model system, and its implications for my argument are, as yet, unclear. It is, however, possible to argue that the extremely large variety of vernacular environments, as already discussed, may provide a 'natural' equivalent by allowing the possible choice of specific characteristics related to the problem or question studied. Note also that on the ST website in Science, there is a Model Organisms category (*Science* 2003a: 1619).
15 When the 2002 Nobel prizes were awarded, a number of reports commented on 'the power of the worm', that the 'worm [was] cast in a starring role for the Nobel prize' (Check 2002: 548). This

nematode, *C. elegans*, has produced insights and findings in such diverse fields as cancer research and genomics. It was selected by Sydney Brenner who was looking for a model system simpler than the fruit fly *Drosophila*, but more complex than bacteria. Because of its short life cycle, genetic simplicity and small size (less than one millimetre long) it has proved extremely useful and is widely used in a variety of subfields. 'You can make major advances in medicine by studying genetically tractable model organisms' (Check 2002: 549).

16 For example, in using *C. elegans* in the study of aging it is emphasized that, since the adult worm has no dividing cells, except in its gonads, it cannot model the important contributions to human aging of impaired cell proliferation in mammals; then, of course, mammalian model systems are used.
17 This corresponds to the range of prehistoric, tribal, traditional, vernacular, spontaneous settlements and popular in the case of 'vernacular'.
18 Announcement of the main topic of a February 2004 conference on Neural Control of Behaviour, *Science* (2003b: 513).
19 This could be extended, e.g. to include Australian Aboriginal rock art, and possibly others (cf. Rapoport 1990a: 75–80).
20 An additional example came to hand while I was writing the final draft of this chapter. In Congo, a country of 50 million people, there are 700 languages and 250 ethnic groups (Lacey 2003). During a recent trip to Ethiopia I discovered that 83 languages and 200 dialects are used by 70 million inhabitants.
21 For example, affect, meaning and evaluation, ergonomics, comfort and culture.
22 Note that, so far the only attribute of vernacular I have added to the original list is the presence/absence and relative importance of the three levels of meaning (Rapoport 1990d).
23 The date is highly significant: the Babylonian conquest. So is the fact that this plan was not used after the Babylonian exile – the culture had changed.
24 As in my previous examples, I rely on a newspaper report. Moreover, these churches are 'popular', in the sense that they are unlikely to be published in architectural magazines.
25 As I have discussed elsewhere, it also leads to lifestyles.

Chapter 11

'Generative concepts' in vernacular architecture

Ronald Lewcock

> *It is clear that even the intentionality of a work cannot help but to defer to secret and hidden motives, to a subconscious stratigraphy.*
>
> (Jung 1922)
>
> *Cuvier's ... 'idea' of the group, or archetype, admitted of endless variation within it; but this was nevertheless subordinate to essential conformity with the archetype.*
>
> (Encyclopaedia Britannica 1911)
>
> *Goethe's conviction of the factual existence of an Urphanomen, an archetypal phenomenon, a concrete thing to be discovered in the world of appearances in which significance and appearance ... idea and experience would coincide.*
>
> (Arendt 1971)

Introduction
Recent arguments for 'archetypes' in architecture came in Gaston Bachelard's *Poetics of Space* (Bachelard 1964). This was followed by the publication of the ideas of Noam Chomsky about the possibility of 'genetic programming' (Chomsky 1972). At this time I noticed unexpected similarities between forms in vernacular architecture scattered over wide regions in many parts of the world. The result

was an article in Paul Oliver's *Shelter, Sign, and Symbol*, 'The boat as symbol of the house' (Lewcock 1975). I further explored the subject in a series of lectures for the Architectural Association, comparatively examining architectures from all over the world.

Subsequent reading in the new literature about the function of the brain and the notions of mind familiarized me with the premise of 'association', with imagination, and creativity (Fodor, Miller and Langendoen 1979; Boden 1990).[1] Intrigued, I studied more seriously the ideas of the cognitive scientists that thought hinges on 'mental models' (Johnson-Laird 1988), with the corresponding conclusion that works of architecture must often be fundamentally conceptual.[2]

The doubts and interest aroused in the subject by these experiences, prompted me to put together some evidence for connections between the 'mental models' or 'concepts' of buildings.[3] The thought had arisen that these might plausibly serve as examples of the presence of 'generative concepts' acting like 'archetypes' in architecture – instances which, if not generic to the built work of all humans, at least seemed to support their functioning in the creative process in certain societies at certain times. I also believe that I have discovered evidence that the link between vernacular architectures and those that are more sophisticated and self conscious often operates at the level of such generative concepts, as I will attempt to demonstrate below.

Generative concepts

Although architectural designs have to respond to practical forces and serve their purpose, they are not, in a strict sense, 'determined' externally. The concept*s* of them must exist in the mind before they are built in reality. And in this way works of architecture naturally reflect the structure of the mental models that formed them.

That works of architecture, like creative works in all fields, are conceptually formed is confirmed by the preoccupation of the arts with imagination, conventionalization, formalism, abstraction, symbolism – the extraordinary, their removal from natural and everyday experiences.

Some cognitive scientists now believe that the human capacity to form such mental structures is genetically programmed. And there is a strong likelihood that a number of capacities of this kind are linked together as a chain capable of giving preferences to certain concepts, including those of space and form, which would mean that they come close to Jung's definition of the archetype (Stevens 1982; Kosslyn 1994).[4]

On a related front: there has been a modern fascination among scholars of architecture with trying to understand the way minds work in conceiving space and form. One of their conclusions paralleled in science, is that *simplification* is a factor in the utilization of concepts and in memory. In 1986 Douglas Graf published an article postulating that most architectural conceptions were based on four components: point, line, plane and volume, which could be combined to give three elementary types of composition: figure, centre and perimeter (Graf 1986). Even earlier, in 1976, Bill Hillier and Adrian Leaman had published a seminal article, 'Space syntax' (Hillier and Leaman 1976), in which they argued that conceptions of form and space could be reduced to an elementary lexicon of twelve combinations of two 'objects' (solid and space, related through properties of continuity or discontinuity, containment and/or permeability, which in turn could be combined in roughly eight ways) and that these were the basic models of thought to which all conceptions could ultimately be reduced.

Do we actually *think* about architecture in the ways described above? Almost certainly the answer is no. What is missing in them is the means of joining a strong 'conviction' to the conception they are setting out to describe. Only if a value is attached to them can concepts attract our attention or have any real meaning for us; they have to carry emotional *commitment*. The way this is affected in the mind is postulated by some cognitive scientists as follows: when experience is checked against existing memory models (concepts), the mind has to have some way of distinguishing between memories of real experiences and memories of dreams or fantasies. The strength of an emotional component is the deciding factor in adding the mark of reality, i.e. 'conviction' or 'commitment' (Mandler 1975; Segal 1983; Warnock 1994; Ellis 1995).

Why should this conviction be so essential in the creative arts? The answer seems to be that through the emotions there comes a sense of 'reality' or 'actuality'. This is presumably what Walter Benjamin meant by the 'aura' of a creative work (Arendt 1971). From this idea it is but a brief step to the deduction that only concepts developed in the mind using the emotional connections of memory, particularly those that are simple and strong, have the potential to form the basis of an architectural or artistic expression capable of carrying the value of commitment. The kind of thing earlier referred to by Schopenhauer as the 'Idea', by Cezanne as his 'little sensation', or by John Stuart Mill as the 'imaginative emotion' (Schopenhauer 1966; Mill 1969; Shapiro 1997).[5]

But how does this work in practice? Recently, scientists have been considering whether the force of emotion might come into play through some kind of association which carries the additional content of meaning. This might depend on social communication and agreement, or it might stem from an individual fascination with certain experiences, or with metaphorical and symbolical models which carry meaning (Restak 1985).

It is the current view that metaphor and symbolism act as devices to link language to visual perception and thus to create deeper memories (Segal 1983). The important point is that this process, or one very like it, is likely to give rise to 'shared' ideas which have the strength of 'archetypes'. Such ideals or concepts would take two forms: (1) those common to all mankind; and (2) those belonging more specifically to one society.

In everyday experience, concepts related to an 'archetype' would be combinations of genetically produced (possibly even genetically inherited) and learned processes. In spite of the shared nature of such concepts, the individual use of a concept is realised in the mind as the result of an emotional trigger.

An example of how this might happen is described by Umberto Eco:

> Let us imagine the point of view of the man who started the history of architecture. Still 'all wonder and ferocity' (to use Vico's phrase), driven by cold and rain and following the example of some animal or obeying impulse in which instinct and reasoning are mixed in a confused way, this hypothetical Stone Age man takes shelter in recess, in some hole on the side of the mountain.
>
> Sheltered from wind and rain, he examines the cave that shelters him, by daylight or by light of a fire (we will assume he has discovered fire). He notes the amplitude of the vault, and understands this as the limit of an outside space, which is (with its wind and rain) cut off, and as the beginning of an inside space, which is likely to evoke to him some unclear nostalgia for the womb, imbue him with feelings of protection, and appear still imprecise and ambiguous to him, seen under a play of shadow and light. Once the storm is over, he might leave the cave and reconsider it from the outside: there he would note the entry-way as 'the hole that permits the passage to the inside', and the entrance would recall to his mind the image of the inside: entrance hole, covering vault, walls (or continuous wall of rock) surrounding a space within. Thus an 'idea of cave' takes shape, which is useful at least as a mnemonic device, enabling him to think of the cave later on as a possible objective in case of rain; but it also enables him to recognise in another cave the same possibility of shelter found in the first one. At the second cave he tries, the idea of that cave is soon replaced by the idea of cave *tout court* – a model, a type, something that does not exist concretely but on the basis of which he can recognise a certain context of phenomena as 'cave'.
>
> (Eco 1980)

Eco is here fleshing out one of Jung's prime archetypes, the cave. Jung describes the next step, the 'construction of a realised model': 'The creative process . . . consists of the subconscious animation of the archetype in its development and formation with the eventual realization of the perfect work' (Jung 1922). Immanuel Kant had put this in more general terms: 'There exists a subjective ground which leads the mind to reinstate a preceding perception alongside the subsequent perception – and so to form a whole series of perceptions' (Kant 1952). Or, in other words, the genesis of architecture appears to come from a natural capacity of the individual brain for association – for lateral thinking; i.e. for making connections between ideas, forms and spaces in the real world and their associations in emotions and symbolism. To this is joined naturally the kind of enjoyment that we have in play, a shared delight which acts to communicate socially to create (the *crucial* term is 'creativity'), a sensation of pleasure: 'space that has been seized on in the imagination cannot remain indifferent space subject to the measures and estimates of the surveyor' (Bachelard 1964).

Arguments are adduced in this chapter that a number of basic concepts shape the thinking of built form in many cultures. The evidence is here examined for the likely development in vernacular architecture of ideas from such elementary concepts as the cave, the covered central courtyard, the walled open space, the hearth, the sheltered space opening to the outside world, and for combinations of two or more of them, sometimes in opposition or in contrast.

Before proceeding to examples there is one other issue that needs to be discussed. There is a notion that a characteristic feature of vernacular architecture is unquestioning conformity to tradition. 'Vernacular architecture does not go through fashion cycles', wrote Bernard Rudofsky (1964). A lifetime of studying the vernacular has convinced me that this is only partly true, like so many ideas about vernacular architecture. Rather, the situation of vernacular building closely parallels that of the middle ages in northern Europe, as discussed in his seminal article on Scholasticism by Erwin Panofsky (1948). In the view of this great scholar there were at that time two influences working on any new building; that of tradition on the one side and conceptual thinking to create innovation and change on the other. What resulted was in some ways a compromise, but one that evolved in a clear direction. The differences between that situation and the freedom of design possible in modern society were twofold. Change was much slower and relatively imperceptible. And vestiges of the original 'traditional' approach survived for many generations.

What I am arguing in this chapter is that in vernacular architecture the underlying presence of a generative concept was so strong that *it* was the essential ingredient that survived throughout the ages. Vernacular architecture is seldom either 'spontaneous', or 'anonymous', but in every case that I have studied in depth involves the agency of highly professional builders who are schooled in principles of design throughout their long apprenticeship; these principles they apply afresh in each new building, using judgement and skill, so that no two vernacular buildings, even if they are contemporary, are ever identical, but vary through a considerable range. Decision making based on deliberation is always a major factor in each building, just as it is in the work of the contemporary architect. In vernacular architecture it is simply the narrower range of choice and the interplay of tradition and change that are different.

The cave as a generative concept

While primitive man took shelter in caves where he could find them, the earliest shelters actually built by man yet discovered were burrows in the silt of wide flood-plains. But these burrows would still qualify as caves, as they are entirely enclosed and have only a single entrance. Apparently the primitive hunters, familiar with cliff caves, having taken up their abode in flat river-side plains, burrowed downwards instead of horizontally, to create 'vertical' caves. For collective security and comfort these sunken dwellings sometimes grew very large, from ten to fourteen feet deep and up to eighty feet long. The oldest so far found, in a river valley in Russia, is estimated to date from 40,000 BC and similar shelters were still being built in northern Europe up to 6,000 years ago (Figures 11.1a and 11.1b). They were clearly roofed: the early examples had mammoth bones around the edge to weight down a stretched roof of animal skins, which was eventually modified by the inclusion of posts and (possibly) horizontal beams of wood, apparently carrying much lighter woven fabrics.

It is from this precedent that the first villages, like Catal-Hüyük in Anatolia, clearly evolved (Figure 11.1c). Built upwards on marshy ground using mud brick covered with plaster, a large number of houses and shrines were packed closely together. Cave-like, each had only one opening to the outside world, a hole in the upper walls or roof, from which descent into the interior was made by means of a wooden ladder.

A similar origin may be ascribed to two of the earliest North American house types, the earth lodge of the Sacramento Valley of California, where the shelter was built dome shaped above ground, and the Pueblo underground dwelling, later transformed as the ritual *kiva*, both of them circular rooms entered through the roofs (Figures 11.1d and 1.11e).

At this point it is appropriate to mention an associated archetypal idea, that of the hearth, which usually became the central focus in these man-made caves. In Gottfried Semper's view the hearth was one of the prime generators of built dwellings (Semper 1851). His ideas are discussed further below.

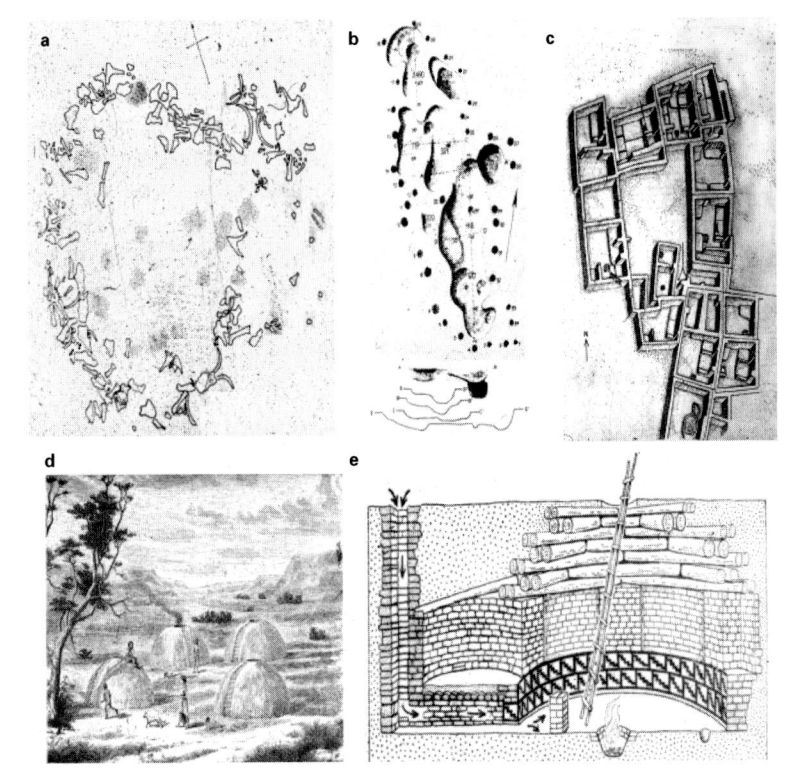

11.1
(a) Middle Palaeolithic sunken tent-stances, c.40,000 BC. Molodovo, South Russia; (b) Sunken tent-house, c.4,000 BC. Koln-Lindenthal, Germany; (c) Part of Neolithic village, c.6,500–5,720 BC. Catal-Huyuk, Turkey; (d) Earth-lodges, c.1,400 AD. Sacramento Valley, California; (e) Reconstruction of the *kiva* of the Great Pueblo phase, Anasazi tradition, with origins that can be traced back to the Hohokam culture, which flourished from the first century AD to AD 1,400.

'Generative concepts' in vernacular architecture

The cave concept was almost always accompanied by the concept of a central hearth needed for the fire, around which the life of the inhabitants was focused. Sometimes there was more than one hearth in a man-made cave, as in the prehistoric underground shelters referred to above. Throughout the Mediterranean there are still scattered descendants of the ancient underground cave house type. The most interesting example, perhaps, was that found by an enterprising Japanese student at the Architectural Association, Tomotsune Honda, who having heard me discuss these ideas, set out one summer to explore the vernacular dwellings of southern Europe and north Africa. Figure 11.2 shows the village of Calcavodo del Param in Spain, although other survivals of the vertical cave tradition exist in other parts of the western Mediterranean, in southern France, in Tunisia and in Libya.

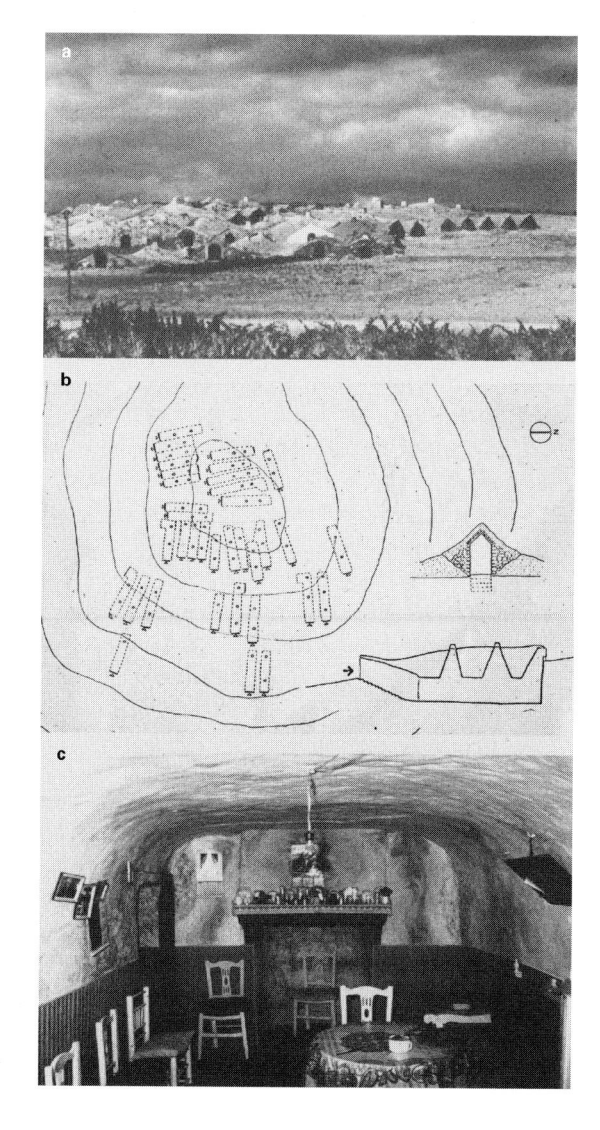

11.2
The underground cave village of Calcavodo del Param, Northern Spain: (a) general view; (b) plan, elevation of the entrance and a section of a typical house; (c) interior of one of the underground houses.

205

Vertical caves were widespread in Asia as well as in Europe, Africa and North America, as is evidenced by living examples from China to Afghanistan. Houses from Hunza in northern Pakistan are characteristic examples. Originally unicellular and strongly constructed in notched logs against the threat of earthquakes, many of these dwellings were subsequently extended upwards so that a second room could be placed above the first, and at the same time horizontal entrances at that level began to be used. The origin of this type of built cave can be traced back at least to the period when Kizil was part of a flourishing culture in Central Asia, AD 350.

Horizontal caves paralleled the development of vertical cave buildings. The earliest man-made versions of such caves have been found by archaeologists in Anatolia and the neighbouring Aegean Islands (Figures 11.3a and 11.3b).[6] From these developed the characteristic throne room type (the *megaron*) of the Cretan and Mycenaean palaces, also evidenced in countless ordinary houses around them.

The Archaic, post-Mycenaean Greek house was in turn a descendant of these. It continued to be characterized by the cave concept (Figure 11.3c). The archaic dwelling was a building totally enclosing a space with a single opening into it for access and light. One of the characteristic late Hellenic house types continued that tradition into Alexandrian times (Figure 11.3d).

At least as early in development as the *megaron* were the stone religious shrines on the islands of Malta, which were constructed as above ground caves entered through man-made cliff faces, with stone seats on either side of the cave entrances (Figure 11.4).[7] To complete the illusion, the rest of corbelled stone domes of each building were covered with earth, apparently planted with grass so that they appeared to be part of the natural landscape. In each, there was a central main cave, lined with two side caves on either side, all of them provided with an altar. That there is a possible symbolism implied by this form is discussed at the end of this chapter.

The practice of building roofs in corbelled stone was widespread in the ancient Mediterranean and was a tradition that continued until modern times. The interiors felt in every aspect like caves. They were devoid of windows and entered in each case through a single doorway. Beneath the *megaron* of Tiryns is the foundation of a great round structure, of which examples have survived with walls of considerable height in Sardinia and other Mediterranean islands. Their ancient descendants were the *tholi* (tombs) of the Mycenaeans and those of the Etruscans. To that ancestry, too, may be traced the much later tradition of the *trulli* in Southern Italy.

Horizontal cave rooms were characteristic of the ancient Maya ceremonial buildings, and of humble dwellings too. Indeed, so persuasive is the cave archetype as a concept of protection that it may be said to underlie most primitive building, from the Arctic igloo to the tropical hut, and from east Asia to the Americas.

In particular, some giant long houses in New Guinea and in South and North America were developments of above ground constructed caves, built in

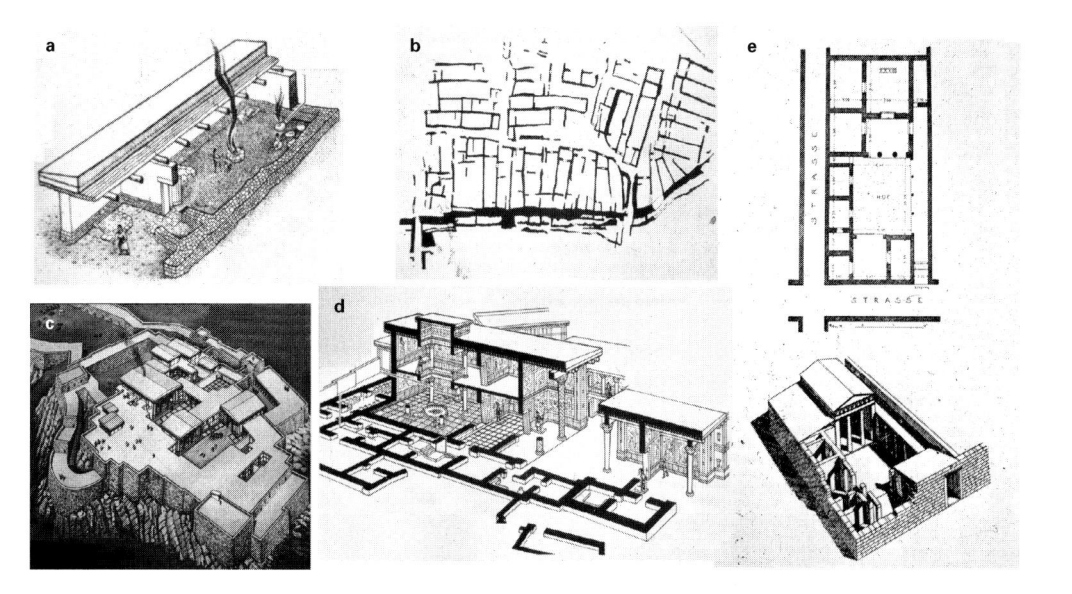

'Generative concepts' in vernacular architecture

11.3
(a) Troy, north-western Anatolia, reconstruction of house of phase 1; 2,700 BC; (b) Thermi, Lesbos, plan of clusters of houses lining streets inside the wall, from Level VC; 2,600 BC (Schachermeys, 1955); (c) Tiryns, Peloponessus, Greece. Reconstruction of the Mycenaean Palace, with smoke shown rising from the hearth in the royal *megaron*, *c*.1,400–1,200 BC; (d) Ptlos, Peloponessus, Greece. Reconstruction of the late Mycenaean *megaron*, with its courtyard and the entrance portico (on the right), *c*.1,400–1,200 BC; (e) Pirene, Asia Minor. Plan and reconstruction of a typical house, *c*.330 BC.

wood and thatch, with, like caves, only one doorway and no windows. The only other small opening was a gap in the ridge to allow smoke to escape.

Other religious buildings which took the form of caves included the early *chaitya* halls of the Buddhists in India. Though many of these were actual caves of enormous size carved into cliff faces, others were freestanding, built-up imitations of them, with few or no windows, each entered by a single door at one end, with a vaulted roof covered by a great curved thatched surface.

11.4
Malta, Bronze Age temple, second half of the third millennium BC: (a) plan and interior; (b) exterior showing the stone seating and the single entrance; (c) two side-chapels seen from the central aisle. Note the beginning of the roof corbelling; (d) Malta, Limestone Idol, second half of third millennium BC. Valetta Museum.

Ronald Lewcock

It may seem that the derivation of all these architectures from such a simple concept as the cave is far fetched and simplistic. But extreme simplicity is the whole point: i.e. the capacity of the human mind to react to strong, simple ideas, and for them to make repeatedly deep impressions in the subconscious throughout thousands of years.

The covered courtyard as a generative concept

Closely related to the cave concept is that of the covered courtyard. That it may be very close is demonstrated in its appearance wherever there are stone outcrops but few natural caves, on the edges of the Arabian desert and other abnormally hot regions. The oldest surviving type of courtyard can be best seen in examples of its relatively recent descendants, which are still clearly cave-related. From these it is but a step backwards to the older central covered courtyards, found in the desert regions. There were examples of these built in ancient Mesopotamia, in ancient Egypt, in nineteenth-century Lebanon and in South Arabia until recently (Figure 11.5).

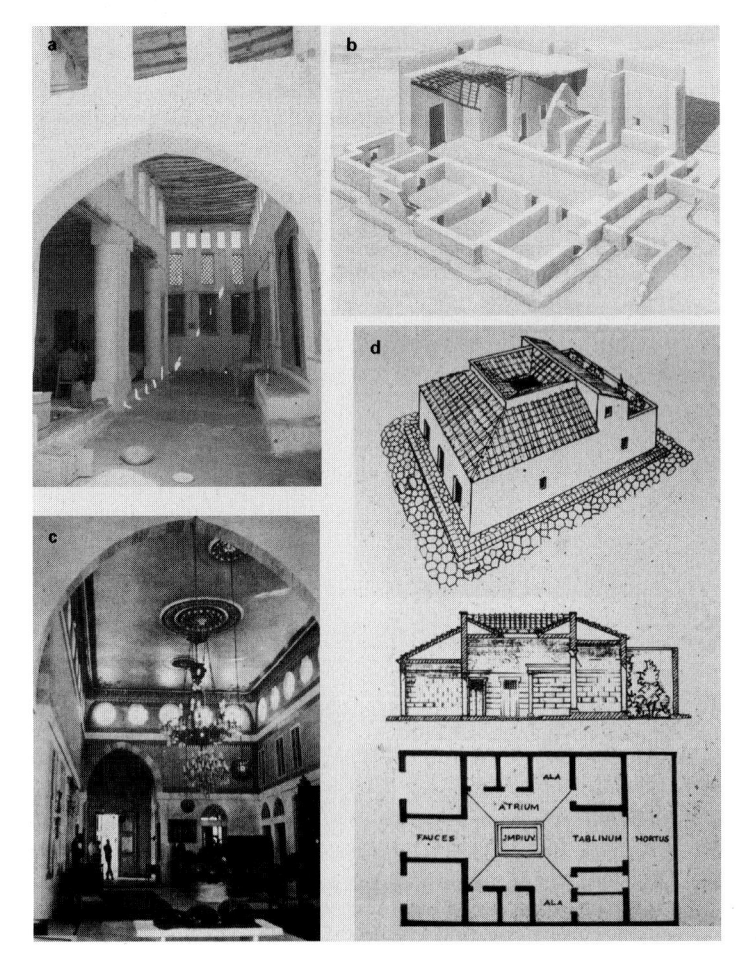

11.5
(a) Zabid, South Arabia. Covered courtyard house, nineteenth century; (b) Tell Madhur, Iraq. Reconstruction of an ʿUbadid house, c.4,500 BC; (c) Beirut, Lebanon. Covered court in Lahoud house, nineteenth century; (d) Pompeii, Italy. Early Roman house, c.250 BC.

'Generative concepts' in vernacular architecture

How the covered courtyard concept developed naturally from the cave archetype is well illustrated in the relationship of the developed Etruscan house, which became the early Roman house (Figure 11.5d) to its earlier prototype, represented by the late Etruscan tomb.

There appears to have been a significant connection between the covered courtyard plan of the eastern Mediterranean and the concept of the mediaeval Venetian *casoni*, and hence with that of the Palladian villa, a striking example of the significance of archetypal descent (Figure 11.6).

The open courtyard as a generative concept

The courtyard open to the sky has demonstrably been one of the earliest protective shelters of man in dry, semi-arid conditions. Like the generative concept of

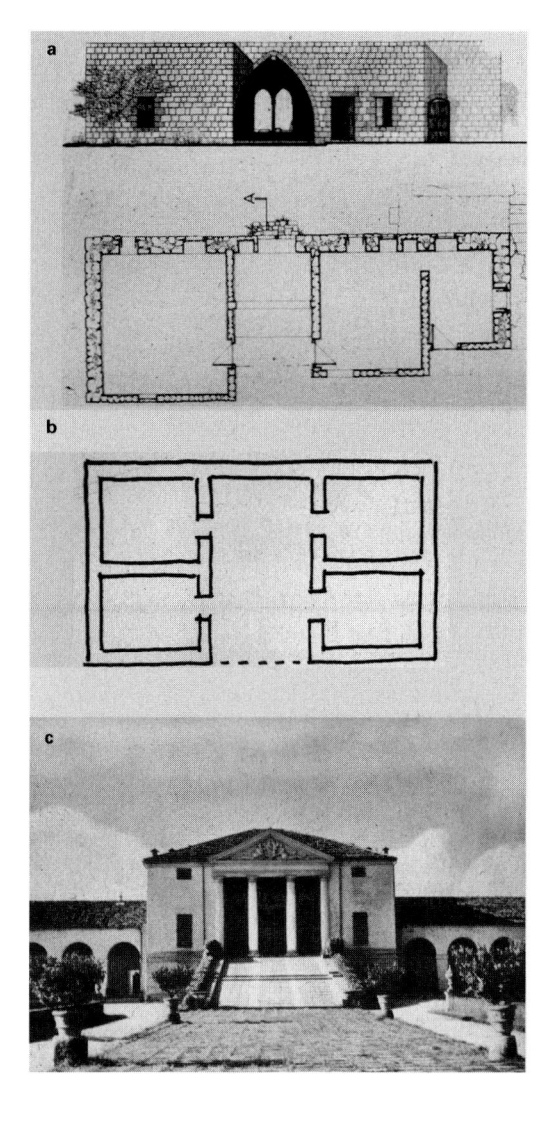

11.6
(a) Beqaa, Lebanon. Covered courtyard house;
(b) Veneto, Italy. Plan of traditional medieval *casoni*;
(c) Veneto, Italy. Villa Emo, Palladio, *c.*1560.

the cave, this open space has slowly acquired sub-divisions of smaller closed volumes within or around it, and thus became the courtyard archetype. The appearance of this concept among the earliest built forms of prehistoric man is well attested in all desert fringe regions, but appears to have emerged somewhat later than the other two primitive archetypes, the built cave and the covered courtyard. One of the finest examples of the use of the open courtyard in early history was another type of developed Greek house of Hellenistic times, that of Delos.

During the period of the Roman republic the Etruscan central hearth was replaced by the *impluvium* in the centre of the floor of the covered courtyard atrium, a trend begun by the opening up of the gap at the ridge that had allowed smoke to escape into a large square or rectangular opening; thus bringing the Roman atrium, though still a covered courtyard, slightly closer to another archetypal concept, that of an enclosure of walls containing a space open to the sky.

It is interesting to note that the developed Roman town house, evidenced by those excavated in Pompeii and Herculaneum, became a combination of these two archetypes, the covered courtyard from the Etruscans, the *atrium*, and the open courtyard, from the Hellenistic Greeks, the *peristyle*. This practice of combining archetypal concepts is one of the traits of architecture in developed civilizations; another example is mentioned below.

The hearth as a generative concept

In the development of prehistoric man, a major milestone was the harnessing of fire, and the attendant possibilities it provided of tenderising food through cooking and of moving to live in cooler climates. Naturally the fireplace became the focus of nascent family and communal life. Means for raising it above the damp ground level and of constructing protection for it against rain and wind followed.

Gottfried Semper (1803–79) was among the first of modern theorists to propose that an important archetype in architecture was thus generated around the notion of the fireplace hearth as the centre of life. In order to protect it, Semper proposed three other essential elements that had to surround it: the raised plinth against dampness, the sheltering roof against rain and screens or walls to create wind breaks (Semper 1851). Semper found prototypes for this archetype in many civilizations, notably in subtropical island climates like those of the West Indies and the Pacific where rainfall and winds were problems, but where threats from wild animals or other humans were not serious enough to justify resort to caves or protective walls (see Figure 11.9a).

The anthropomorphic analogy as a generative concept

Anthropologists have discovered that an anthropomorphic metaphor underlies the concept of built form in many cultures. The use of the metaphor of the human body as a spatial concept is perhaps less an archetype than an archetypal way of thinking. What Northrop Frye calls 'the imaginative limit of desire . . . the content of an infinite and eternal living body which, if not human, is closer to being human than to being inanimate' (Frye 1957).

'Generative concepts' in vernacular architecture

11.7
(a) India. Characteristic *purusha mandala*, used for ritual meditation and rituals; (b) Thanjavur, India. Brahadeswara temple, eleventh century; (c) Characteristic South Indian village temple. Note the pedestal for the navel stone and the mast in the forecourt.

Ancient Hindu belief identified the shrines of gods as manifestations of their physical forms, so that the parts of Hindu Temples today are popularly read in anthropomorphic terms: the shrine as the upper body, the navel represented by an upturned stone lotus on a pedestal in front of the doorway, and the organ of reproduction by a symbolic mast in the centre of the forecourt (Figure 11.7). In a similar way, the plans of the early temples of Malta are seen by some scholars as, expressing through metaphor, an anthropomorphic concept (cf. Figure 11.4).

Throughout Africa the influence of the anthropomorphic archetype is widely evidenced. This is particularly clear in West Africa in the architecture of the Dogon people (Biederman 1987). Their traditional houses are built with thick mud or unbaked brick walls, and each consists of a number of circular or cylindrical rooms clustered around a courtyard. The rooms have different functions and together make up the dwelling. Usually, only one doorway gives access to each room, but small openings in the roof let in light. In the eastern areas the room units adopt rectangular forms. The research of Griaule (1949 and 1975), Griaule and Dieterlen (1965), Lagopoulous (1975) and others has shown that an anthropomorphic archetype is frequently given expression in the names of elements within a Dogon compound, or in the conception of the whole compound (Figure 11.8).

Many other architectural archetypes have been proposed. One frequently referred to is the animal skin or canvas tent, while Bachelard discusses the 'cellar' or catacomb; the 'garret' or hidden space in roofs; the 'shell'; and the 'nest', among others (examples of the shell concept are the hard white polished plastered or marble volumes favoured in Arabia and northern India. I suppose its ultimate expression is the Taj Mahal. Nests are man-made bowers such as those created in mediaeval wall-beds, Victorian window alcoves or Ottoman *diwans* like that in the Sultan's pavilion at Topkapi) (Bachelard 1964).

More important for our purposes is the generative concept of the 'primitive hut', already referred to above, which proved fundamental to the revival of Greek architecture, whether by Palladio or the eighteenth-century neo-classicists. In the brief discussion above I have focussed on the cave and its related forms, not only because the evidence suggests that these seem to have been the earliest primordial architectural concepts, but because they were in any case included as basic primitive concepts alongside the 'primitive hut' by such authors as Quatremere de Quincy (1999), a fact that is often overlooked when classicism and neo-classicism are discussed.

Mimi Lobell (1983) has proposed another kind of generative concept, the simple configuration of lines or

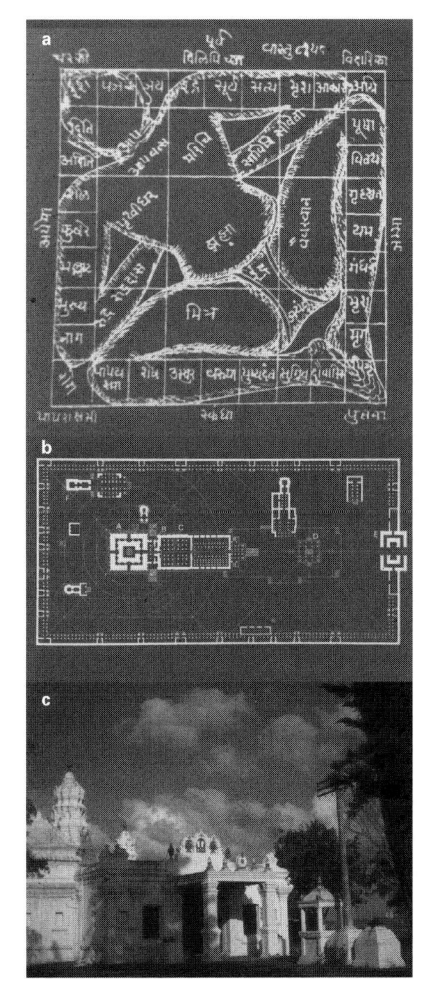

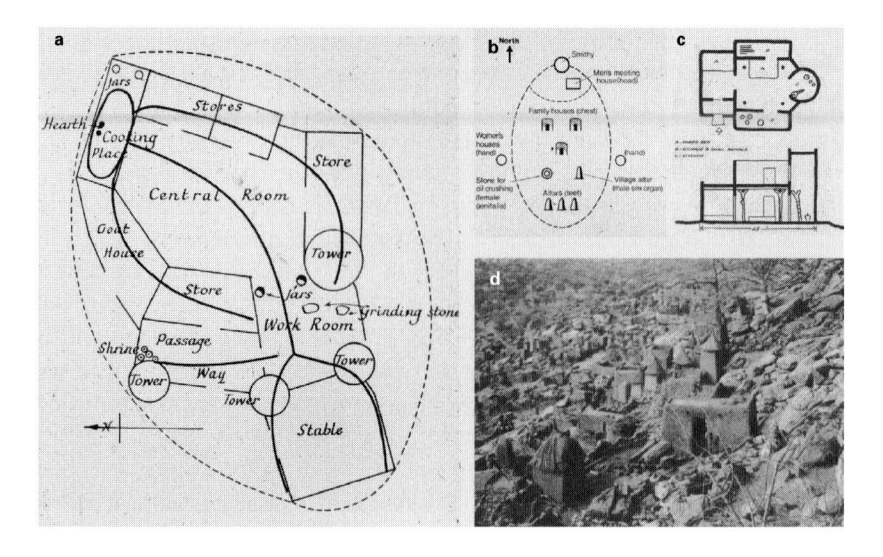

11.8
Mali, West Africa: (a) Dogon family house plan with superimposed anthropomorphic conceptual diagram; (b) Dogon village diagrammed, with the parts identified anthropomorphically; (c) Dogon house. Plan and section; (d) Bandiagara. Dogon village.

axes, as in the spiral, axial cross within a square, the circle, the triangle (or pyramid), the radiating axes of a star and the grid. But these seem too abstract to carry the kind of conviction discussed at the beginning of this chapter.

Doubtless many more 'archetypal' concepts than those suggested above can be proposed. Who is to say the number that human experience and conception can embrace? But an essential characteristic is that they have to be extremely simple.

Architecture has the capacity to speak to the deepest level of our being. In conclusion, we might pause to consider for a moment the extent to which concepts like those discussed here might still underlie traditional architecture today. And what is the extent to which they might be present in the work of our best designers (Mario Botta, for instance, has said of his own home, 'I would like my house to be a reminder of a cave')?

There are at least two possible sources for the phenomena of similarity I have noted above. One explanation is that of the school of Goethe and Jung; that is the archetypal generative concept. But the other distinct possibility is the continuing preference in the minds of successive generations for *simplification*, with the result that the same simple concepts may be arrived at in differing circumstances, and from different sources. But in either case the central issue remains: is this one of the generative ways in which minds think about architecture? And if so, how pervasive is this phenomenon? Then we must ask: did concepts like these become more elaborate and evolve in complexity with the development of civilization? We have already seen instances when one concept was played out against another. What about the extension by combinations of

'Generative concepts' in vernacular architecture

groups of such concepts into settlements, towns and cities? These now seem to me to be promising areas for future investigation.

Generative concepts and the vernacular in the twenty-first century

Carl Jung believed that generative concepts recharge ideals. For him, a created work is both mediation and the regeneration of old values:

> By turning its back onto the imperfection of the present, the artist's aspiration withdraws until it reaches in his subconscious, the primordial image which can compensate most effectively the imperfection and partiality of the contemporary spirit. It takes hold of this image, and from the deepest subconscious it draws it closer to the conscious, modifying its shape to render it acceptable to the contemporary person, according to his capacities.
>
> (Jung 1922)

It will have become apparent from their strength and resilience that the presence of such essential 'archetypical' generative concepts in local and regional vernacular traditions is likely to persist into the foreseeable future. They also constitute a resource that has been drawn upon in the past by original architects, as Frank Lloyd Wright did in his Prairie houses, combining two generative concepts: the archetype of the hearth (duly set in the centre of the building under the shelter of the overhanging roof) with the concept of the open road, or the axis out into the countryside, to produce a dichotomy of representational qualities (Figure 11.9).

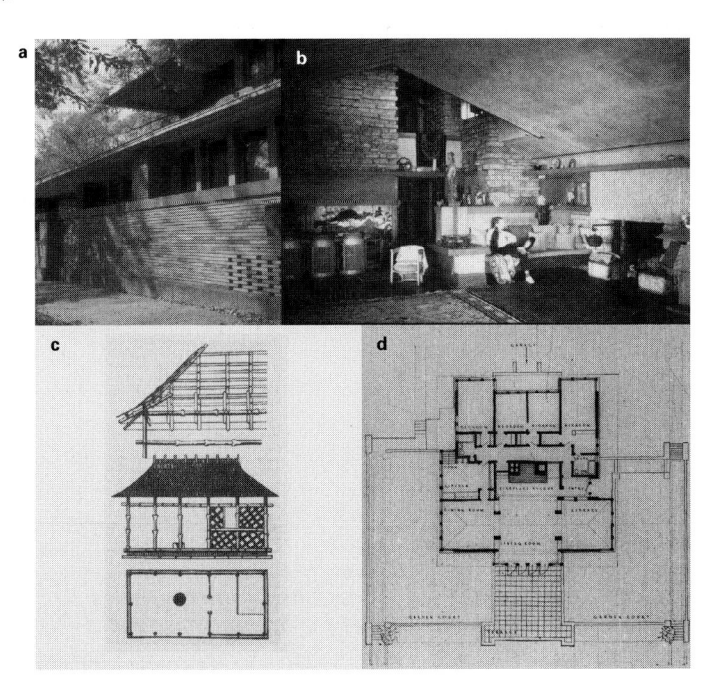

11.9
(a) Buffalo, New York, North America. Martin House, 1905; (b) Wisconsin, North America. 'Taliesin East'. Fireplace in living room, c.1910; (c) The Indian hut on display at the Great Exhibition, 1851; (d) Oak Park, Illinois, North America. Cheney house plan, 1905.

213

Ronald Lewcock

Bachelard's provocative interpretation, quoted below, of the way in which the 'primordial image', the generative concept, 'work[s] its effect on the mind' suggests that such generative concepts are still with us, and will continue to exert a profound influence in the future:

> The relation of a new poetic image to an archetype lying dormant in the depths of the unconscious is not, properly speaking, a causal one. The poetic image is not subject to an inner thrust. It is not an echo of the past. [But] through the brilliance of an image, the distant past resounds with echoes, and it is hard to know at what depth these echoes will reverberate and die away.
>
> (Bachelard 1964)

Notes

1. Boden reviews several computational algorithms which seem to simulate the creative/generative process, which he calls 'genetic algorithms'. See also Kant (1952), Restak (1985), Dissanayake (1988) and Bronowski (1978).
2. Johnson-Laird (1988) proposed the term to offer a descriptive model of deductive reasoning. Mental model theory assumes that reasoning begins with comprehension of the premises in syllogistic and conditional problems. This comprehension results in a representation of the problem, which is the mental model. See Giere (1992).
3. 'Concept: an abstraction or a general notion . . . Some concepts may be powerful thinking tools even when they are not at all fully understood' (Gregory 1998).
4. Stevens (1982) and Kosslyn (1994) deal with storage of visual information and our perception of space; Biederman (1987) introduces a structural theory of representation of which the basic unit of representation is called a geon, meaning a geometrical icon. He hypothesises that the geon originates in the primitive long-term memory.
5. The distinction between 'knowledge' and 'conviction', or between a statement of fact and belief in it, was the source of 'Moore's Paradox', which is briefly discussed in Edmonds and Eidinow (2001).
6. That even earlier examples must once have existed further east is evidenced by the houses of the Ubaid (c.5500–4000 BC). These were courtyard houses derived from caves and will be discussed in the next section. Of course, the Paleolithic huts, even those made of branches, which are covered and entered only on one side, functioned conceptually as horizontal caves.
7. There were also below-ground caves hollowed out in the natural rock, but these existed somewhat later.

Chapter 12

The future of the vernacular
Towards new methodologies for the understanding and optimization of the performance of vernacular buildings

Isaac A. Meir and Susan C. Roaf

> Dwellings are built to serve a variety of functions, but one of the most important is to create living conditions that are acceptable to their occupiers, particularly in relation to the prevailing climates. Buildings do not control climate, which, apart from the wind or sun shadow that they cast, remains largely unaffected. But from within the dwelling can modify the internal climate, even though it is affected by the external conditions. The materials that are used, the forms they take, the volumes they enclose, and the services that are installed may all contribute to the 'micro-climate' that the house generates. This is not always precisely what the occupants require in temperature, ventilation or relative humidity.
>
> (Oliver 2003: 130)

Isaac A. Meir and Susan C. Roaf

Introduction

At the threshold of a new century, and faced with the potentially devastating impacts of climate change and the end of the 'fossil fuel age', questions are increasingly being asked on 'what types of buildings will be most resilient in the face of such challenges'.[1] Instinctively, many people's response is that traditional vernacular prototypes are best adapted to the different climates they occupy and therefore are better suited to provide sustainable prototypes for a future without cheap energy, far more so than energy expensive high-tech building types. But is this right? Here we question this response, in a search for genuinely resilient building types that will safely house people in the changing circumstances of the twenty-first century.

Historical, traditional and vernacular housing prototypes have been considered as inherently adapted to the constraints of the natural environment. A reliance on such deterministic assumptions has often led to misinterpretation of facts, wrong conclusions regarding appropriate technologies and design solutions in general, and in particular those relevant to low-cost housing for developing countries. This chapter analyzes a number of generic types of housing common around the Middle East and the Mediterranean, and assesses their actual performance vis-à-vis different low-tech upgrade and retrofit strategies. A number of methods and techniques were employed, including monitoring, modelling, numerical analysis, simulation and infra-red thermography. Investigations included different building technologies and materials, morphologies and details, under different arid conditions typical of the Middle Eastern and Mediterranean climatic regions, with a view to exploring methods of optimizing the performance of vernacular prototypes to provide resilient buildings for the twenty-first century.

Performance stereotypes of the vernacular

It has become rather common to encounter in architectural publications statements advocating the study of vernacular and historical housing prototypes as a base for environmentally conscious design. One such typical statement is the following:

> Temperatures (within Nabatean buildings) were controlled by proper construction of the walls. These were made three layers thick and were hermetically sealed on the outside and the inside with a porous insulating layer between. In addition, all openings of the living rooms faced south and west in order to benefit fully from the sun. Slot like windows placed below the ceiling facilitated ventilation but prevented, at the same time, the penetration of dust. In this way temperatures were always much higher in the rooms in winter and considerably lower than the heat on the outside of the building in the summer. The extremely small courts around which were grouped the living rooms only helped in this matter.
>
> (Negev 1980: 30)

The future of the vernacular

This, of course, is only an illustration. Similar examples are numerous and can be encountered throughout professional and academic literature describing different building types, systems and materials; among them internal courtyards, windcatchers, vaults and adobe. Traditional urban layouts in the Middle East and the Mediterranean are usually dense and built around courtyards, some internal and fully enclosed, others attached and/or semi-enclosed (see Figure 12.1). It has been assumed and very frequently noted that this almost homogeneous dispersal of this building type implies an inherent microclimatic advantage. However, the matter was never thoroughly investigated, although in some of the few cases monitored results were contradictory to the theory.

It might be assumed that geometry, proportions and orientation could affect the thermal properties and behaviour of such spaces, and that their function might have been other than, or at least not solely that of climatic amelioration. Studies such as those by Rapoport (1969) suggested the existence of an obvious discrepancy between the environmental or climatic requirements of, and the behavioural patterns to be accommodated in, such spaces. This discrepancy is one explored in this chapter.

The densely built urban form typical of Middle Eastern and Mediterranean regions mentioned above created ambivalent microclimatic conditions. On the one hand it created narrow and winding streets and alleys with a very specific and often ameliorated microclimate, especially due to the shading of surfaces (Pearlmutter 1998). However, on the other hand this densely built-up form suffers from a reduced cooling potential both due to the restricted ventilation potential and due to the reduced Sky View Factor (SVF) of narrow spaces, which reduces outgoing long-wave radiation. This last parameter depends on the proportions of the open space, namely its width and height. The narrower a space in relation to its depth (or the height of its built-up edges), the smaller the proportion of sky dome 'viewed' by the surfaces, the less efficient the heat exchange between these surfaces and the sky dome. To compensate for the poor ventilation and the reduced SVF, many traditional settlements were aligned to ensure that the

12.1
Internal courtyards.
Left: St Gerasimos Monastery (Deir Hajle), near Jericho;
right: Patio de los Leones, Alhambra.

prevailing winds were efficiently channelled through the streets and alleys to flush the heat from them, and cool their users by convection and evaporation as appropriate (Bonine 1979).

Another way in which traditional architecture dealt with this problem was through the use of windcatchers. Such scoops or towers protrude above the densely built urban form and catch air flowing at higher velocities compared to that within the urban open spaces. This air is cooler than that within the city. However, at particular times of the day or year it may be at higher temperatures than those inside parts of the buildings such towers serve. In various cases the thermal and aerodynamic properties of the windcatchers have been discussed in a non-scientific way and a number of misconceptions regarding their actual function have developed over the last fifty years (Roaf 1990). Typical of these misconceptions is that the notable height of the windcatchers of Yazd in Iran, resulted from climate influences alone (see Figure 12.2). In fact, their apogee in terms of height and elaboration of design and detail was largely influenced by the British–Chinese opium trade of the mid-nineteenth century, spurred on by the opening of Hong Kong as a major British port in the South China Sea. In this trade, the Yazdi farmer and merchant flourished, trumpeting their wealth and social elevation with a plethora of 'high' towers, not unlike the civic towers that sprouted around late Mediaeval and Renaissance cities in Europe (Roaf 1989). The sophisticated design, and enhanced height and climatic performance of these towers, resulted largely from the socio-economic conditions that created them.

Traditional building technologies and structural systems appear often to be viewed by architects as little more than aesthetic statements. However, the notion of copying vaults and domes as simple morphological emblems or attributes of 'desert', Middle Eastern or Islamic architecture carries with it the danger of misunderstanding and misinterpreting the underlying reasons for the evolution of particular construction systems, and their attendant detailed designs and performance characteristics, in the climates and environments in which they developed. Alternatively, the automatic justification for their existence and continued use, based on their perceived climatic advantages alone is also dangerous because it denies us the benefits of re-interpreting (understanding) rather than re-using (copying) the technology. It also relies on the veracity of those perceptions of performance. It is important to remember that under the considerable climatic constraints of the regions discussed in this chapter, wood is a scarce and expensive commodity.

Key performance issues within the vernacular

The readily available building materials in dry lands are soil and stone, both very poor in tensile strength. Thus, the structural systems developed through a long trial and error process, stretching over some nine millennia, were those that took advantage of the compressive strength of these materials. The dominant roofing systems in many parts of the Middle East are consequently vaults and domes, arches and sometimes 'latent' arches in the form of corbels. The question that should be asked is how such roofing systems actually work climatically, in relation

12.2
Windcatchers on the village of Nausratabad outside Yazd, Iran.

to their form and details, but again, as with the windcatchers, surprisingly little research has been done until fairly recently.

Soil has been used since the earliest excavated dwelling settlements dating back to the seventh millennium BC, with different techniques such as adobe bricks or rammed earth cast in moulds. Systems and techniques of building in both soil and stone have undergone long periods of detailed development and adaptation, building on the characteristics of the properties of the materials in each region.

Structural issues were key to the evolution of the crude building prototypes, with their characteristic systems and forms, such as the arches and vaults that emerged in response to the need to provide adequate shelter in these harsh climates. However, climatic considerations may often have been of secondary importance, after behavioural, economic and cultural influences, in the evolution of the detailed design of these buildings. Assuming an inherent climatic suitability or superiority of materials and form in vernacular buildings may be misleading, as demonstrated when their excessive thermal mass, and minimal, often unglazed fenestration, more or less necessitated by traditional constraints, are considered vis-à-vis the climatic conditions of different drylands and deserts.

These issues become of special importance when inspected through the looking-glass of climate change and energy sources depletion. The former has been extensively discussed in recent years and in spite of a first reluctance to identify climate change as a planning and design issue, the close observation of changing climate patterns has shown the urgency needed in relating to them. Today it has become obvious that local and global changes are occurring at an accelerating pace, often observed as extreme droughts, storms, heat waves, cold spells, floods and aridization. Although the general trends seem to be those of global warming, it is anticipated that local cases of cooling may occur, such as in the case of the UK. Whereas technology was considered in the past to be a panacea, it has become obvious that current practices have severe limitations for a number of reasons. First and foremost, it has been shown that the current rate of energy consumption typical of industrialized countries, will eventually reach a forced decline due to the limitations of fossil fuel resources (Bartsch and Mueler 2000). At the same time, it has been shown that such intense and unrestricted fuel consumption has direct influences on climate, environment, health and the well-being of people (Jean-Baptiste and Ducroux 2003; Odum and Odum 2004).

In industrialized countries buildings are the biggest consumers of energy, using about half of it, invested mainly in heating, cooling, lighting, ventilation and movement. A vicious circle has been created in which such excessive use of energy adversely influences the built-up space and its microclimate, thus creating a higher demand for more energy to be invested in the buildings and their services (Santamouris *et al.* 2001; Steemers 2003). It is obvious that this is not the case in the less developed countries, where often the majority of the population is still rural, using traditional buildings and technologies, and having very limited access to fossil fuels or the building services technologies so widespread in the industrialized countries. For those masses of people the only obvious viable solution is the upgrade of traditional and vernacular buildings. However, for this to be possible it is necessary to study critically the actual behaviour of such buildings, their morphology, technology and systems.

Tools and methods

If we are to be able to really understand, and learn from, the extent to which traditional dwellings are suited to their natural environments, we need clear and systematic research methods for doing so. This paper outlines a programme of research applied to the houses of the desert regions of the Middle East and drylands of the Mediterranean, designed to enhance our understanding of exactly how these buildings work in their climates.

The paper presents a range of methods used in this project, including theoretical and fieldwork studies undertaken by the Desert Architecture and Urban Planning Unit of Ben-Gurion University in the Negev Desert, Israel, and theoretical, simulated model studies undertaken within the Energy Efficient Buildings programme at Oxford Brookes University, UK. The parametric studies described here include *in situ* monitoring; 1:1 scale model monitoring; infrared thermography; thermal and daylight simulations; and numerical analysis (see Figure 12.3). In several cases, 1:1 scale models were used to calibrate simulation tools. Investigations included different building technologies and materials, morphologies and details, under different arid conditions typical of the Middle Eastern climatic regions, as well as semi-arid and drylands conditions typical of many Mediterranean areas. Indoor climate was analysed vis-à-vis visual and thermal comfort. The results have been used to assess heating loads, and these were used to estimate the probable indoor air quality and environmental implications due to the use of combustible fuels under poorly ventilated conditions.

The future of the vernacular

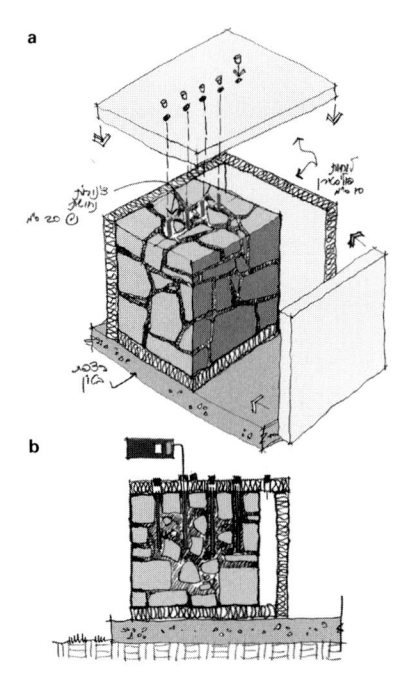

12.3
(a) 1:1-scale masonry models built and monitored under real conditions to allow simulations calibration; (b) Drawing showing copper pipes embedded in the walls at 20 centimetres intervals to allow temperature measurements; (c) Models with lime mortar (left) and dry masonry with mud and rubble core before

In situ surveys, data collection and literature reviews indicated that, although the overall number of specific house types of the region was relatively large, these could easily be grouped under generic types. Such grouping allowed the creation of a relatively limited, and manageable, number of prototypes common throughout the Middle East and the Mediterranean (Canaan 1932–3; Ragette 1980; Khammash 1986; Hirschfeld 1995). These house prototypes included various forms, morphologies and geometries, materials and technologies, geo-climatic location and dispersal, details and variations. A comprehensive study of such dwellings would eventually need *in situ* monitoring, but the initial requirement was for the development of a methodology that would allow researchers to gain an overall understanding of the performance of the generic prototypes in a range of locations and climates.

First the methodology would have to provide appropriate strategies and solutions to overcome constraints such as limited access to potential sites, partial preservation of buildings, monitoring in occupied buildings, security and safety, and a large number of parameters. The solution chosen was a multi-partite protocol based on the following procedures and methods:

1. classification of types in time and space/climate
2. monitoring of available/accessible case studies
3. construction of scale and full size physical models
4. monitoring under real conditions
5. use of model monitoring results for the calibration of simulation programs

221

Isaac A. Meir and Susan C. Roaf

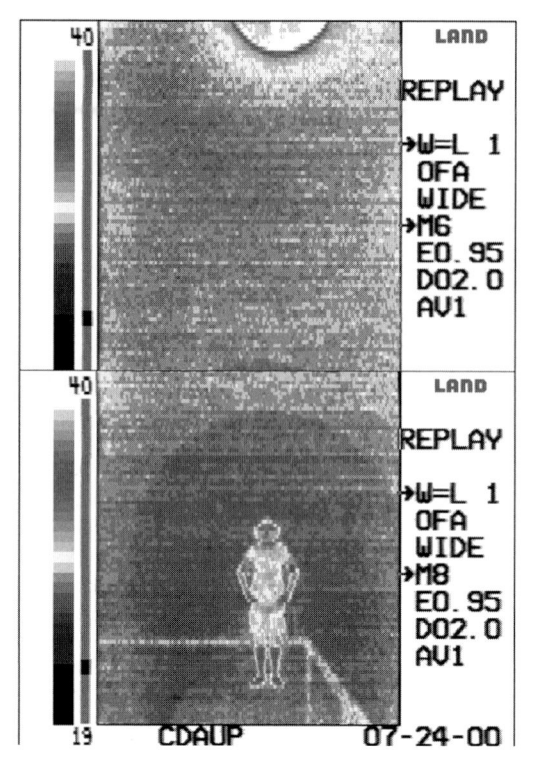

12.4
Roman bathhouse dome and interior, Avdat, Negev Highlands. Left: thermal image showing surface temperature variations at summer (July) noon; right: structural details – lighter stone hue indicates reconstructed part of the dome.

6 simulation of case studies and variations
7 infrared thermography for the verification of simulation results
8 numerical modelling combining parameters of simplified physical and simulated models
9 parametric studies of simulated upgrade and retrofit.

The prototypes examined up to the preparation of this chapter, and the procedures and protocols followed, are summed up in Table 12.1. They represent hundreds of runs of simulation models (primarily with simulation software Quick/Easy and Toolbox, but also Ecotect, Daylight and others), and hundreds of days of *in situ* surveying, documenting and monitoring. Some of the results have been discussed and analysed, many of them summed in a number of papers submitted by graduate students as part of their assignments. Many of these have already been published as partial, type or site-specific studies (Meir, Pearlmutter and Etzion 1995; Pearlmutter and Meir 1995; Meir 2000 and 2002; Meir, Mackenzie Bennett and Roaf 2001; Meir and Gilead 2002; Meir and Roaf 2002; Peeters and Meir 2002; Runsheng, Meir and Etzion 2003a and 2003b). (See Figure 12.4.)

Results and discussion

The results have been interesting both from the research point of view and from the educational one. Most historical and vernacular prototypes are, by nature, of

The future of the vernacular

Table 12.1 Types and parameters used in this study, and research activities undertaken so far

	Coastal	Lowlands	Highlands	Deep valleys
Building type				
Subterranean (cave/complex/earth integrated)		▦▦ ▮	▦▦ ▮ ▬	▦▦ ▮
Conventional 1 storey	▦▦ ▣	▦▦ ▮ ▣	▦▦ ▮ ▣ ▬	▦▦ ▣
Conventional 2 stories	▦▦ ▣	▦▦ ▣ ▬	▦▦ ▮ ▣	▦▦ ▣
Conventional 3 stories	▦▦ ▣	▦▦ ▣	▦▦ ▣ ▬	▦▦ ▣
Building material				
Adobe	▦▦ ▣	▦▦ ▣	▦▦ ✕ ▮ ▣	▦▦ ▣
Stone	▦▦ ▣	▦▦ ▣	▦▦ ✕ ▣ ▬	▦▦ ▣
Light: woven fabrics, reeds, etc.	▦▦	▦▦	▦▦	▦
Light: reused fabrics, metal sheets, etc.	▦ ▣	▦ ▮ ▣	▦ ▮ ▣	
Plan				
Square	▦▦ ▣	▦▦ ▣	▦▦ ▮ ▣	▦▦ ▣
Rectangular	▦▦ ▣	▦▦ ▣	▦▦ ▮ ▣	▦▦ ▣
Wall section and details				
2 layers dry ± plaster (mud/lime)	▦▦ ▣	▦▦ ▣	▦▦ ✕ ▮ ▣	▦▦ ▣
2 layers + mortar ± plaster (mud/lime)	▦▦ ▣	▦▦ ▣	▦▦ ✕ ▮ ▣	▦▦ ▣
Fenestration				
Door only	▦ ▣	▦ ▣	▦ ▣	▦ ▣
Door + window (various orientations)	▦▦ ▣	▦▦ ▣	▦▦ ▣	▦▦ ▣
Door + windows (various orientations)	▦▦ ▣	▦▦ ▣	▦▦ ▮ ▣	▦▦ ▣
Roof				
Flat	▦▦ ▣	▦▦ ▫ ▣ ▬	▦▦ ▫ ✕ ▣	▦▦ ▣
Vault	▦▦ ▣	▦ ▫ ▣ ▬	▦▦ ▫ ✕ ▣ ▬	▦▦ ▣
Dome	▦▦ ▣	▦ ▫ ▣ ▬	▦▦ ▫ ✕ ▣ ▬	▦▦ ▣
Light pitched	▦▦ ▣	▦▦ ▣ ▬	▦▦ ▫ ▣	▦▦ ▣
Courtyard				
Enclosed	▦	▦ ▦ ▮	▦▦ ▮	▦▦
Semi-enclosed	▦ ▦	▦ ▦ ▮	▦▦ ▮	▦▦
Adjacent	▦ ▦	▦ ▦	▦▦ ▮	▦▦
Portico	▦ ▦	▦ ▦ ▮	▦▦	▦▦
Peristyle	▦ ▦	▦ ▦ ▮	▦▦	▦▦

Legend: ▦ historical; ▦ contemporary vernacular; ▫ numerical modelling; ✕ physical modelling; ▣ computer simulation; ▮ monitoring; ▬ infrared thermography.

high thermal mass, with very limited fenestration area, usually unglazed. These properties make them very inert in relation to ambient daily fluctuations. However, this extreme inertia is counter-productive due to the inability of such structures to take advantage of solar gains in winter and of night cooling by cross ventilation in summer, primarily, but only, due to their limited fenestration size. Thus, the construction technology and building types traditionally considered to be, by default, adapted to the environmental constraints proved to be uncomfortably hot in summer and uncomfortably cold in winter for most of the hours of the day. The thermal performance of such buildings proved to be better in highland and mountain regions rather than the lowlands and more humid coastal plains. No significant differences were found between stone masonry and adobe construction.

Curved roofs were found to perform thermally better than flat ones by promoting more comfortable indoors. The geometric advantages of such roofs were originally investigated experimentally by Pearlmutter on test cells with negligible mass roofs (Pearlmutter 1993). Numerical modelling undertaken within the broader framework of the research described here showed similar results for massive curved roofs modeled under arid conditions (Runsheng, Meir and Etzion 2003a and 2003b). Results showed that a domed or vaulted roof absorbs more solar radiation than its corresponding flat roof. This increases with the increase of the half-dome or vault angle, but is insignificantly affected by the climatic characteristics and latitude of the location. However, the main reason for improved indoor conditions under a curved roof is exactly this enlarged surface area, which allows for more heat to be dissipated at night through radiation and convection. This, of course, limits the suitability of such geometry to areas with clear night skies, typical of continental and especially highland deserts, but not necessarily coastal ones.

Minor differences were found between vaults of different orientations. To quantify such discrepancies, numerical modelling was performed to estimate the insulation absorbed by vaulted and domed roofs, based on angular dependence of absorptance and solar geometry. A north–south facing vaulted roof was found to both reduce the solar heat gain of buildings in summer months and increase solar heat gain in winter months, compared to identical vaults facing east–west; the greater the proportion of area exposed to the sun, the smaller the amount of beam radiation that will be absorbed by a curved roof. Furthermore, results showed that even if absorptance were assumed to be constant this would affect the total solar heat gain of the roofs studied by less than 4 per cent (Runsheng, Meir and Etzion 2003a and 2003b). The role of the seemingly negligible windows positioned on the upper part of gable walls in such structures proved to be of very significant importance in the overall behaviour of the structures, when ventilation was applied to the model and the simulations.

The worst indoor temperature conditions measured and simulated were within buildings with light roofs. Shading was found to have a favourable effect in summer. However, having said that, it is important to mention that over fifty per cent of the thermal loads of buildings originate in the roofs and therefore

12.5
Monitored and simulated stone building on the Negev Highlands, with different roof details and configurations.

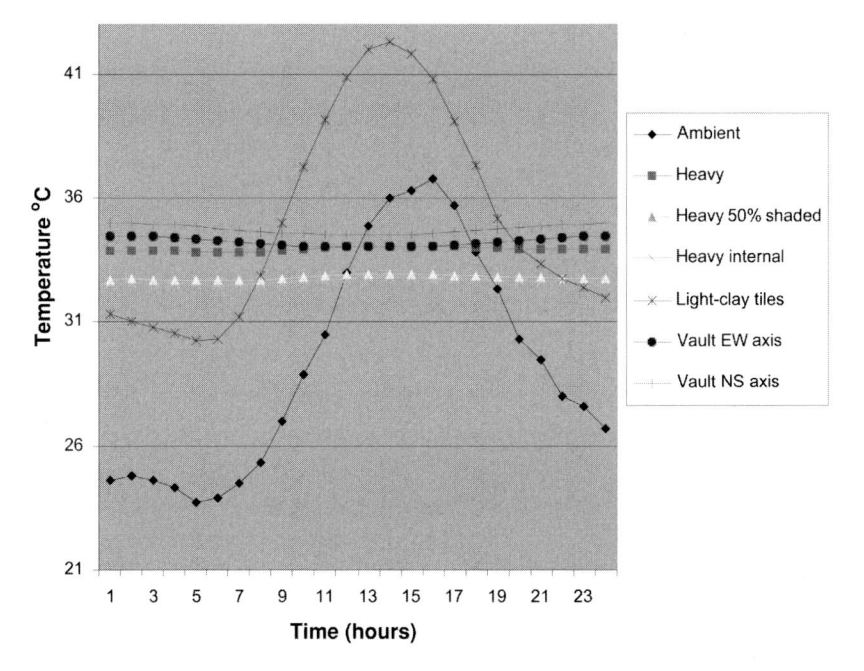

insulating a flat roof will have a much more profound effect than replacing it with a curved one.

Fenestration in such structures is typically negligible. In many cases it is limited to one opening only, a door. Where windows do exist they tend to be very small, a fact dictated by the construction technology and materials, as mentioned previously. Traditionally such apertures were unglazed, often totally blocked in winter to avoid heat losses, thus also minimizing daylighting, while not solving entirely the problems of infiltration. This lack of fenestration and proper glazing materials did not allow the structures monitored and simulated to take advantage of the otherwise advantageous ambient conditions and temperature differences. In cases simulated with modified fenestration and operation regimes the improvement was limited, again due to excessive thermal mass.

One such case monitored and simulated is a stone structure typical of the Roman period in the Negev Highlands in Israel (see Figure 12.5). It has an entrance door and four narrow and high slots (15 by 85 centimetres) located at about 4 metres above the floor in the wall perpendicular to that of the door. Both measured and simulated temperatures (for different orientations and operation modes of the windows and door) indicated a marginal rise of indoor temperature in comparison with an identical room simulated without windows. This marginal rise of approximately 0.5°C was similar in winter and summer, and may be attributed to air movement due to buoyancy differences and wind (Meir 2002).

Energy input for heating in winter, a necessity in most drylands, turned out to be a significant burden. The most common sources of energy are firewood, dried dung and agricultural residue. Whereas the use of the former is considered one of the main contributors to desertification in semi-arid regions (Loevenstein, Berliner and Keulen 1991; Sauerhaft, Berliner and Thurow 1998), the use of all three has serious health implications, especially when they are burned within confined and poorly ventilated spaces, where the pollutants produced (some of which may well be carcinogenic) make respiratory complications as high as a thousand times more likely than when such biofuels are burned outdoors (Environmental Protection Agency 1995; Nazaroff, Weschler and Corsi 2003).

In many traditional settlements cooking is still done outside the main living quarters, either in a separate structure or in the open-air courtyard. However, winter heating is still a problem, since in many cases it is based on an open fire. Although no studies were done within this research regarding particles and pollutants common in the stone, mud or concrete buildings, or tin huts and tents, heated in this way in winter, the data available through parallel investigations point to some very disconcerting facts. Indoor air exposure to suspended particulate matter increases the risk of acute respiratory infections, one of the leading causes of infant and child mortality in developing countries. In Asia, such exposure accounts for between half a million and one million excess deaths every year. In sub-Saharan Africa the estimate is 300,000–500,000 excess deaths. Around 30–40 per cent of cases of asthma and 20–30 per cent of all respiratory diseases may be linked to air pollution in some populations (World Health Organisation 1999). Such cases may also be linked to poorly heated indoor spaces which encourage the growth of moulds whose spores are often allergenic (Smith 2003).

Fenestration alterations and enlargement, and roof insulation and/or shading were identified as vital for the improvement of thermal performance, especially when combined with a reasonable thermal mass, i.e. lower than that described above. A variety of insulating materials and details were investigated, among them various recycled materials. Simulation results showed that the 'resistive' insulation (that which resists or stops the passage of heat through a wall) plays a significant role only when the typically high thermal mass (the 'capacitive' insulation relying on the density and thermal capacity of the envelope to retard the heat flow) is reduced (Roaf, Fuentes and Gupta 2003). It was also demonstrated that the shading of heavy flat roofs can have a significant effect, lowering indoor temperatures by up to 3°C under certain conditions. Lightweight roofs, such as tile roofs already common in the Hellenistic, Roman and Byzantine periods, or, even worse, today's corrugated sheet metal roofs common in many developing countries, have an extremely negative effect on indoor temperatures, both in summer and in winter. Shading of flat roofs may have been a common practice in the past and can still be seen in Middle Eastern and Mediterranean villages, where vines provide summer shading, or where temporary shading is provided by 'transient layers' such as tobacco, peppers and other agricultural produce dried on the roofs. Fabrics may also have been used in the past, as indicated by details identified on the parapets of roofs.

These results showing poor indoor conditions explain the phenomenon of 'intra-mural migration'; i.e. the use of different parts of traditional housing

The future of the vernacular

prototypes for different parts of the year or the day, and especially the habit of sleeping on rooftops, balconies or in patios, where the summer night conditions may be significantly better than the indoor ones (Rapoport 1969). The educational aspects of this project, as it developed, were very important, too. Many of the students participating in different modules had been aware of the advantages of thermal mass in desert climates. Their intuitive reaction to the poor indoor conditions indicated by simulation was the addition of more thermal mass, which proved at best to have no effect. As a result, students participating in the project developed a much more realistic and practical attitude toward traditional building technology and details, and appropriate methods for the improvement of indoor climate and energy conservation. In most cases it was realized quite early on that part of the excessive thermal mass could be effectively replaced by insulation in

Table 12.2 An example of the simulation of heating and cooling loads of construction types, building parameters and modifications, in a highland desert climate – in this case, a mud building ('Size' indicates the estimated peak load demand for heating and cooling equipment; 'Units used' indicates the annual energy consumption)

Amman mud		Summer			% of people comfortable at max temp.	Winter			Cool plant Size KW	Cool plant Units used KWh	Heat plant Size KW	Heat plant Units used KWh
		max	min	ave		max	min	ave				
AMud1	Base	23.4	22.5	23	100	4.9	4.7	4.8	0.1	33.5	1.7	2807.2
AMud2	Thicker roof	23.4	22.5	22.9	100	4.9	4.7	4.8	0.1	31.6	1.8	2851.4
AMud3	Thicker walls	23.4	22.5	22.9	100	4.9	4.6	4.8	0	7	1.4	2325.4
AMud4	Thinner roof	23.4	22.5	22.9	100	4.9	4.6	4.8	0.1	38.2	1.7	2796.8
AMud5	Thinner walls	23.9	23.1	23.5	100	5.2	4.9	5.1	0.2	75.2	2.5	4015.3
AMud6	Glass north window	23.4	22.5	23	100	5	4.7	4.8	0.1	36.7	1.8	2865.8
AMud7	Window in south	24.1	23.1	23.1	100	5	4.7	4.8	0.2	67.4	1.8	2870.9
AMud8	Internal insulation	24.2	22.1	23.2	100	5.2	4.5	4.8	0.2	52.8	1.4	2203.9
AMud9	Intermediate insulation	23.4	22.5	22.9	100	4.9	4.6	4.8	0	5.6	1.8	2711.4
AMud10	External insulation	23.4	2.5	23	100	4.9	4.6	4.8	0	9.8	1.2	1834.2
AMud11	Vent N window, E door	23.7	23.3	23.5	100	4.9	4.7	4.8	0.1	34.9	1.7	2740.1
AMud12	Vent W window, E door	23.5	17.1	20.1	100	4.9	4.7	4.8	1.7	129.3	1.7	2738.1

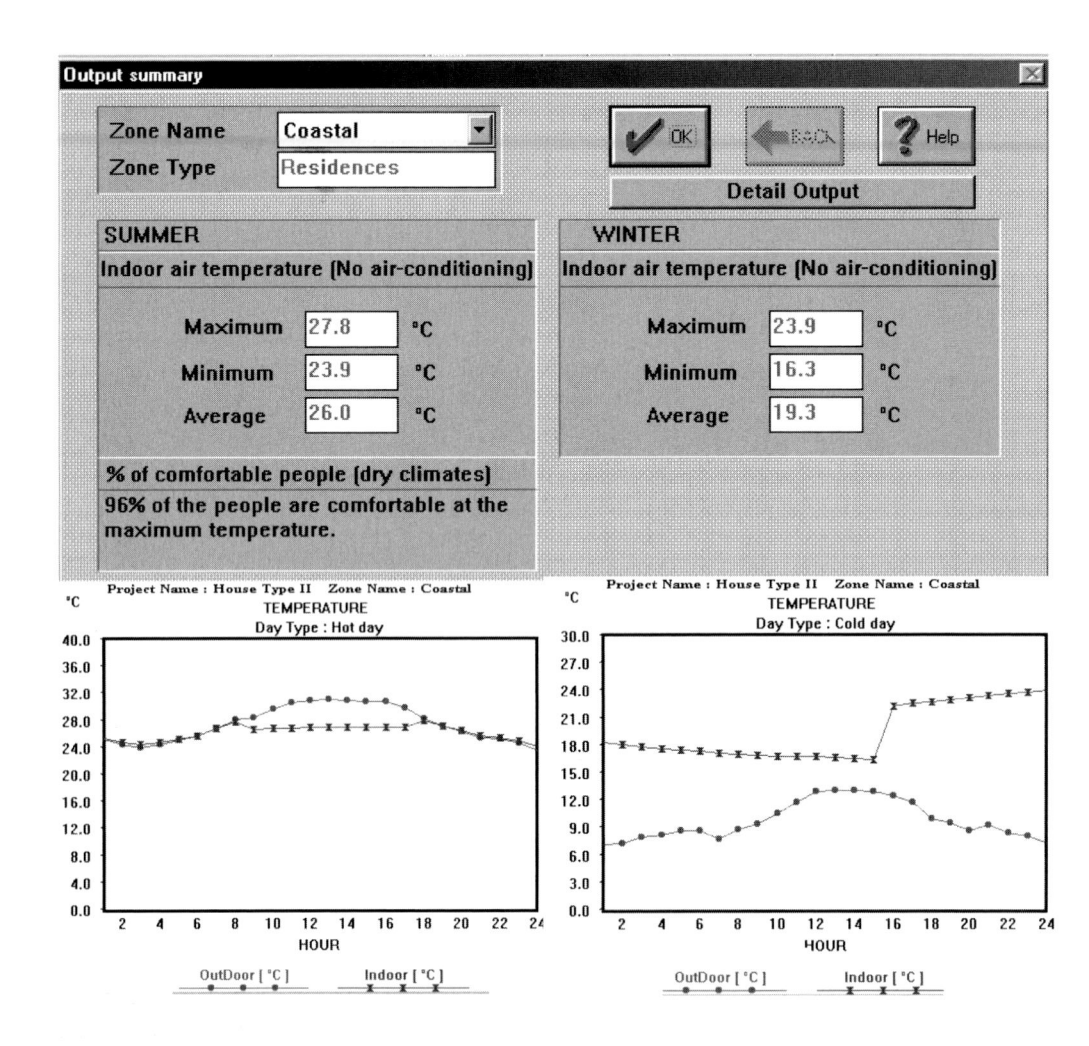

the centre or outside of compound wall constructions, with significant improvements to wall and building performance as a result (Meir, Mackenzie Bennett and Roaf 2001). The study also demonstrated that although summer cooling needs may be an important issue for certain arid regions (especially the hot continental valleys and the humid coastal plains), winter heating is vital for most arid regions, extremely so for the highlands and mountains.

Conclusions and implications

This ongoing study continues to highlight the need for the systematic research of historical, traditional and vernacular building types and technologies if we are to truly understand the living conditions of the past and present in such buildings. Such an understanding is vital if we are to improve living conditions for millions of people within a sustainable development framework (Roaf, Horsley and Gupta 2004b). This, in turn, should take into account commonly available materials, construction methods and know-how, the opportunity for simple improvements and

12.6
Building thermal performance improvements achieved through the introduction of compound wall constructions in simple houses in the Middle East, from an MSc study using the Building Toolbox model, by Rajat Gupta at Oxford Brookes University.

alterations at realistic costs, and particularly the possibility of enhancing comfortable indoor conditions with minimum auxiliary energy input.

Traditional and vernacular architecture is based on locally available materials and cultural dictates, which have given birth to interesting building forms and types, systems and details. Those should be studied and understood, yet in a critical way which will allow the true assessment of their interaction with climate vis-à-vis a rising demand for better, more comfortable and healthier indoor environments. The continuation of current, poor, practices within the vernacular building stock is worrying both from an environmental and a health point of view. Issues of poor indoor air quality and uncomfortable conditions are being further exacerbated by the growing numbers of people living in inadequate housing, and by the changing climate. Current research has established to a great degree of certainty the connection between urbanization, fossil fuel use, deforestation and land degradation, and unsustainable production processes on the one hand, and climate changes and unpredicted climatic extremes on the other hand. Such extreme climatic events, which are becoming more common in recent years, are leading to more uncomfortable indoor conditions and the accelerating processes of desertification, witnessed in the last few decades, in turn driving ever more people into ever poorer dwellings.

One of the important outcomes of this study, so far, stems from the counter-intuitive results of the monitoring and simulation studies. It is such 'intuition' stemming from 'common knowledge' and theoretically 'thoroughly established' historical paradigms that cause misconceptions and assorted problems, not least among NGOs and development organizations operating in developing countries, many of which are defined as deserts. Such misconceptions have given birth to housing units with massive walls and lightweight sheet metal roofing, as bad a solution as one could possibly conceive.

The majority of people in the world live in 'vernacular' buildings, evolved from antecedents in the distant past, and continually evolving into those buildings that will house future generations. Changing circumstances drive the evolution in buildings, just as the apogee of the opium trade in Hong Kong resulted in the great windcatchers of Yazd. The changes we face today are those of increasing populations, depletion of resources, pollution, rising prices of finite fossil fuels and climate change (Roaf, Crichton and Nicol 2004a), each of which individually would be a reason for investing in improving the vernacular paradigms, but put together add up to a compelling imperative for change. Applied building performance studies such as those outlined above, can generate the understanding not only of how such buildings work today, but also of how they can be modified to optimize their performance in the future. This can be done with minimal economic and environmental costs, as some recent research is aiming at demonstrating, either by using local traditional solutions (Esteves *et al.* 2003) or by using innovative processes based upon traditional building methods, such as those using recycled or re-used waste materials (Garcia Chavez 2004).

We believe that, using these and other methods, every student in every school of architecture around the world should be taught to understand how local traditional buildings work in relation to the local cultures, climates, environments and economies within which they provide shelter. In so doing students will be able to acquire a deeper understanding of how buildings really perform, and can be improved. Such knowledge will provide an essential foundation from which to build the more resilient, and regionally appropriate, buildings that are essential if we are to survive the exigencies of the twenty-first century.

Note

1 Numerous individuals, groups and institutions have contributed so far to this ongoing research. Parts of the parametric studies were undertaken by graduate students at the Department of Architecture of Oxford Brookes University, UK (2000–1), and the Albert Katz International School for Desert Studies, at the Blaustein Institute for Desert Research, Israel (2001–4). Infrared thermography processing was done in co-operation with Wolfgang Motzafi-Haller. Partial results were summed in a paper co-authored by Meir *et al.* (2003). Access to and work in archaeological sites in Israel was facilitated by the Israel Nature and National Parks Protection Authority. The help of these and many others is kindly acknowledged.

Chapter 13

Architectural education and vernacular building

Howard Davis

Introduction
This chapter is about the idea that architecture schools might play a formative role in the education of people who will help guide the production of the vernacular architecture of the future. The academic study of vernacular architecture is important to this, but needs to contribute to a new kind of professional education, that sees professionals in quite a different way than contemporary architectural education does. In this new way, professionals are humbler, draw their knowledge from the situations in which they find themselves, and have deep respect for other professionals, builders, clients and members of communities. This requires an altogether new attitude towards professional education.

There is an obvious contradiction that needs to be resolved. On the one hand, healthy vernacular architecture, an architecture that does not inhibit the honest expression of people's lives and cultures, has traditionally developed through common, culturally embedded knowledge of building, along with the political and economic ability to put that knowledge to use. On the other hand, formal education has traditionally promoted an attitude toward professional expertise that seems opposed to the idea of a shared, embedded knowledge. Is it possible to imagine formal education being helpful to the production of a healthy vernacular architecture? I will argue that the answer to this question is 'yes'. A new vernacular architecture for the twenty-first century must include the worldwide production of some 40 million buildings a year. This need should be met partly through the education of people who can help guide the complexities of building production in ways that lead to buildings that allow the life of people and their communities to flourish, and in ways that minimize negative environmental impacts.

These challenges differ greatly from place to place. In the 'developed' world, for example, both expertise and control of the building process have been taken away from ordinary citizens. Most people have been reduced to consumers, buying or renting products over which they have little control. In these places, developers for whom the bottom line is profit, and planners, who work within systems of bureaucracy, control the shape of cities and houses, and prevent the 'from the ground up' kinds of processes that might lead to more human results (Davis 1999).

It is however in the rapidly growing cities of the 'developing' world, ranging from Mumbai to Lagos to Lima, that the issue is perhaps most potent. In these cities, informal settlements do not lack people with expertise. What these communities lack is power, which resides in outside institutions like governments, NGOs and banks. These institutions often get in the way of the communities' ability to use the expertise that already exists within them. Within this view, the people who need appropriate education are not in the settlements, but in these currently unhelpful institutions.

But additionally, even though communities need to control what happens inside them, people in communities can benefit from more and different kinds of expertise than they already have. This is a difficult argument to make, because many communities and organizations maintain the view that professional expertise of any kind is unhelpful to them. They see professionalism as akin to colonialism, and can often point to experiences where professionals attempted to exert their own will on situations in which such an attitude was inappropriate. This argument has been most cogently put forward by John Turner, and I have talked to community leaders in India, South Africa, Egypt and Nepal who could cite specific projects where professionals did not understand the realities of the communities and made serious mistakes in the projects (Turner 1976). But this chapter challenges the view that professionals cannot work alongside and within communities in helpful ways. Communities and educational institutions may come to see the advantages of both inside and outside knowledge and experience. Indeed, vibrant cultures have often maintained their life by assimilating knowledge from outside themselves. The critical thing is that they not be overwhelmed in the process.

In both the 'developed' and the 'developing' world, there is a need for cities and communities to use holistic thinking while at the same time allowing the vernacular to develop from the grassroots. The issue is largely one of *process* rather than of *form* or *style* (Alexander 2003). With such a view, the quality of buildings, which represents an intimate reflection of the lives and dreams of communities, will only emerge through the cultivation of processes that allow communities themselves to take responsibility for their own buildings.

Let us assume that architecture schools can do more in educating people to help take care of the built environment as a whole, including the 98 per cent of buildings in the world that are vernacular buildings, rather than only educating people with a primary orientation toward the two per cent of buildings that are not. This assumption, however, is only helpful if it does not include the idea that architecture schools should be primarily responsible for such education. Indeed, one of the principal fallacies that lies behind formal, university education in architecture since its inception in the nineteenth century is that architects should dominate the building culture. The disparity between that view and the reality of the way buildings get built has meant not only that architects have not had nearly as much control as they want, but more significantly that the influence that might rightfully be theirs has been seriously diminished (see Gutman 1988 and Larson 1993).

So, although this chapter is primarily about the education that happens within architecture schools, architecture schools should not be seen as the dominant players in the maintenance and generation of knowledge about the built world. A healthy building culture will include a variety of institutions that play complementary roles and the members of which will operate with a healthy respect for each other. In addition to architecture schools, these institutions include the regular apprenticeship that happens in building practice, craft schools, schools of planning and urban infrastructure, the training and raising of political awareness that happens in community based organizations and NGOs, government sponsored research institutes, the research and development arms of building product manufacturers, and probably others. This diversity of institutions will never (and probably should not) operate in a way that is completely without conflict, as a certain amount of conflict will help to sharpen the position of each institution. But a higher level of co-ordination is clearly needed.

It is only within such a scenario that architecture schools can be justified at all in having a strong role to play in the maintenance and regeneration of the vast bulk of buildings that make up our everyday world. Within such a scenario, which is based on a shared set of intentions, architecture schools will have the responsibility for training professionals in a way that pays respect to their counterparts in other professions, trades and institutions.

Vernacular architecture as part of a healthy building culture, and the role of professionals

What is the character of a healthy vernacular architecture for the new century? What kind of processes are required to produce it? And what kind of institutions are required to support these processes? It is only after these questions are answered that we can deal with the issue of the kind of education that might be helpful.

As vernacular architecture always has been, healthy traditions of vernacular architecture for the twenty-first century will be different in different places, and will emerge out of the lives of people and groups they belong to. These will be new traditions that serve people's lives well, and that elevate their spirit. But they may not look very much like the vernacular architecture that to large extent preceded industrialization and that co-existed and declined during the centuries of industrial culture. This older vernacular architecture consisted of the hundreds of traditions that make up the bulk of the entries in Paul Oliver's *Encyclopedia of Vernacular Architecture of the World* (Oliver 1997a). They are largely rural traditions, based on pre-industrial techniques and materials.

The situation now is different, not least because most people are now living in cities rather than in rural settings. Although in cities there may be a persistence of long-standing cultural patterns as expressed in room arrangements or settlement patterns (such persistence is often seen both in informal settlements and in the culturally based use of space in modern buildings), the most ubiquitous modern vernacular looks much the same in different parts of the world. Many new areas of cities, particularly in the 'developing' world, consist of buildings of some kind of mixed concrete and unit masonry construction, with pre-manufactured windows and doors, covered with plaster, stucco or thin tiles. Other places, particularly in the west, are characterized by suburbs based on an American model, with repetitive houses built by developers. Each of these contemporary manifestations has a reality which, quite realistically, is not about to disappear.

The problem that communities are faced with, therefore, is not the restoration of historic vernacular patterns, but instead the transformation and extension of the existing built world, in ways that support the lives of those communities and the lives of people in them. The forces of modernism, industrialization and globalization have irrevocably altered the built landscape, and those same forces cannot be wished away in favor of some idealized pre-industrial vernacular. Instead, it is the context of the present built landscape and the institutions that produce it that will be transformed into a new vernacular and new processes that make it.

Nevertheless, this new vernacular architecture, if it is to serve people's lives well, needs to have a series of characteristics that have deep similarities to some of the vernacular architecture of the past. It is the shepherding of this transformation in appropriate ways that represents the biggest challenge to communities and professionals all over the world, and to the institutions that support them and that help train people to be a part of them. These characteristics are as follows:

With respect to buildings:
- Building types are expressive of, and supportive of, local place and local culture. They serve people well, and they persist in their meaning to people. They have the capability to touch people deeply. At the same time, they are appropriate to people's contemporary lives, and evolve as they need to. They are not necessarily like the vernacular types of the past.
- Buildings and their arrangements with respect to each other are supportive of community, understanding that communities may not be as homogeneous as they were in the past.

- Buildings and communities are culturally and environmentally sustainable.
- Buildings have individual uniqueness. Buildings of the same type take on differences from each other that come from differences in site, inhabitants and builder.

With respect to the processes that make these buildings:
- Buildings are largely a product of local culture. This is not to say that external institutions, processes, or materials are not involved. What it does mean is that the motivation to build, and decisions about what and how to build (and how best to utilize external institutions) is taken by people who are most affected and involved. Decisions are taken at as local a level as possible.
- Within the local culture, knowledge about building is as widely shared as possible. Even if they are not themselves builders or architects, many people understand buildings and can deal with them articulately. This shared knowledge includes rules, techniques and procedures that make sense.
- The process of building, and the ongoing creation/re-creation of a city or town, is primarily a process of growth rather than one of design/creation. This is not to say that design is not important in particular instances. But it needs to be seen as part of a larger, natural process that allows the actions of thousands of individual players to be co-ordinated. The result of such a process may be predictable as far as overall type is concerned, but unpredictable in its actual, detailed form.
- The process of building is primarily one of craftsmanship rather than manufacture, where craftsmanship is understood as an activity which allows the decisions about the making of a building to be vested in the person who actually makes it.

And with respect to the institutions of the building culture (banks, builders, organized client groups, materials suppliers, architects, regulatory agencies, etc.) that can support these processes:
- Institutions are locally based and accessible. The people who lead them have a stake, beyond a purely financial one, in the quality of the built outcome.
- Economically, these institutions do not take money out of the community but recycle it within the community in ways that regenerate the community's economy and body of knowledge about building.
- Institutions operate with shared overall goals about the future of the community.

Building cultures that operate in the ways described above, and that produce buildings in quantity that have deep human meaning, do not exist in many places today (Davis 1999). There may be a role for professionals in the creation and maintenance of such building cultures, but these need to be different kinds of professionals than most architects are today. These are professionals who understand that their roles are not ones of total control; who are willing to be guided by situations rather than force their will onto situations; who are genuinely respectful of people who are playing other roles in the building culture; who see buildings as needing to take their place in a larger whole rather than as individual expressions of ego; who see their roles as helping to heal and improve the built world as it is experienced by many people and who know the importance and delicacy of community processes.

Such professionals, working towards a healthy building culture that can take care of the ninety-eight per cent of building that today's architecture students do not expect to deal with, will not be educated through a system that is very much like today's. Formal education needs to be fundamentally different. And to the extent that formal education is useful at all, it cannot be expected to provide all the expertise needed, and it cannot set up an ideological opposition with traditional forms of knowledge that are embedded in the lives of communities. Instead, it will allow all forms of expertise to co-exist, draw on and ultimately enrich each other.

Architectural education today

Before the advent of the formal architectural profession, people who designed buildings had a much closer relationship to the activity of building than they do today. Building types were more commonly understood, so even the word 'design' had a different meaning. The design of buildings was a necessary adjunct to construction, and some building craftsmen, who learned their craft through apprenticeship, also learned how to draw plans as a part of their job. Eventually, some people, such as building craftsmen, surveyors and 'gentlemen architects', began to do design exclusively, and an ultimately unbridgeable split emerged between architecture and building. But even then, architecture was learned through apprenticeship.

The system of contemporary architectural education in universities is a relatively recent invention. It developed first during the nineteenth century, partly as a way to control entry to the emerging formal profession (Abbott 1988; Woods 1999). Architecture schools today have their origins in a series of institutions which still persist in the schools in the form of procedures and attitudes towards building and the building professions.

Architectural apprenticeship in the nineteenth century, through which pupils entered architectural offices and learned directly from experienced architects, has come down to us in the system of direct, one to one instruction in studios that still exists in most schools of architecture (Crinson and Lubbock 1994). Apprenticeship first co-existed with and then was followed by the French *Beaux Arts* system, which set the content for many architecture schools in the world for several decades beginning at the end of the nineteenth century. Competitions and pressure of time were central to this system, and still are critical in many architecture schools (Carlhian 1979). The *Beaux Arts* system had little to do with local concerns, and design problems were the same at many different schools all over the world. The *Beaux Arts* system was followed by that of the *Bauhaus*, in which innovation and creativity rather than adherence to established norms were valued. And finally the universities themselves provided the context for academic pedagogical models in the form of lectures and non-studio classes. They are now exerting pressure on architecture departments to carry out research according to standard academic models.

The present system is an amalgam of these origins, not necessarily optimized for learning architecture in today's world, and certainly not optimized to educate professionals to be helpful in the emergence of a healthy vernacular architecture (Boyer and Mitgang 1996). Although the present system of architectural education includes some positive features (which will be mentioned later in the chapter) its pedagogy is based on at least five unhelpful assumptions:

1. There is the assumption that professional expertise is better than grassroots knowledge, and that the architect must, almost at all costs, maintain dominance over the builder as well as the client and the client's community. This assumption, which is taught more implicitly than explicitly, helps educate the architect as someone who is fundamentally opposed to his or her culture, and leads to future professional relationships that cannot foster the kind of sharing of cultural rules that characterize traditional vernacular architecture.

2. There is the assumption that the 'star designer' is more valuable than other members of the profession, much less contractors and builders who are not even a part of it. This marginalizes students who want to be involved with architecture but not as star designers, and pushes pedagogy and curricula in ways that marginalize most other roles.

3. Connected to the idea of the 'star designer', there is the assumption that the 'object building' is more worthy of study than the ordinary house. Many design projects in architecture school deal with monumental or highly symbolic public buildings (museums, libraries, cultural centres of one kind or another) or unusual private buildings (galleries, artists' studios) rather than housing, shops and workplaces for ordinary people.

4. There is the assumption that the future has little to learn from the past. Certainly, architectural history is taught in architecture schools. But another basic premise of design teaching is that every problem is new and needs to be solved from first principles, and that the direct use of precedents is to be discouraged. This is quite different from processes of traditional architecture, in which the repetitive use of types helped to give coherence to complex, collective configurations like cities and settlements.

5. There is the assumption that learning can happen by drawings, away from buildings, construction and communities. Learning is expected to happen in the studio in front of the drawing board and now the

computer, and in classrooms. Theory predominates over reality. There is the occasional field trip to view a construction site, or study trip to Italy or Finland. But by and large, the idea that present day reality itself might be the best teacher is not present.

These assumptions do not support the production of a healthy vernacular architecture, and contemporary architectural education is therefore not the answer.

Characteristics of a new system of architectural education

A system of architectural education that will be a positive force in the creation of the built world as a whole, will first of all recognize a variety of different roles, ranging from policy-makers to local community leaders to architects, planners, developers, builders and engineers. It will seek to help these various players speak a common language and to understand each other's concerns. It will recognize the importance of both common goals but also the inevitability of conflict and contradiction.

It will include a variety of content. Academic content will include the architectural and social nature of vernacular building, the character of processes of growth of settlements and cities, and the political and economic frameworks of governance and communities. Practical content will include building construction, the needs of communities and training in techniques necessary to work on the ground, with buildings and people in communities. This content will include, as much as possible, work with communities themselves, mixing people of different backgrounds and professional goals. It will incorporate modes of learning, akin to those of medical training, in which students are learning from direct experience.

And it will be forward looking and optimistic, not afraid of new technologies or emerging frameworks of thought, within the context of helpful intentions. Specifically, it will differ in two critical ways from the present system: in its attitude toward professionalism and its attitude towards technology. It will also respect certain features of the present system: the idea of integration of subject matter, and learning through problem-solving.

Attitudes toward professionalism

A new system of architectural education will differ in professional attitudes and the nature of architectural knowledge. The importance of reality (real places, actual communities, the complex processes of society, 'bricks and mortar') must be central to the new architectural education. This is fundamentally different from the present system, in which the most basic attitude toward teaching design is indicative of a way of thought that itself controls how professionals act in the world.

In today's schools, one of the most basic things taught in design is the importance of inventing and imposing a *parti*, or simple conceptual diagram of a building, very early, and 'sticking with it' during the development of the project

(Herdeg 1983). The importance of the *parti* is an indication of the importance that architecture schools give to a certain kind of abstract thought. By its very nature, school is removed from communities and from the real human processes that take place in them. Architecture school deals with abstraction and with idealizations of reality. In architecture school qualities such as 'clarity', 'economy of means' and 'simplicity' are valued, along with buildings that at first glance appear to have them. And indeed, twentieth-century architects saw such qualities in vernacular architecture, and used such interpretations as justifications for their own work. Le Corbusier for example, even as he clearly learned from vernacular architecture, also idealized it, extracting what he saw as formal qualities of Platonic purity (Le Corbusier 1987).

But the order that exists in communities and in the processes of building that take place in them is different from the simple order that sometimes appears on the surface, and that architects idealize. Communities are vibrant, multilayered, political and complex. Because formal education deals so much with abstractions, the reaction in architecture school to a messy reality is often that communities need to be 'straightened out', that they need order where little appears to exist, that their roughness and irregularity is a problem to be solved rather than a reality to be respected on its own terms.

This kind of value judgment made about communities comes from a view of looking in from the outside, from a place where simple and clear conceptual models are valued. The simple clarity that is valued as an academic construct is transformed first into a particular kind of interpretation of an actual situation, and then into a paradigm for action. So developing an attitude of respect for the complexity of living and changing communities, which may include looking at communities from the inside, requires a conceptual shift in architectural thought, and is not simply a matter of exposure or awareness.

And from a practical point of view, it results in new ways of teaching and learning. It means that ambiguity and even contradiction need to be accepted and respected. It means that learning experiences in communities, with members of communities, may be as important as learning experiences in the classroom or studio. It means that awareness needs to be developed of the full range of economic, social and institutional forces that shape communities, including architectural ideas but not restricted to them. And it means that the classroom or studio itself may be seen as a community, and not only as a locus for individual learning.

The result of such educational experiences will be more thoughtful practitioners, for whom there are fewer easy answers, and who will have more willingness to delve into the depths of issues. These professionals will also see the value of other players in the building culture and be more willing to engage them as equals.

Attitudes toward technology

A system of architectural education that is useful to the production of a healthy vernacular architecture will not shun advanced technology. Instead, technology will be regarded as a way of enabling both the widespread dissemination of knowledge and the production of uniqueness and variety among buildings.

In architecture schools today, vernacular architecture is seen as being archaic or even 'primitive', made crudely with hand tools. Of course, this is a superficial view, since one of the things that characterizes all living cultures is the adoption of available technology as it becomes available and useful. It follows that in today's world, the use of advanced technology can support the production of successful vernacular buildings and settlements.

Indeed, in contemporary cultures where 'traditional' vernacular architecture is still prevalent (largely in various marginalized cultures in the world), the use of advanced technology is as welcomed as it is in cultures that are more economically powerful. There are numerous examples outside the world of building that start to define a new context that embraces technology. One is the use of the mobile phone as the kernel of an economic development strategy among women's groups in Bangladeshi villages, and the Grameen Bank lending money for such initiatives. Another is the use of the internet among farmers in Africa to provide the most up to date information about weather conditions that affect crops. And yet another is the development of web pages among Rajasthani village craft co-operatives to market their products. There is a more widespread use of the web for numerous local initiatives ranging from health care to education.

It may be argued that such initiatives are in fact destroying the cultures they are purporting to support. But the use of advanced technology is simply part of the natural evolution of cultures, and it is therefore critical to strike the most effective balance between cultural identity and use of global technologies. We are now witnessing the re-assertion of regional identities in the face of globalization, along with the weakening of national boundaries. It seems more reasonable to imagine that advanced technology will allow cultures and their emerging vernacular traditions to survive and transform, than to imagine that by ignoring technology it might be possible to freeze cultures in time. The examples cited above are indicative of a new context of advanced technology that does not respect cultural or national boundaries. This will spread and affect building activity itself in a helpful way.

Consider, for example, that in South Africa and other places, there are initiatives to use the satellite technology of Geographic Positioning Systems (GPS) in concert with computer based Geographic Information Systems (GIS) to identify and record irregular property boundaries in informal settlements that have evolved organically (Barry *et al.* 2002). This is often prohibitively expensive to do with conventional surveying techniques. But when done, it legitimizes lot definition, thereby facilitating both legal title and the use of the lot as collateral for loans. It allows traditional cultures to continue to use their own forms of land division, in a way that does not prevent their participation in external governmental and financial institutions. Although there is still a great 'technology gap' between the 'developed' and the 'developing' worlds (Glaser 2004), these initiatives are seeds that will grow as that gap gets smaller.

Or consider the use of computer programs which allow for a direct connection between the three dimensional visualization of an object or building, and its construction (McCullough 1996). Such programs, which already exist, allow lay people to participate directly in the design of their own houses, in a way that does not require the interpretation of a professional. There are also modern manufacturing processes in which the goal is products that are uniquely tailored to individuals but at costs that are no greater than mass produced, repetitive things. Other existing computer programs help to do materials take-offs, cost estimating, and management of labour and materials. Such programs could allow for the dynamic process that makes vernacular buildings to be entered into with as high a level of confidence as the process in which costs are controlled through drawings, specifications, and bids.

Many of these processes are still rough in their specifics. What they all do, however, is use technology in ways that allow for local control in situations that had developed in ways in which long-distance, global control was typical. And they all have obvious connections to education, where technology may be put in service of a field of study that had heretofore been seen as 'pre-technological'.

This represents a real difference from the way technology is now seen in architecture schools. Right now, the study of digital technology has two purposes. The first is a purely vocational one, to prepare students for work in a profession in which computer screens have almost completely replaced drawing boards. The second is a creative one, to generate innovative forms that are intended mainly as provocations or critical reactions to what is seen as the status quo. In a sense, the first and second purposes are diametrically opposed to each other. And up until now, neither of them seems to have had very much to contribute to the design of buildings that are carefully shaped to their clients or to their places in the world. That role has been assumed by architects whose process of design has not yet been really enhanced by the use of digital media.

But as that begins to change, architectural education will change as well. The use of digital technology in ways that help to solve real problems of building design, production, and settlement layout will have a number of effects, not the least of which is that it will make the study of vernacular architecture the investigation of innovation rather than the study of something archaic.

Attitudes toward 'traditional' architectural education

At the same time we recognize how current assumptions in architectural education need to be challenged, it is important not to 'throw out the baby with the bathwater'. For all of its problems, the pedagogy of architectural education has aspects that remain important: the idea of learning through problem-solving; the potential it has for demonstrating the importance of the integration of subject matter; the human energy that may be present in the design studio. These may all be valuable models for learning how to operate in professional or community situations that are enormously complex.

Furthermore, the emergence of new and useful typological models, which can be replicated again and again in the vernacular, will depend to some extent on the inventiveness of architects. In most traditional cultures, vernacular types emerged slowly, as conditions changed slowly. In contemporary society, the emergence of coherence in the built environment will not be the automatic result of present, largely market-driven processes of typological dissemination and transformation. Architects can help with this. This will happen not through the architect's normal role of control over processes of design and construction: in some form, local, community-based processes must take care of this. But architects have an essential role in the design of prototypical buildings that may be transformed and replicated through community processes, and in the design of buildings and building types that are compelling and beautiful enough to enter into the vernacular. Within a building culture in which innovation may happen through changes that come about by builders, or communities, as a result of real need, the architect's role is as an innovator whose work will become useful not because he or she designs many buildings, but through the gradual acceptance and dissemination of his or her ideas. And the studio culture which exists in most architecture schools can continue to foster the inventive spirit that may be needed for such innovative work.

Consider for example the role of architects in the design of housing in the twentieth century. Most of the innovative designs of architects were done within paradigms of mass production, removed from communities and people's desires. The transformation by residents of Le Corbusier's project at Pessac is perhaps the classic example (Boudon 1972). But most housing in the 'developed' world is built by developers, without much influence of architects, using typological models that are based on the financial bottom line. But suppose architects were to insert themselves a little more squarely into existing production processes, and at the same time develop models that are closer to people's needs and the health of communities. Then, their work could have a much more widespread effect: necessarily transformed over time, but acting as important seeds for processes that will ultimately be community-based ones.

Good 'traditional' architectural education may have an important role in the emergence of a healthy, new vernacular. This can only come about, however, if the useful processes of design innovation can be separated from the unhelpful idea that architects must exert ultimate and complete control over the emergence of the built world. Then, the innovation that architects are so good at will take its rightful place within the natural vernacular processes of the evolution of types and techniques.

Conclusion: the separation of expertise from power

In the same way we need to think of buildings as the product of a building culture that has many players, we need to think about education as a system that has many different modes of learning and training, all of them necessary within a common intention of improvement. This system will include schools within which theoretical advances are being made, and it will include community based training programs in which skills are being transmitted. It will include the education of architects who will be exemplary designers of new housing, and the education of different architects who do not have a strong role in the design of buildings at all. Such a system is not based in a single institution, but it is a co-ordinated network, with at least some shared intentions and the recognition of the varied strengths of the different institutions inside it. Architecture schools, which in the nineteenth century began to assume the attitude that the professionals trained in them needed to take complete responsibility for decisions about buildings, need to give up that attitude.

There are, in the world today, schools, curricula and individuals who are working in ways that are helpful. Some of this is happening inside formal architectural education, and some of it outside. Some schools bring together traditional disciplines of architecture, planning and landscape architecture, recognizing the integrated nature of the built environment. There are new programs in urban infrastructure and the urban built environment, recognizing the need to address the forces that affect the context within which the vernacular gets built. There are schools of traditional craft and construction, outside of universities, recognizing the limitations of architecture schools in dealing with the realities of building. Studios and courses within architecture schools place students in communities, and have them work directly on real problems, with people who can speak to them first hand. Public programs to educate citizens about architecture and their architectural heritage exist in cities in various places in the world.

Although these initiatives are the seeds of positive change, they have not yet changed the overall character or perceived relevance of architectural education. Such a radical change represents a change of fundamental attitude, along with a recognition that formal architectural education cannot do it all. In the end, we are talking about educational systems that are interwoven with large and diverse building cultures, that themselves contain the means of training and the transmission of knowledge, just as traditional cultures did. Such educational systems do not stand apart from their building cultures. They recognize the diversity of roles within the building culture and the varied means of education and training that are necessary to support those roles, and they insist on respect and empathy among the different players.

I teach a regular course in vernacular architecture to architecture students. Some of these students are working toward 'alternative' careers in low-cost housing or community development, but most of them will enter the mainstream architectural profession. It may be argued that for these students, learning about vernacular architecture is an unnecessary 'extra' in a curriculum that is based largely around questions of contemporary practice. However, the content of the course is not as important as the attitude that is necessary to understand the built world well. That is a fundamentally professional attitude, in which students are asked to step into the shoes of people and groups very different from themselves, to see other worlds as much as possible from the inside. This is not only the stance of an anthropologist; it also needs to be the stance of any professional who seeks to work with any client in a way that is empathetic and helpful. Such a stance is one which respects people's realities and which does not seek to impose, through a position of power, a different reality. Such a stance recognizes that power needs to reside with the client, and that professional expertise must be carefully separated from professional dominance.

A new educational system that is dealing more generally with the production of the vernacular environment, and that can be helpful to the success of that environment, needs to do exactly the same thing. Educational institutions like architecture schools are faced with the choice of becoming relevant to the large majority of people and cultures or remaining within the elitist margins of the changing societies of billions of people. Their relevance, like that of the institutions they serve, depends on the extent to which they can separate expertise from power.

Chapter 14

Educating architects to become culturally aware

Rosemary Latter

Introduction
In the UK and throughout the world the role of the architect is affected by social, cultural and technological change and global economic imperatives. In this critical period for architectural education, both students and the profession must address some of these changes, for the sake of the future sustainability of the planet. It is often said that half of the world's population lives in rapidly expanding urban areas, and that this is where the principal efforts to mitigate poverty should be made. But the 'pull' to the cities is linked to the 'push' away from the villages due to lack of support for regional rural economies, so contributing to the problems implicit in rapid urbanization (Oliver 2003). Most international aid effort is focused on the towns and cities, sometimes leaving the other rural half of the population unsupported. However, if and when the needs of traditional, rural societies are addressed, sensitive interaction by well informed agencies is crucial if the villages are to prosper, retain their communal identity, and increase the likelihood that they will be sustainable. Unless supported in a culturally appropriate manner by policy-makers, how will the increasing billions of people who inhabit these areas of the planet be able to use their own considerable resources? As the global population grows, how will those who may be involved in managing the built environment acquire the knowledge and skills needed for a culturally sensitive approach to development? The specific needs of individual social, cultural and religious contexts will be disregarded if standard solutions are universally applied

by global financial institutions, and the architects employed by them. Many emerging economies are looking towards western practices for models of large-scale planning and high-rise iconic buildings. The global industries of banking or tourism may require such skills, but the application of this approach to housing and community building projects will almost certainly not address the more specific local needs. Most of the world's rural housing is built by owner-occupiers, and is likely to continue to be so in the foreseeable future. However, as pressure on settlements increases they must also be appropriately planned and serviced by supportive agencies, aware of regional traditions, rather than be subject to the imposition of standardized solutions that may meet physical needs, but which disregard cultural identity.

Unless they limit themselves to designing iconic buildings with solely engineering and economic criteria, architects must respond not only to physical and environmental determinants, but also to cultural contexts. Many examples exist of development interventions that have been made in the recent past without knowledge or consideration of the culture in which they are set, particularly in the field of post-disaster housing. Large funding institutions such as the Red Cross and UN agencies now have a better understanding of the negative, wasteful consequences of a culturally ignorant approach to such projects, but problems still remain. Architects with awareness of varying cultural contexts will be in a powerful position to influence and advocate sensitively on behalf of communities coping with such interventions. A knowledge and understanding of the vernacular architecture of the world can only be helpful in achieving the aim of producing more sensitive practitioners.

Among professionals of the built environment, architects often have characteristics which make them adept at confronting such situations. They have a wide education in the humanitarian aspects of architecture. They have practical experience of problem solving and they are often extremely able lateral thinkers and creative spirits with a genuine social conscience. They are a potentially invaluable resource for confronting development problems. Unfortunately, the profession has become less inclined to pursue culturally sensitive practice and, recently, educational curricula in the UK have become increasingly narrow in focus. Students often come to university without any clear concept of their own culture in any deep sense, let alone with any insight into other world views. Without discussion of the anthropological aspects of their work they are likely to continue their education into a professional elite with more or less uniform values. But the younger generation of architects should be encouraged to confront the challenges and implications of globalization on traditional societies and their buildings, and a few are seeking to widen their prospective architectural remit.

The significance of vernacular architecture in education

Many people ask for a definition of vernacular architecture, discussed in other texts (e.g. Bourdier and AlSayyad, 1989; Oliver 1997b), and are unsure about the significance of studying it as part of an architectural training. This chapter describes some of the issues covered in a Masters course in 'International

Studies in Vernacular Architecture' (ISVA).[1] These range from policies of conservation to disaster resistant measures in housing, from technology transfer to the ethical problems of intervention. Rather than a specialist subject, as it may at first appear, the course in ISVA seeks to broaden the perspective of students and to open up areas of philosophy and anthropology omitted from most other courses. The examination of such issues should constitute an essential part of architectural and environmental education. The Royal Institute of British Architects' *Criteria for Validation* (RIBA 2003) sets out the minimum levels of awareness, knowledge, understanding and ability that students of architecture must attain at key stages in the process of qualifying as an architect. They include a section on cultural context stating that:

> students will demonstrate within coherent architectural designs and academic portfolios understanding of . . . The interrelationship between people, buildings and the environment and an understanding of the need to relate buildings and the spaces between them to human needs and scale.
>
> (RIBA 2003: 8)

The study of vernacular traditions throughout the world is of crucial importance if this aim is to be achieved. An inter-disciplinary education, connecting fields such as architecture, social and development work, geography and anthropology, is essential for architectural students in tackling the complexities of the world today. This is in contrast to much architectural practice in western countries, where culture is often narrowly defined as a special interest in the 'arts'. Any wider or deeper meaning of culture is barely discussed in schools of architecture, so it is not surprising that vernacular architecture and architectural practice seem disassociated. It is important to explain the field to colleagues as one of crucial importance to the future of the built environment in the world, not, as often presumed, as just the backward looking study of thatched cottages. The students themselves are sometimes unsure of what they are starting at the beginning of their course and only realise its relevance later in their careers.

A past student, who now works as an architect in a London practice, wrote to us several years after completing the course:

> My experience from the ISVA course has been particularly appropriate for my current project and other work I have been doing in the office . . . I am really glad that I chose the ISVA course. It certainly has proved most relevant to the work I have done since.[2]

Similar comments have been received from students as far apart as Germany and Venezuela.

So far, over eighty students from all continents have attended the course and in their feedback at the end of the year they consistently say how beneficial this multicultural aspect has been to their learning experience.

Participants' backgrounds and professional education contribute to the variety of exchanges from disciplines such as architecture, anthropology, teaching and social work, or experience with aid agencies. The gender and ethnic diversity of each cohort is balanced when compared with most other architectural courses, with over fifty per cent female and forty per cent non-white constituency. The reasons for this are unclear, but we have debated possible causes in an effort to encourage this trend.

Students draw on their own backgrounds for many of their assignment presentations, although they are gradually encouraged to research and examine other cultures, using a comparative method. Often it is the personal experiences and skills of their fellows that illustrate an academic point far better than a text or a lecture. A short verbal presentation at the beginning of the academic year serves as a way of letting the course staff and the students' peers know a little more about a student's background, ideas and expectations, including educational or working experience, country or region of origin and the types of vernacular buildings encountered. This serves as a useful point of reflection at the end of the course from which to evaluate personal progress.

Approximately half of the student intake is of direct entry Masters students who come from other countries, at considerable expense, to study the taught course and write the dissertation. They are usually clear about their field of study and how they want it to inform their future careers. The other half of the cohort are architecture diploma students, mostly from the UK or Europe, who are able to follow a 'special route' of ISVA within the post-graduate diploma course, leading to a qualification of RIBA part II. These students are sometimes less sure about vernacular architecture, as it is not commonly encountered in undergraduate programmes in schools of architecture. They often have an intuition about the course, and its relevance to their education may only become evident later in practice.

What all architectural students should learn

When deciding on their route through the architectural education system, students have to rely largely on the course literature for their information. The following two quotations are taken from the ISVA course handbook and may be termed a 'credo', describing the philosophy of the teaching and learning agenda:

> Architectural responses to building needs must be developed that respect local traditions rather than apply universal design solutions irrespective of cultural diversity and differing aspirations.

> We believe it to be essential for the welfare of the environment and of present and future generations that much more should be known about vernacular traditions, so that they may be supported all over the world and so that architect-designed buildings can relate to them.

> (Latter and Oliver 1997–2004)

The course objectives underpin this 'credo' and inform the subjects studied in the programme, and excerpts from the course handbook are included in this chapter where they relate to the assignments. The assignments in each subject area are set to enable students to research topics and apply seminar and lecture themes to their own learning. In order that this chapter is not merely a description of the programme, but also an explanation of the teaching aims, the course objectives are demonstrated by one or two examples of the assignments set on each topic. Some of the assignments are repeated every year, although student responses to their work and staff feedback are a constant stimulation and help us to change and develop assignments in detail or substance.

The reading list for the course is extensive and includes the *Encyclopedia of Vernacular Architecture of the World* (Oliver 1997a) and *Dwellings: The Vernacular House World Wide* (Oliver 2003). While these, and other texts (e.g. Upton and Vlach 1986; Bourdier and AlSayyad 1989; Turan 1990; Glassie 2000), have established and consolidated the subject of vernacular architecture as an academic field, the course that delivers this learning has not been published but exists as disparate texts, including the course handbook, the assignment descriptions and student feedback. A version of the course is described in this chapter, and although edited, may provide interest and inspiration to other educators of future architects, concerned about cultural sustainability.

The development of disciplinary and professional transferable skills is integral with the assignment and individuals' capabilities develop in exchange with fellow students. Increasing employability after study, practice in these skills encourages teamwork and reflection, as exercises examine the students' own inherent strengths and weaknesses in a supportive learning context. Assessment by the staff, peer group and self, encourages high-level skills such as forming judgments and evaluating work. The curriculum is augmented with talks on recent research, and field work, and illustrated by a unique slide collection of over 40,000 images collected over fifty years of travelling and studying indigenous peoples and their buildings. This rich archive demonstrates the significance of social structures and cultural behaviour for vernacular architecture, using examples from every continent, and encourages a comparative analysis of the forms, uses and meanings of buildings.

Coursework includes poster presentations, design exercises, field notebooks, essays, seminar papers, structured debates and finally a dissertation. By making presentations of work, particularly in poster form, concepts and information can be explained both graphically and verbally, which is helpful to a class in which English is a second language for approximately half of the group. The discourse within the student group stimulates thought, enables comparative study, practices presentation skills and also discourages plagiarism. There are always plenty of attractive posters to display at the end of the year exhibition, and many computer and graphic skills are gained during their production.

Throughout the year students undertake approximately sixteen assignments of varying kinds before beginning the dissertation. This enables them to learn not only the curriculum but also to develop their own personal knowledge and skills, gradually identifying the area of study that they wish to pursue, and to define the methodology best suited to demonstrate their hypotheses.

Some questions and exercises

In order to describe the diversity of student learning rather than focus on the delivery of teaching, a selection of several assignments given to the students follows.

Throughout the seminar series aspects of anthropological theory and research that bear upon shelter and settlement are discussed. The students must demonstrate that they have an understanding of cultures and their buildings encountered in diverse regions, climates, traditions and economies, and an awareness of the learned behaviours by which cultures express values and identities in building. An appreciation of the architectural expectations and needs of individuals, families and cultures is essential if the vernacular traditions of a society are to be understood. Several of the assignments relate largely to the 'Anthropology of Shelter' module, and one of these has been successfully run every year since the start of the course.

By looking at kinship systems it is possible to appreciate how the physical form of a building or settlement is organized around the family structure. The best way to grasp the techniques of kinship diagrams is for students to draw one, either of their own family with themselves as *ego*, or of another closely related family. With one exception, all the students have chosen to draw their own families and have discovered a wealth of information to analyse. At the start of the project its purpose is somewhat unclear to the students and many of them wonder what this might have to do with their architectural careers. By the completion of the exercise, when they may have gone home for the holidays to interview the keeper of family history, they are enthusiastic about making their presentations. Relating the members of the family to the buildings in which they live or lived illustrates many issues in a personal and powerful way. Patterns of migration, inheritance customs, family structures, property ownership and occupational changes are among the subjects that become very evident. Subjects have included migration of young men in Botswana within extended families, matrilineal descent in Japan and compound settlement relationships in Nigeria. The impact on families of economic migration in Thailand, Indonesia and the Philippines has been in contrast with the early twentieth-century Jewish Diaspora in the US. Post-Second World War social transformation in Britain explains changes of occupation, housing, manufacturing, education and equality for women. All of these relate to the buildings and settlements that the students call home, and the group dynamic of the participants is enriched by their knowing much more about each other as people. The course objective underpinning this assignment aims to provide 'a substantial examination of those aspects of culture which have been and are the subject of anthropological research and their significance for the ethnology of vernacular architecture' (Latter and Oliver 1997–2004).

The 'Philosophy and Theory' seminar series introduces principal philosophical theories relevant to the study of vernacular architecture, such as structuralism and semiotics. The module as a whole sets out to give 'an introduction to the philosophy of vernacular architecture studies and its bearing on other aspects of theoretical and philosophical developments related to architecture and culture' (Latter and Oliver 1997–2004). It aims to increase the student's capacity to apply theory to problem-solving, to synthesize information and to form proposals, whether as design approaches or written theses. Assignments are designed to encourage creative participation in examining the relationship between person and world within the vernacular environment, and include the development and examination of one's personal 'credo', together with a critical review of one of the books on the reading list, done as a peer assessment with fellow students.

The 'credo' is the essence of the course, and we ask the students to critically evaluate and re-write the two phrases mentioned earlier in this chapter, according to their own thoughts, while considering the nature of the term 'essential'. To write down in concise terms and within a specific word-count the basis and motivation of their approach to vernacular architecture, is a challenging task. However, the ability to express and convey views clearly is particularly important to this aspect of the course. This assignment takes place at the beginning of the course and helps the students to position themselves within a philosophical framework, and allows re-assessment and reflection during their period of learning and into practice. During the poster presentation of their work, students are encouraged to offer positive criticism of each statement; these can be revised and re-written as understanding of the subject increases throughout the year.

In her first term, a student produced this excerpt from her poster:

> We are losing the diversity of distinctive cultures due to universalism, globalization and the hegemonic pressure of western influence . . . I believe that only the understanding of local vernacular contexts can help stave off the distortions of globalization. Architects should be culturally aware, flexible, but above all else should understand the environment in which they work.[3]

The credo is presented to the group for discussion, and debating the work develops an ability to make thoughtful judgements on the philosophical issues surrounding vernacular architecture. This method of poster presentation is used for many of the assignments, enabling the issues to be put concisely in both verbal and visual terms and improving communication skills. Scholarly argument and methods of discourse such as 'antithesis' help students understand other points of view and challenge prejudice, so that they may develop objectivity in their thinking. Theoretical concepts must be communicated in uncomplicated language to students coming from many diverse countries; good practice for those who will work in development contexts with people who do not appreciate the jargon of the architect.

The ability to evaluate the impact of design interventions on vernacular architecture traditions is crucial for architects involved in proposals or policy that will have huge impacts on traditional societies that may be far removed from

14.1
Debating in the studio.

decision-making processes. In one structured debate the students were asked to research, and take up arguments for and against, the architectural response to a re-settlement project for those people displaced by the Yangtze River Dam, to be funded by the World Bank. The approach to the re-housing as proposed by a London firm of structural engineers was reported in the British architectural press. The design of the terraced units was compared with the Georgian Nash terraces of houses in the city of Bath and was to be based on the precedent of the Chinese shop–house model. The steel framed units were to be mass-produced in one of the large dockyards of Beijing (Middleton 2001).

Although this was a very short article about an extremely complex problem, there was no reference to the cultural traditions or knowledge of the hundreds of thousands of people to be relocated, or any use made of their indigenous building skills. Perhaps, in reality, the engineers had no other choice but to see the problem in purely technical terms, given the immediate political and time constraints of re-housing large numbers of people, but as a teaching vehicle this issue provoked some stimulating thoughts. The students were encouraged to examine their position on some of the cultural, political, ethical, economic and practical aspects of this proposal. Half of the students were asked to take positions in the argument that were antithetical to their own moral and ethical stance, so that they could understand the issues facing both the government, the engineers and the inhabitants in this troubling issue. By considering the points of view of the interested parties, professional or laymen, local or foreign, and then clarifying the implications of undertaking the proposed strategy when confronted with a demand for mass housing, the complexity of the issue and the importance of supporting local culture became evident. Several students suggested alternative modes of practice that, in this instance, were likely to have successful long-term outcomes for the people being relocated. For example, one student proposed:

the breakdown of the large resettlement program into a series of smaller locally based programmes with involvement of both the local community and the refugee community. Improving the employment prospects and creation of a sustainable local economy for the resettlement refugees would help to create a long term and sustainable economy enabling people to settle down and build their own houses.[4]

Coming at the end of the taught programme, this project gave students an opportunity to reflect on what they had learned on the course as a whole, and apply a critical perspective on the ethics of professional intervention to a real problem, developing an 'awareness of the ethical problems of "intervention", of the practicality of participatory planning, and the nature of rapid change in technologically developing nations' (Latter and Oliver 1997–2004).

Priorities for architects today and in the future

Architects working in development contexts will need to comprehend the cultural significance of the vernacular if they are to support and complement it. The course indicates the nature of development in vernacular contexts in the UK and regions across the world, facing transformations related to urbanization and globalization. It examines how the intentions of development practitioners, builders and users traditionally matched skills and resources within recognizable traditions, and discusses the issues involved in facilitating building processes within the community. We discuss how traditions are 'handed down' or transmitted and diffused, and how they may be sustained, supported and developed appropriately to meet changing requirements. In this way, learning about vernacular architecture is put into the wider context of development. We are aware that vernacular architecture studies have been seen as concerned only with the past, and many local history courses cover the built environment. However, in the ISVA course the students are presented with problems that confront communities and practitioners in the future. We know that the knowledge of patterns of space use in specific cultural contexts and an understanding of the relationship of vernacular architectural traditions to the reinforcement of regional identity will aid sensitive and sustainable development in the context of increasing globalization. 'An understanding of economic systems (pastoralism, agriculture, production, trade, etc.) and their relevance to the vernacular, seen in their international and transcultural environmental contexts', is crucial to enable students to develop these skills (Latter and Oliver 1997–2004).

Where possible, such as in the 'Design in a Vernacular Context' module, current issues are brought into the curriculum. Projects have ranged widely and are not always 'design based', and include a report on supporting home building in the development context of South Africa, 'Accommodating Cultures' (Latter and Oliver 2003). The report was prepared for ClaroMundi, a for-profit, socially responsible financial services organization whose aim is to facilitate finance for low-income housing applicable to multiple cultural, environmental and economic contexts. ClaroMundi's idea was to bridge the gap between local retail finance institutions, such as microfinance institutions, village banks and savings organizations, which need medium-term funds, with international socially responsible investment funds. They were seeking to use standardized quality systems to combat the corruption and inefficiency that had been a feature of the large-scale and centrally organized housing projects of previous decades. Initially we were asked to develop a building performance specification for a low-cost house which could be guaranteed for a ten year mortgage period. The idea for a standardized, mass produced unit supplied universally was anathema to our approach and although an unlikely partner for us, the opportunity for dialogue and research with a venture capitalist proved stimulating for staff and students.

The students researched case studies in Africa to illustrate why some previous low-income housing projects had succeeded in the longer term and why others had failed. Issues of standardization were discussed, alternative approaches compared and a critical viewpoint on institutional and professional intervention encouraged. Local planning contexts, homes for people, measurements of success, social values, material values and cultural values were factors which resulted in the students proposing a different paradigm to the one initially suggested by ClaroMundi. As summarized by one student:

> Fundamental is the inclusion of environmental and social concerns to the capitalist system of development that has principally economic concerns at its heart ... In order to be sustainable buildings must therefore fit into a wider environmental and social context, yet still be economically viable.[5]

Western, conventional architectural or planning education does not prepare professionals for meeting many of the problems they may encounter in new housing projects in other cultures. These may relate to resources, materials, skills training and above all respect for culturally differing needs for accommodation. We devised a short course programme intended to supplement the ISVA course for those who undertake the organization of work in this field, should the demand for such skilled practitioners be required, in 'preparation for the practice of architecture, planning or social work in development, with respect for, and the sustainability of vernacular traditions in a period of globalization' (Latter and Oliver 1997–2004).

The ability to make judgements on issues affecting the sustainability of vernacular architecture and an understanding of the forces impacting on the built environment in current and anticipated future contexts, is crucial to a designer or planner wherever in the world they are to practice. The assignments require the students to synthesize information and to form proposals for the support of vernacular traditions, resolving conflicting concepts of regional identity and modernity. The village of Haddenham in Buckinghamshire is subject to such pressures to considerably expand its housing stock. An ISVA group was invited by local people to make proposals for a sensitive brownfield site in the centre of the village adjacent to the conservation area. We were glad to have another 'real world' place for the students to study and the wider issues were particularly relevant to current government policy revisions and to giving the students 'knowledge of the problems and possibilities in the conservation, maintenance and protection of traditional buildings in natural materials, and the sustaining of traditional wisdom and skills' (Latter and Oliver 1997–2004).

Development plans and impacts on the village were examined in the context of the Barker Report (Barker 2004) into the crisis facing British housing. Brownfield and greenfield sites were assessed and the traditional patterns of the village were analysed. The role of the professional architect in relation to the community was considered, to develop an understanding of the challenges facing design teams working within vernacular settings. Village appraisal and design statements were discussed. Parish council members were consulted and proposals for housing and infrastructure, designed to be rooted, but not stuck, in the past, were developed. The students presented their work to members of the parish council and others, giving them the experience of explaining their ideas outside the school environment and to an audience unfamiliar with the language of architectural discourse. The feedback was positive and we have been asked by the community to continue with the projects in the following academic year, endorsed by the students' positive learning experience.

Several of the students commented that this was one of the hardest projects they had ever done, resolving complex but subtle issues. Although on the studio walls of the Department of Architecture the schemes may have appeared elementary, an examination of the issues underlying the 'design appearance' reveals the real issues confronted. Unfortunately, students and tutors are often reluctant to tackle this kind of low key housing project, preferring to focus on iconic building types, often illustrated with abstract computer modelling techniques. This exemplifies the current debate concerning the position of architectural education, whether it is training for practice or a place to experiment with conceptual design projects. This dichotomy is not helpful and there should be place for both in schools of architecture. If we do not equip students with the skills to confront some of the issues of house-building, then the regional identity of many villages in the south-east of England and beyond, will be lost to the volume house-builders operating within bland planning guidelines, to the detriment of people and places.

Social motivation underlies the learning, and 'lived experience' is valued above the abstract ideas in another course module. 'Vernacular Resources and Technology' discusses the range of available resources and skills employed in vernacular buildings around the world, and aims to develop an understanding of the technical, cultural and environmental factors involved in the building process. The seminar series gives the students a broad knowledge of materials and techniques used in constructing and maintaining vernacular buildings in differing parts of the world. Assignments are set to extend the students' understanding of the effects of changes associated with resource scarcity, sedenterization, permanence, the cash economy, industrialization and urbanization on vernacular building technology. They are individually assessed, but related, and gradually culminate in a design project incorporating all aspects of the taught module. A precedent study of a particular trade skill, including the tools, methods and culture of working, together with the range of materials used, refers to an existing structure. A subsequent essay topic relates the precedent study to changes in resources, technology and economy that have led to changes in work practice. Practical fieldwork is carried out at the Centre for Alternative Technology in Machynlleth, Wales. A field diary with sketches, notes and photographs records the materials and processes experienced at the time, and is used as a resource book for further design development in the studio. The experimental ambitions of the practical work and the material and human resistances encountered, together with the accommodations made while trying to achieve those ambitions, are recorded. This exercise encourages reflection on design ideas or insights into working practice developed during the Wales modelling exercise, to improve the understanding of the dynamic interplay between material and human agency involved in the culture of making vernacular building. The opportunity to build slate walls, or green wood structures, or straw insulating panels, is often a new experience, encouraging students to learn for themselves in a direct way rather than through the teacher in a relative way. They invent ways of fixing and joining found materials and change from the 'what do I want this to look like?' approach to design to a 'what can I do with this material?', questioning and experimenting. This is a subtle but powerful shift in thinking that not all students, or architects, ever experience. It gives insight into alternative modes of practice in the field of vernacular architecture and 'an understanding of the resources and materials used in vernacular building, and of the structural systems, techniques and details employed in diverse regions and environmental conditions' (Latter and Oliver 1997–2004).

14.2
Building a shelter from found materials at the Centre for Alternative Technology, Machynlleth, Wales.

Applying theory in fieldwork

The last part of the course that I would like to describe in some detail is the 'Field Workshops'. These, and the more extended 'Field Studies', have been an integral part of the course since its inception and give insights into building traditions and how they satisfy needs, employ available resources and accommodate local conditions. An essential tool of the anthropologist, knowledge and understanding of appropriate methods of field research applicable to a specific situation, can be studied in texts, but enhanced only by practice. It satisfies the objective of 'practical experience "in the field" in the study of vernacular architecture in socio-cultural contexts, including surveys and technical skills' (Latter and Oliver 1997–2004).

Dartmoor, in the county of Devon on the south-western peninsula of England, is exceptionally rich in archaeological sites. It affords opportunities to study various kinds of settlement, from farm and hamlet to village and town, including Stiniel, Poundsgate and Chagford. On Dartmoor the history of vernacular traditions is clearly evident. The availability of materials, access to water, methods of farming and raising of animals, means of communication, religion and other contextual and cultural factors have been expressed in settlement and building. The field trip offers opportunities to observe a number of different settlement patterns, and to consider these in relation to the landforms, climate, economies and ways of life of people over thousands of years.

Coming right at the beginning of the academic year, the arrival at the Bronze Age settlement of Grimspound makes a startling contrast to the hustle

14.3
Drawing in Scorhill stone circle on Dartmoor.

Rosemary Latter

14.4
Study group in the valley of Valldigna, Valencia province, 2001.

and bustle of university enrolment. The stone rows, hut circles and stone circles at Fernworthy, Bellever, Merrivale and Shovel Down, together with the stone circle at Scorhill, astound and enthral students from all over the world. An element of culture shock and the necessity of travelling, living and cooking together for four days with new acquaintances, usually coalesces the ISVA cohort, the more confident students supporting those who may be far from home for the first time.

Participants stay in a converted 'longhouse' on a Dartmoor farm, and are taught to measure and draw the buildings. The origins and development of the longhouse forms one theme of the trip, with visits over the moor to the site of an ancient settlement (over a thousand years old) at Houndtor, the village of Widecombe and existing longhouses at Bonehill. Identification and analysis of relevant data are formulated into hypotheses based on the field research, which may be synthesized to form proposals on conservation issues. Students discuss and assess the ideas of colleagues, and work together in the formation of proposals, collaborating with local inhabitants and within local constraints. All the skills are transferable to future fieldwork situations and the field log book of sketches, notes, photographs and ephemera provides a structured way of recording information. The logbook is assessed, and detailed feedback is given to help the students improve on subsequent trips.

These assignments give a glimpse into the work of the ISVA Masters students, but do not cover the full range of the modules offered in the programme. The following gives a brief outline of some of the other assignments covered in the course, to give an idea of the diversity of topics and approaches used to enrich the students' understanding of the subject. In the 'Natural Hazards and Traditional Building' module, students can study how the construction, form and settings of traditional buildings and the planning and disposition of settlements have affected their ability to resist extreme climatic or geological hazards.

It enables students to develop an understanding of natural hazards and how they affect buildings, and gives them awareness of the way builders have responded to such threats, as well as 'knowledge of the significance of climate and natural hazards with regard to environmentally appropriate building and secure housing, and measures to record and achieve them' (Latter and Oliver 1997–2004). Another module provides students with an introduction to the relationship between 'Climate and Vernacular Buildings' and a working knowledge of the tools used to measure this relationship, using short projects to apply the knowledge to a design problem in a vernacular context (Latter and Oliver 1997–2004). The 'Cultural Geography' module shows how physical structures reflect a complex interaction between the physical environment and the evolving culture of social groups in many parts of the world. Lectures, seminars, tutorials and fieldwork relate to research work on the *Atlas of Vernacular Architecture of the World* (Oliver and Vellinga forthcoming) and live research projects where possible. Students investigate how places are identified and marked in vernacular traditions, and what kinds of rites and rituals or forms of worship are observed in their construction, celebrated within them, and even defined their form. This provides an opportunity for students to try to understand the beliefs and value systems of a different culture, and to investigate how these are expressed in its vernacular architecture, as well as an 'awareness of the spatial and environmental impact of culture and the implications of the cultural geography of societies in relation to dwellings and settlement in all continents' (Latter and Oliver 1997–2004).

Extending the project based nature of the 'design' assignments, a European Union 'Culture 2000' funded project was undertaken with three other universities, and resulted in an exhibition and a publication (Latter and Oliver 2002). Based on fieldwork, the students devised conservation recommendations in the valley of Valldigna in the province of Valencia, identifying the prime characteristics that made the villages remarkable, and assessing the impact of and potential for tourism. Historic significance, location and appearance were considered, together with the assets and weaknesses from a conservation or tourism point of view, and the possible conflicts of interest which these two provoke. Proposals for change or development and recommendations for repair, restoration, preservation or demolition were illustrated with drawings and photographs, and combined into a broad policy document presented to the Fundación de Jaume II el Just as a catalyst for decision and action by local people. In the same context students from Spain and Germany produced some very diverse solutions; a stimulating learning experience in its multi-cultural nature.

The research and writing of the dissertation follows this intensively taught period, bringing together concepts of research theory, methodology and practice introduced from the beginning of the course. Students identify issues of criticism and judgement in the use of sources and the construction of arguments in the presentation of their ideas in seminar groups, and many of the assignments provide opportunities to develop research and writing skills. The choice of dissertation subject often develops some aspects of the student's interests already studied, building on a cross-cultural, inter-disciplinary, comparative and discursive theme. Some topics have been a focused study of a particular issue bearing upon an individual field of interest. We encourage inter-disciplinary and cross-cultural enquiry, extending the knowledge gained by studying the taught course. Coming at the culmination of the Masters course, the dissertation acts as a bridge between the Masters programme and career progression into professional life. Topics have included subject areas as diverse as tradition and transmission, practical and theoretical issues, development paradigms and those with a historical or geographic focus.[6]

All of the modules either teach or practice disciplinary and professional skills alongside transferable skills, which make the students attractive to a range of employers. Students have gone on to work in aid agencies, architectural firms, conservation, teaching, research and publishing. So although not strictly vocational, the course aims to support the career aspirations of the students.

Concerning teaching

Although comparative studies in international vernacular architecture have yet to become embodied in the curriculum of architecture schools, the time has come for those enlightened managers and educational policy makers to support such programmes. Although not necessarily focused on vocational skills for training architects as in a twentieth-century model, it is an essential part of architectural and environmental education in the twenty-first century in order to ensure that future generations' cultural and physical needs are met. At a time when the representations of other cultures are largely polarized by the media, an appreciation of the validity of cultural norms different from 'the western model' is crucial to a more sensitive approach to global architectural practice. In any future policies of a multinational scale and with an increasingly globalized and industrialized building industry, a generation of architects will be in a better position to assist with planning and housing issues, having had some education in vernacular architecture.

Notes

1 ISVA Masters course run at the Department of Architecture at Oxford Brookes University, UK. This chapter is less an academic treatise than a summary of a teaching philosophy, based on my experiences of being taught by Paul Oliver in the 1980s and, since 1997, of developing and running this course with him. It draws on the course documents and my own observations. Students have been quoted in the text; however, I did not intend to single out particular people, but have rather tried to illustrate a particular point. The many students who have taken the course have all contributed greatly to the first post-graduate, cross-cultural course of its kind in the world and this chapter is written with thanks to my colleagues associated with the course, Professors Ian Bentley, Michael Lloyd, Maurice Mitchell and Paul Oliver, and with the support of the Department of Architecture at Oxford Brookes University.
2 Philip Knudsen, 2002.
3 Ayaka Takaki, 2001.
4 Martin Nissim, 2000.
5 Argus Gathorne-Hardy, 2003.
6 Masters Theses titles include: 'The House as a Reflection of Status in Sumatra' – Yvonne Widjojo; 'The Character of the Traditional Betawi House' – Puteri Cempaka; 'Adjustment and Sustainability of Chinese Shop-houses in South-East Asia in the Contemporary World' – Widya Sujana; 'Temples

on Wheels, Hindu and Shinto Religious Portable Architecture' – Viviana Vivanco; 'A Study of the Thai and Isan Kitchens in Thailand' – Pratima Nimsamer; 'Communication and Safety: Floods in the Capital Region of Sudan' – Rania Dagash; 'Dedicated Training Programs: The Crucial Link in the Appropriation of Sustainable Housing' – Melissa Malouf; 'Notes on Building Tools: Continuity and Change' – Wael Sabry; 'The Windrush Spring Settlements' – Degang Wang; 'Oxfordshire Almshouses: Architecture, Use of Space and Setting' – Juan Fernando Bontempo; 'Cave Dwellings and Ghorfas of Southern Tunisia' – Martin Nissim; 'The Houseboats of Pin Mills' – Anthony Reid; 'Tradition: Life and Meaning in Rock Cut Cappadocia' – Korinna Adrianopoulou; 'Japanese Homes in British Houses' – Sawako Furuno; 'Travellers: Changing Contexts in Irish Nomadic Culture' – Anna Hoare; 'Understanding the Adaptation of Bugis Vernacular Architecture – A Semiotic Approach' – Yenny Gunawan; 'The Small House Policy and Its Impact on the Identity of a New Territories Village in the Hong Kong SAR' – Tiffany Gibbs; 'Cultural Sustainability and Development: Drukpa and Burman Vernacular Architecture' – Regina Lim. These titles are intended to illustrate the range and depth of subjects covered by students, and the theses are kept as a resource for students in the archive of the Oxford Institute for Sustainable Development: International Vernacular Architecture Unit, adding to the knowledge on vernacular architecture.

Afterword
Raising the roof

Paul Oliver

Although I have been invited by the editors to write a conclusion to this remarkable collection of papers, produced by several of the most authoritative researchers and writers in the field of vernacular architecture, I have to admit that a 'winding up' is scarcely possible when so many themes have been examined, so many approaches brought to them, and so many issues raised. As the categories in which the editors have placed the contributions clearly indicate, these can be grouped in a number of ways, ranging from the spatial intuitions that emanate from the emergence of *homo sapiens* to the projected applications of the analyses of fundamental behaviour in relation to the vernacular forms. They range from conservative adherents to a religious faith to family relationships in suburban houses; from speculations on the future of vernacular studies to the description of a course which sought to define a route to the future. Consideration of the applications of the analysis of suburban family life within a range of dwelling types may be regarded in the context of future design projections.

    If the diversity of themes and approaches may leave some readers wondering what is left to be done, for many others this extensive range of approaches and studies reveals a spectrum of possibilities for future research. In some instances the direct application of such research may be difficult to ascertain, but in others the routes have been opened for the application of vernacular studies to housing needs and policies in the twenty-first century. This is not to say that all approaches to vernacular architecture studies have been introduced here, nor that their significance for the future has been fully defined; none of the contributors would claim this.

    In the light of these illuminating studies on the subject of the future of vernacular architecture, not only have many facets been revealed, but also a number of problems exposed. Some lie in the experiences of many of its serious advocates and students, especially in the identification, recognition and active encouragement of the overall subject itself. That the field is vast is evident even in the range of building types and functions chosen to illustrate the issues raised by contributors to the collection. This is not intended to suggest in any way that these traditions are inapplicable to the term 'vernacular architecture', but rather to indicate the diversity of building types and traditions that are now perceived to be encompassed by it. Of considerable importance are the many lessons that are still to be learned from the study of vernacular traditions. Many of these may reflect the nature of the way of life, of different cultures – nomadic or sedentary, rural or urban, or poised in transition between these and other contexts. They may be related, or separate and distinct, but the cultures in all continents whose building requirements and traditions have yet to be the subject of focused studies, number in thousands. Every culture yields more information and, if we are receptive, greater understanding of their built structures, the identification of these with particular functions, the organization of spaces and their relation to social patterns.

Given the stability of some cultural features and the mobility of others, their significance in terms of age, gender, occupation, status and belief may be observed and, to a degree, comprehended. Although these are closely related to the economies of the cultures concerned, as well as to their societal structures, they raise issues concerning the forms that they may take in the future. An important but largely overlooked aspect of this is the study of the singularities and the commonalities shared by otherwise different building traditions of diverse societies, and how these have developed, diverged or converged over time. Comparative research has scarcely begun; until the bases of such comparisons have been identified in research, studies will continue to be largely of isolates.

Traditions and sustainability

Aspects of the natural and the built environment may be determined or constrained by their connotations within some cultures, while others may have stimulated the evolution of building traditions over generations. Like the study of the languages of humankind, each may reflect specific perceptions and awareness of the world their creators inhabit, and the world of habitation that they have in turn created. We may experience and take pleasure in the spaces that have been created within and without the buildings, but we are not necessarily sharing this, especially when the culture whose vernacular tradition we have chosen to study may have no words for space and form, for tradition and sustainability – and hence no concepts of them. Personally, it's my view that research into vernacular architectural traditions, while intense in specific instances, still leaves large areas awaiting serious examination and understanding. 'Such as?', a reader may ask.

Such as, for example, the ecological implications of housing demands in the next half century. The impact on the environment, and on natural resources, has to be considered with care, bearing in mind the massive increase in numbers of dwellings, quite apart from that in numbers of service centres, stores and social buildings which growth in the population will entail. Fundamentally, a high proportion of the materials employed in building traditions around the world are replaceable, or renewable. It is widely accepted that earth, in many forms, whether as mud, rammed earth, adobe or fired brick, is the most extensively used traditional material. While its uncontrolled use could impact upon the soil available in the land for agriculture, for the growing of grasses for pasture, a serious consideration of the soil types, levels and geomorphology could determine what could be used for building that would not adversely affect agriculture.

Original research is necessary for this.

Similarly, the development of tree plantations, the establishment of woods and forests of indigenous, not introduced kinds, with a view to their being used for regional building over a span of three or four decades, would be of practical value. The ecologies of indigenous buildings in many parts of the world, in both their positive and their negative aspects, need to be examined in order to eliminate unfounded assumptions and to provide the basis for policies for their management and long-term planning for consumption and renewal. Corresponding to this is the utilization of the land for agriculture and pasture that population growth will also demand, in addition to the stresses, so painfully evident, carried by the lack of reliable water supplies. In this respect, studies in salination as a result of natural forces and human agencies are necessary. This is not to say that they have not been made; they have. But their bearing on the water supply, and the siting and erection of desalination plants, are rarely, if ever, considered as issues in development housing.

Simply summarised, these areas of research can be considered as 'quantitative', but there are also important aspects of 'qualitative' research that need to be pursued: for example, the issue of aesthetics. Many of us, perhaps all, have been drawn to the study of vernacular architecture by the experience of it. We may regard it as beautiful, as simple, as inspiring, or however we may wish to express our response to it, or to a specific tradition. What we are responding to, in itself, needs examination: whether we *perceive* the values that we strive for in sophisticated 'architecture', such as proportional relationships or the juxtaposition of primary forms; or whether we are *projecting* these values upon it. The implications of the aesthetics of many cultures relate not only to the forms and sensory experience of the buildings, but also to the visual and tactile qualities of the materials used and, in some societies, the decoration employed: the motifs, visual symbols, mouldings and geometric or amorphous shapes, let alone the cosmological and anthropomorphic connotations that their plans, structures or the spaces that they generate, with which those who inhabit them identify.

But there is another aesthetic issue that I consider to be extremely important: namely, how does a vernacular tradition express the aesthetic of the culture concerned? What are the qualities in the building or its details that are valued by its builders, its craftspeople and its users? Are these shared by all members of the culture concerned, or by just a few; are they articulated verbally, or only by their production? This brings a further serious issue into the discussion, for the safeguarding and promotion of skills is fundamental to the future of vernacular architecture. As yet there are many skilled craftspersons working in materials of all kinds, who are to be found engaged in building in every continent. But their numbers are declining, as the attractions of electronic media have their effect upon the younger generations. Encouragement and support of the crafts needs to be given, including the passing-on of acquired expertise with regard to materials and their working, tools as well as techniques, as well as the transmission of knowledge related to forms and functions, strengths and weaknesses, and means to ends.

I feel obliged to raise an issue that I have written about in the past, but on which there still seems to be remarkably little new research, namely, the means by which traditions in building are passed on to succeeding generations, and how change is effected and assimilated within such traditions and their transmission. In particular, I believe that a greater understanding is vitally necessary – an understanding of the perceptions and aspirations of young people in cultural regions throughout the world, who will be facing the problems of housing themselves and their families in a dozen years or so. External intervention in the means and methods of the transmission of traditions raises ethical considerations. To what extent a tradition needs to be sustained, and the degree to which change can be introduced or may evolve over time is another major factor, given the dramatic changes that have taken place within the first few years of the twenty-first century. It may be concluded that the maintenance of a tradition no longer has a place, but it is fundamental to development that continuity is ensured. It is essential to human nature that the past, the present and the future, whether perceived as essentially the same, as linear, as cyclic or unfolding in time, form a continuum; one that few would wish to see interrupted or concluded.

It can be argued, with validity in my view, that vernacular architecture is the time-honoured, truly sustainable architecture that, in its multitudinous manifestations, has evolved over the centuries, changing or adapting when necessary to variable environments and the nature of family and social growth. Sustainability (from the Latin, *sus tenere*, to uphold) is not simply a condition, as 'sustainable vernacular architecture' might suggest; it implies and requires active involvement and support. In a period of dramatic change on an unprecedented scale – technological, electronic and social – the sustaining of an architectural tradition is much more than its conservation – also a major issue that should not be overlooked. Sustainability, in the face of massive innovations and the introduction of new means of communication, transportation, production and marketing, not excluding their control and restriction by tariffs and subsidies, applies to all aspects of living of which the vernacular is a substantial part.

Vernacular in the 'globalizing' world

In contradiction to this view is the argument that we now live in an age which is witnessing the early stages of globalization, and that the distinctions between the world's cultures will cease to have relevance. They will eventually become unrecognizable as they are subordinated to the exploitation and manipulation of the world's resources, both material and human, by a small number of superpowers roughly corresponding to the world's continents. In broad terms, this may be so, but I am of the opinion that the loss of identity, both of the individual and of the group, will lead to a striving for the reclamation of cultural identity. This is already evident in the revival of indigenous languages, and of aspects of clothing, and may well extend to other forms of material culture, of which the largest and most significant is that of buildings. Whether I am right or wrong in this speculation I do not know, but I am convinced that the problems of housing and building in the immediate future have to be faced.

A major issue in the book is the future of such building. Some authors do not anticipate that vernacular architecture, as we know it today, will survive to the end of this century. That may well be true when considered in relation to the traditions of many African and Asian countries in their unique forms as documented and illustrated in numerous books. It may not be true as far as informal, self-built settlements are concerned, even if there are severe measures taken leading to their destruction on a massive scale – something which has occurred at intervals in several countries. Even if we assume that the average number of people occupying a dwelling unit is six (and this is a low figure in many countries), some 500 million units will have to be built in little more than four decades, to meet the accommodation needs generated by the growth in the global population of 3 billion, predicted by the middle of the twenty-first century. It is by no means clear how such demands are to be met without reliance on the self-builders and community-builders who still construct by far the majority of the world's dwellings. Labour for the erection of such scarcely imaginable numbers of buildings is unlikely to be found, or paid for, by a large proportion of the world's nations – which may be used to justify factory-made and mechanically erected mass housing units on a vast scale.

'Mass housing', with its patronizing, dismissive connotations of 'the masses', with little concern for the size of communities, their distinct values, ways of life and specific housing or other needs, will undoubtedly appeal to construction firms intent on exploiting the housing shortage to their commercial advantage. It may also commend itself to governments and national agencies that would welcome any measure that would release them from the responsibility of developing culturally appropriate housing. But 'mass housing' is a likely concomitant to globalization which, as a concept, is inimical to the specific identities of the world's distinct cultures. An equally large problem is the demand on land which such housing would make. The customary alternative of 'high-rise' multi-storey buildings, with their considerable structural and service requirements, including elevators, is unlikely to meet these substantially increased needs, even when considered away from the evidence of their past anti-social records.

Obviously, the expense involved in the construction of hundreds of millions of new homes, let alone the costs in improved servicing, maintenance and possible eventual replacement of the existing structures, is scarcely imaginable. An unavoidable expenditure will arise with the growing need for materials and resources; even today, after more than a decade of attempting to ameliorate the suffering occasioned by its limited availability, fresh water supply on a regular daily basis is still denied to hundreds of millions of people. The failure to achieve this much publicized objective is indicative of the problems that will be encountered when necessary services must be provided, including sanitation and waste disposal, electricity or gas supply, lighting and ventilation, heating or cooling depending on the climatic circumstances, and the provision of appropriate transport infrastructures.

We may well ask how all this is to be financed? To a large extent the financial implications of 'sweat equity', or the balancing of the owner-occupier's physical efforts against the cost of paid labour, could go far to meet the expenses, which could be otherwise directed to effective servicing. This may seem exploitative, but it may be realistic; it still costs, but many nations have committed a percentage of their gross national product to the alleviation of poverty, and, it is to be hoped, the rehabilitation of many communities in the world's peoples. If the end of poverty is a declared intention, the end of the impoverishment of rural lands, created by the ruthless exploitation by multinational corporations, should be a high priority. But not at the expense of the lack of culturally appropriate, well-serviced domestic dwellings and settlement clusters.

As far as building itself is concerned, the basic requirements are building sites and title to them, and the resources, materials and relevant technologies with which to construct homes. These requirements apply in all continents and in all regions where there is the need for human habitation. Resources for building, whatever the scale, draw heavily on natural materials, many being transformed into concrete, brick, steel or glass by technologies that consume vast quantities of fossil fuels, minerals and water. By comparison with the exploitation and destruction of the forests of Brazil or Indonesia, the Asian sub-continent, and many other parts of the world, the use of natural materials in vernacular architecture is likely to represent a small proportion of the volume consumed as a whole. I cannot state what that proportion may be, but this is just one of the many areas of research that need to be undertaken. Moreover, the structures are seldom of a temporary nature, but are often utilised for generations, even centuries, as countless examples demonstrate.

Whatever the forms that housing may take in thirty, fifty, a hundred years time, human settlement will remain a fundamental issue. Some will doubtless argue that people will adapt to change, and future generations will accept what has been and is being built, having known no other. Yet, as cultural distinctions become blurred in the urban haze, and as populations large and small, national and cultural, become aware of an increasing loss of both group identity and personal individuality, I fear serious reactions may occur. These may take the form of new settlements, resulting, perhaps, in the neglect or destruction of others; they may lead to the occupation of some lands and the reclamation of more, although the rising of sea levels with global warming may make this problematic. To those involved in development work and studies much of what I have written may seem obvious, and painfully obvious indeed it is; except, apparently, to those with the power to make a difference. There is little indication that the majority of these matters are receiving any serious international consideration by politicians.

Afterword

Educating for the future

It is possible that the only way in which the housing of the world's peoples will become a matter of serious attention among politicians – and educationalists, anthropologists, architects and other professionals – will be through its inclusion in general education. It is one of the ironies of education that even professionally designed architecture of the Western world is not represented in the work of the classroom; or, if it is, then it is only as an adjunct to the art class. For the most part, people enter the world of designed architecture and accept it, grumble about it, enjoy it or reject it – but still use it. Concern for the dwellings and buildings of peoples in other parts of Europe, let alone other continents, is negligible. This ignorance of the relationship of buildings to the cultures that produce them inevitably leads to the assumption that Western architecture meets the aspirations of the world's peoples, and provides the solutions to their building problems and requirements.

An assumption on the part of the governments and the populace that the study of the world's buildings and vernacular architecture is a part of the education of professional architects who are preparing to design for the future, would not be unreasonable. Except for the fact that it is not the case. A number of the papers in this collection are by educationalists who run courses in schools of architecture, but they are in the minority. Education in architecture and other related disciplines, including anthropology, ecology, geography and history, pays little or no attention to the buildings of the world's peoples, except occasionally for the grand manner 'architecture' designed by architects; but certainly not to the vernacular.

I have remarked before, that this is a field without a 'discipline', by which I mean 'a branch of instruction, a department of knowledge' (*Shorter Oxford English Dictionary* 2002). I notice that this has been echoed by one or two contributors. In some respects it is an advantage, as the frontiers of the subject are open to expansion and creative thought. But in terms of validation and recognition, including the employment of its graduates, a course that is not a 'recognised discipline' is at a considerable disadvantage. Most of those who work in the field and are involved in education are aware that the subject is 'multi-disciplinary'; in other words, one that involves (or should involve) participants from a wide range of studies: architecture, building science, anthropology and sociology, planning, climatology and cultural geography, conservation, planning and development studies, and several other specialist fields ranging from economics to ecology, ethnology to ethics. How can this be possible? Ideally, by the establishment of substantial courses, professionally acknowledged and serviced by specialists from such fields, leading to national (preferably international) accreditation. It is a problem that has still to be addressed, but it is one for which there has to be an answer, if appropriately trained graduates are to be prepared who can help nations, governments, NGOs, cultures and communities to meet the challenges of settlement and housing in the coming decades.

Bibliography

Abbott, A. (1988) *The System of Professions: An Essay on the Division of Expert Labor*, Chicago: The University of Chicago Press.

Abernethy, F.E. (1985) 'Folk art in general, yard art in particular', in F.E. Abernethy (ed.) *Folk Art in Texas*, Dallas: Southern Methodist University Press.

Abrahams, I. (1988) 'Some booths I have known', in P. Goodman (ed.) *The Sukkot/Simhat Torah Anthology*, Philadelphia: Jewish Publication Society.

Abu-Lughod, J. (1992) 'Disappearing dichotomies: First world–third world; traditional–modern', *Traditional Dwellings and Settlements Review*, 3 (2): 7–12.

Afshar, F. and Norton, J. (1997) 'Developmental', in P. Oliver (ed.) *Encyclopedia of Vernacular Architecture of the World*, Cambridge: Cambridge University Press.

Ahmed, A.S. (1982) 'Nomadism as an ideological expression', *Economic and Political Weekly*, 3 July 1982: 1101–6.

Akbar, J. (1988) *Crisis in the Built Environment: The Case of the Muslim City*, Singapore: Mimar Books.

Alcock, J. (2001) *The Triumph of Sociobiology*, New York: Oxford University Press.

Alexander, Ch. (2003) *The Process of Creating Life* (Book Two of *The Nature of Order*), Berkeley: Center for Environmental Structure.

Alexander, Ch., Ishikawa, S. and Silverstein, M. (1977) *A Pattern Language*, New York: Oxford University Press.

AlSayyad, N. (ed.) (2004) *The End of Tradition?*, London: Routledge.

Altman, I. (1975) *The Environment and Social Behavior: Privacy, Personal Space Territory, Crowding*, Monterey, CA: Brooks/Cole.

Alward, W.L.M. (2003) 'A new angle on ocular development', *Science*, 299 (5612): 1527–8.

Amerlinck, M-J. (2001) *Architectural Anthropology*, Connecticut and London: Bergin & Garvey.

Ames, K.L. (1977) *Beyond Necessity: Art in the Folk Tradition*, New York: W.W. Norton for the Henry Francis du Pont Winterthur Museum.

Arendt, H. (1971) 'Introduction', in H. Arendt (ed.) *Walter Benjamin*, München: Piper.

Arias, E.G. (ed.) (1993) *The Meaning and Use of Housing*, Aldershot: Avebury.

Aronson, D.R. (1980) 'Must nomads settle? Some notes towards policy on the future of pastoralism', in P.C. Salzman (ed.) *When Nomads Settle*, New York: Praeger Publishers.

Arreola, D.D. (1984) 'House color in Mexican-American barrios,' paper presented at Conference on Built Form and Culture Research, Lawrence, KS, October.

—— (1988) 'Mexican-American housecapes', *Geographical Review*, 78 (3): 299–315.

Asad, T. (1970) *The Kababish Arabs*, London: C. Hurst.

Ashmore, J. et al. (2003) 'Diversity and adaptation of shelters in transitional settlements for IDPs in Afghanistan', *Disasters* 27 (4): 273–87.

Asquith, L. (2003) 'Spatial construction: An evaluation of space use and claim in family homes', unpublished PhD thesis, Oxford Brookes University.

Austin, M.R. (1997) 'Maori', in P. Oliver (ed.) *Encyclopedia of Vernacular Architecture of the World*, Cambridge: Cambridge University Press.

Aysan, Y., Clayton, A., Cory, A., Davis, I. and Sanderson, D. (1995) *Developing Building for Safety Programmes: Guidelines for Organizing Safe Building Improvement Programmes in Disaster Prone Areas*, London: Intermediate Technology Publications

Bachelard, G. (1964) *The Poetics of Space*, 2nd edn, New York: Orion Press.

Bachman, C.G. (1961) *The Old Order Amish of Lancaster County*, Lancaster, PA: Pennsylvania German Society.

Bibliography

Banning, E.B. and Byrd, E.F. (1987) 'Houses and the changing residential unit: Domestic architecture at PPNB 'Ain Ghazal', Jordan,' *Proceedings of the Prehistoric Society*, 53: 309–25.
Barker, K. (2004) *Delivering Stability: Securing Our Future Housing Needs.* Online. Available HTTP: http://www.hm-treasury.gov.uk/media/053/C7/barker_review_execsum_91.pdf.
Barkow, J.H., Cosmides, L. and Tooby, J. (1992) *The Adapted Mind: Evolutionary Psychology and the Generation of Culture*, New York: Oxford University Press.
Barnett, R. (1977) 'The libertarian suburb: Deliberate disorder', *Landscape*, 22 (3): 44–8.
Barry, M., Roux, L., Barodien, G. and Bishop, I. (2002) 'Video-evidencing and Palmtop computer technology to support formalising land rights', *Development Southern Africa*, 19 (2): 261–71.
Barth, F. (1961) *Nomads of South Persia*, Boston: Little, Brown & Company.
Barton, H. and Tsourou, C. (2000) *Healthy Urban Planning*, London: E. & F.N. Spon.
Bartsch, U. and Mueller, B. (2000) *Fossil Fuels in a Changing Climate: Impacts of the Kyoto Protocol and Developing Country Participation*, Oxford: Oxford University Press.
Bauman, R. (1972) 'Differential identity and the social base of folklore', in A. Paredes and R. Bauman (eds) *Toward New Perspectives in Folklore*, Austin: University of Texas Press.
—— (1983) 'Folklore and the forces of modernity', *Folklore Forum*, 16: 153–8.
—— (1992) 'Performance', in R. Bauman (ed.) *Folklore, Cultural Performances, and Popular Entertainments: A Communications-Centered Handbook*, New York: Oxford University Press.
Becker, J.S. (1998) *Selling Tradition: Appalachia and the Construction of an American Folk, 1930–1940*, Chapel Hill: University of North Carolina Press.
Bedaux, R., Diaby, B., Maas, P. and Sidibe, S. (2000) 'The restoration of Jenné, Mali: African aesthetics and Western paradigms', in International Committee for the Study and Conservation of Earthen Architecture and the University of Plymouth (Centre for Earthen Architecture) (eds) *Terra 2000: 8th International Conference on the Study and Conservation of Earthen Architecture*, London: James & James.
Bedaux, R., Diaby, B., Maas, P., Sidibe, S. and Keita, M.K. (1995) *Plan de Projet: Rehabilitation et Conservation de l'Architecture de Djenné (Mali)*, Leiden: Bamako.
Ben-Amos, D. (1984) 'The seven strands of "tradition": Varieties in its meaning in American folklore studies', *Journal of Folklore Research*, 21: 97–131.
Bendix, R. (1989) 'Tourism and cultural displays: Inventing traditions for whom?', *Journal of American Folklore*, 102: 131–46.
Berland, J.C. and Salo, M.T. (1986) 'Peripatetic communities: An introduction', in *Nomadic Peoples: Special Issue on Peripatetic Peoples*, 21-2: 1–4.
Bernstein, B. (1970) *Class Codes and Control: Theoretical Studies Towards a Sociology of Language*, London: Routledge & Kegan Paul.
Bernstein, F. (2002) 'A critic takes the Catholic church to task for its architecture', *New York Times*, 7 September 2002.
Biederman, I. (1987) 'Recognition by components: A theory of human image understanding', *Psychological Review*, 90: 115–47.
Blanton, R.E. (1994) *Houses and Households: A Comparative Study*, New York: Plenum.
Blasdel, G.N. (1968) 'The grass-roots artist', *Art in America*, 56: 24–41.
Blier, S.P., (1987) *The Anatomy of Architecture: Ontology and Metaphor in Batammaliba Architectural Expression*, Cambridge: Cambridge University Press.
Boden, M.A. (1990) *The Creative Mind: Myths and Mechanisms*, London: Sphere.
Bonine, M. (1979) 'The morphogenesis of Iranian Cities', *Annals of the Association of American Geographers*, 69 (2): 208–23.
Boudon, P. (1972) *Lived-in Architecture: Le Corbusier's Pessac Revisited*, Cambridge, MA: MIT Press.
Bourdieu, P. (1977) *Outline of a Theory of Practice*, Cambridge: Cambridge University Press.
—— (1990) *The Logic of Practice*, Cambridge: Polity Press.
Bourdier, J.-P. and AlSayyad, N. (eds) (1989) *Dwellings, Settlements and Tradition: Cross-Cultural Perspectives*, Lanham: University Press of America.
Bourdier, J.-P. and Minh-ha Trinh, T. (1996) *Drawn from African Dwellings*, Indiana: Indiana University Press.
Bourgeois, J.-L. (1989) *Spectacular Vernacular: The Adobe Tradition*, New York: Aperture Foundation.
Boyden, S. (1987) *Western Civilization in Biological Perspective: Patterns in Biohistory*, Oxford: Oxford University Press.

Bibliography

Boyer, E.L. and Mitgang, L.D. (1996) *Building Community: A New Future for Architectural Education and Practice*, Princeton: The Carnegie Foundation for the Advancement of Teaching.

Brackman, B. (1999) 'Remember the grotto: Individual and community', in B. Brackman and C. Dwigans (eds) *Backyard Visionaries: Grassroots Art in the Midwest*, Lawrence: University Press of Kansas.

Brackman, B. and Dwigans, C. (eds) (1999) *Backyard Visionaries: Grassroots Art in the Midwest*, Lawrence: University Press of Kansas.

Bronner, S.J. (1979) 'Concepts in the study of material aspects of American folk culture', *Folklore Forum*, 12: 117–32.

—— (1985) 'Researching material folk culture in the modern American city', in S.J. Bronner (ed.) *American Material Culture and Folklife*, Ann Arbor: UMI Research Press.

—— (1986a) *Grasping Things: Folk Material Culture and Mass Society in America*, Lexington: University Press of Kentucky.

—— (1986b) 'Folk objects', in E. Oring (ed.) *Folk Groups and Folklore Genres*, Logan: Utah State University Press.

—— (ed.) (1992) *Creativity and Tradition in Folklore: New Directions*, Logan: Utah State University Press.

—— (1998) *Following Tradition: Folklore in the Discourse of American Culture*, Logan: Utah State University Press.

—— (1999) 'Cultural historical studies of Jews in Pennsylvania: A review and preview', *Pennsylvania History*, 66: 311–38.

—— (2000a) 'The meanings of tradition: An introduction', *Western Folklore*, 59: 87–104.

—— (2000b) 'The American concept of tradition: Folklore in the discourse of traditional values', *Western Folklore*, 59: 143–70.

—— (2002a) *Folk Nation: Folklore in the Creation of American Tradition*, Wilmington, DE: SR Books.

—— (2002b) 'Questioning the future: Polling Americans at the turn of the new millennium', *Prospects: Annual of American Cultural Studies*, 27: 665–86.

Bronowski, J. (1978) 'The imaginative mind in art', in J. Bronowski (ed.) *The Visionary Eye*, Cambridge, MS and London: MIT Press.

Brown, D.E. (1991) *Human Universals*, Philadelphia: Temple University Press.

Brown, P.L. (2002) 'Megachurches as minitowns', *New York Times*, 9 May 2002.

Bucko, R.A. (1998) *The Lakota Ritual of the Sweat Lodge: History and Contemporary Practice*, Lincoln and London: University of Nebraska Press.

Bunge, M. (1998) *Philosophy of Science*, vol. 1, New Brunswick, NJ: Transaction Publishers.

Bunimovitz, S. and Faust, A. (2000) 'Ideology in stone: Understanding the four room house', *Biblical Archaeology Review*, 28 (4): 32–41, 59.

Burnham, P. (1975) '"Regroupement" and mobile societies: Two Cameroon cases', *The Journal of African History* 16 (4): 577–94.

—— (1979) 'Spatial mobility and political centralization in pastoral societies', in L'Equipe Écologie et Anthropologie des Sociétés Pastorales (ed.) *Pastoral Production and Society*, Cambridge: Cambridge University Press.

Burns, G.L. (1991) 'What is tradition?', *New Literary History*, 22: 1–22.

Cain, A., Afshar, F. and Norton, J. (1975) 'Indigenous building and the third world', *Architectural Design*, 4: 207–24.

Canaan, T. (1932–3) 'The Palestinian Arab house: Its architecture and folklore', *Journal of the Palestine Oriental Society*, 12: 223–47, 13: 1–83.

Cannon, H. (ed.) (1980) *Utah Folk Art: A Catalog of Material Culture*, Provo: Brigham Young University Press.

Cantwell, R. (1993) *Ethnomimesis: Folklife and the Representation of Culture*, Chapel Hill: University of North Carolina Press.

Cardinal, R. (1972) *Outsider Art*, New York: Praeger.

Carlhian, J.P. (1979) 'The École des Beaux-Arts: Modes and manners', *Journal of Architectural Education*, 33 (2): 7–17.

Carroll, S.B. (2003) 'Genetics and the making of Homo sapiens', *Nature*, 422 (6934): 849–57.

Carroll, S.B., Grenier, J.K. and Weatherbee, S.D. (2001) *From DNA to Diversity: Molecular Evolution and the Evolution of Animal Design*, Oxford: Blackwell.

Bibliography

Carter, T. (1997) 'Folk design in Utah architecture, 1849–1890', in T. Carter (ed.) *Images of an American Land: Vernacular Architecture in the Western United States*, Albuquerque: University of New Mexico Press.
Casey, E. (1996) 'How to get from space to place in a fairly short stretch of time', in S. Feld and K. Basso (eds) *Senses of Place*, Santa Fe, NM: School of American Research Press.
Casimir, M.J., Lancaster, W. and Rao, A. (1999) 'Editorial', *Nomadic Peoples* 3 (2): 3–4.
Chatty, D. (1980) 'The pastoral family and the truck', in P.C. Salzman (ed.) *When Nomads Settle*, New York: Praeger Publishers.
Check, E. (2002) 'Worm cast in starring role for Nobel Prize', *Nature*, 419 (6907): 548–9.
Chomsky, N. (1972) *Language and Mind*, New York: Harcourt Brace Jovanovich.
Cierrad, I. (ed.) (1999) *At Home: An Anthropology of Domestic Space*, New York: Syracuse University Press.
Clottes, J. and Lewis-Williams, J.D. (1998) *The Shamans of Prehistory: Trance and Magic in the Painted Caves,* New York: Harry Abrams.
Cohen, D.H., Yamaguchi, A. and Spengler, J.D. (2003) 'A bio-regional approach to environmental building: A case study of the KST house', in R.J. Cole and R. Lorch (eds) *Buildings, Culture and Environment: Informing Local and Global Practices*, Oxford: Blackwell.
Collins, T.K. (1996) *The Western Guide to Feng Shui: Creating Balance, Harmony and Prosperity in Your Environment*, Carlsbad: Hay House.
Corsellis, T. and Vitale, A. (2004) *Transitional Settlement Displaced Populations*, Cambridge: Martin Centre.
Crinson, M. and Lubbock, J. (1994) *Architecture, Art or Profession? Three Hundred Years of Architectural Education in Britain*, Manchester: Manchester University Press.
Daniels, T. (1981) 'A Philadelphia squatter's shack: Urban pioneering', *Pioneer America: Journal of Historic American Material Culture*, 13: 43–8.
Davis, H. (1999) *The Culture of Building*, New York: Oxford University Press.
Davis, I. (1978) *Shelter After Disaster*, Oxford: Oxford Polytechnic Press.
—— (ed.) (1981) *Disasters and the Small Dwelling*, Oxford: Pergamon Press.
—— (1985) 'Shelter after earthquakes', unpublished thesis, University College London.
—— (2001) *Location and Operation of Evacuation Centres and Temporary Housing Policies: Global Assessment of Earthquake Countermeasures*, Kobe: Hyogo Province.
De Saussure, F. (1972) 'Course in general linguistics', in R. and F. DeGeorge (eds) *The Structuralists from Marx to Lévi-Strauss*, Garden City, NY: Doubleday.
Dewhurst, C.K. (1984) 'The arts of working: Manipulating the urban work environment', *Western Folklore*, 43: 192–202.
—— (1986) *Grand Ledge Folk Pottery: Traditions at Work*, Ann Arbor: UMI Research Press.
Dewhurst, C.K. and MacDowell, M. (1983) 'The conduit tile buildings of Grand Ledge, Michigan', *Pioneer America: Journal of Historic Material Culture*, 15: 91–104.
Dillehay, T.D. (2002) 'Climate and human migrations', *Science*, 298 (5594): 764–5.
Dissanayake, E. (1988) *What is Art For?*, Seattle and London: University of Washington Press.
Dovey, K. (1999) *Framing Places, Mediating Power in Built Form*, London: Routledge.
Downing, F. and Fleming, U. (1981) 'The bungalows of Buffalo', *Environment and Planning B*, 8. Online.
Dugatkin, L.A. (ed.) (2001) *Model Systems in Behavioral Ecology: Integrating Conceptual, Theoretical and Empirical Approaches*, Princeton: Princeton University Press.
Dyson-Hudson, N. (1972) 'The study of nomads', in W. Irons and N. Dyson-Hudson (eds.) *Perspectives on Nomadism*, Leiden: E.J. Brill.
Eco, U. (1980) 'Function and sign: The semiotics of architecture', in G. Broadbent, R. Bunt and C. Jencks (eds) *Signs, Symbols, and Architecture: Architecture as Communication*, New York: Wiley.
Edmonds, D. and Eidinow, J. (2001) *Wittgenstein's Poker*, London: Faber.
Edwards, B. and Turrent, D. (2000) *Sustainable Housing: Principles and Practice*, London: Routledge.
Edwards, J.D. (1971) 'An analysis of vernacular architecture in the Western Caribbean', Paper presented at the 1971 meeting of the Southern Anthropological Society, Dallas, April 1971.
Ellen, R.F., Parkes, P. and Bicker, A. (eds) (2000) *Indigenous Environmental Knowledge and its Transformations: Critical Anthropological Perspectives*, Amsterdam: Harwood.

Bibliography

Ellis, R.D. (1995) *Questioning Consciousness: The Interplay of Imagery, Cognition, and Emotion in the Human Brain*, Amsterdam and Philadelphia: J. Benjamins.

Encyclopaedia Britannica (1911), London: Encyclopaedia Britannica Ltd, 14: 17.

Ennis, M. (2002) 'Small stuff', *Texas Monthly*, April 2002. Online. Available HTTP: http://www.texas monthly.com/mag/issues/2002-04-01/art.php.

Ensminger, R.F. (1992) *The Pennsylvania Barn: Its Origin, Evolution, and Distribution in North America*, Baltimore: Johns Hopkins University Press.

Environmental Protection Agency (1995) *The Inside Story: A Guide to Indoor Air Quality*, (United States Environmental Protection Agency and the United States Consumer Product Safety Commission, Office of Radiation and Indoor Air (6604J) Document # 402-K-93–007), Cincinnati, OH: National Service Center for Environmental Publications. Online. Available HTTP: http://www.epa.gov/iaq/pubs/insidest.html.

Essex County Council (1973) *A Design Guide for Residential Areas*, Chelmsford: Essex County Council Planning Department.

Esteves, A., Ganem, C., Esteban, F. and Mitchell, J. (2003) 'Thermal insulating material for low-income housing', in W. Bustamante G. and E. Collados B. (eds) *Rethinking Development: Proceedings of the 20th PLEA International Conference*, vol. 2, Santiago de Chile: Escuela de Construction Civil, Pontifica Universidad Catolica de Chile.

Evans, C. and Humphrey, C. (2002) 'After-lives of the Mongolian yurt: The "archaeology" of a Chinese tourist camp', *Journal of Material Culture*, 7(2): 189–210.

Evans, D. (1982) *Big Road Blues: Tradition and Creativity in the Folk Blues*, Berkeley: University of California Press.

Fathy, H. (1970) *Construire Avec le People: Histoire d'un Village d'Égypte, Gourna*; trans. Y. Kornel, Paris: Sindbad.

—— (1973) *Architecture for the Poor: An Experiment in Rural Egypt*, Chicago and London: The University of Chicago Press.

Feld, S. and Basso, K. (eds) (1996) *Senses of Place*, Santa Fe, NM: School of American Research Press.

Ferris, W. (1974) 'Don't throw it away: Folk culture and our dwindling resources', *Yale Alumni Magazine*, 37: 19–24.

Filarete, A.P.A. (1965; 1st edn 1465) *Treatise on Architecture*, New Haven: Yale University Press.

Fodor, J., Miller, G.A and Langendoen, D.T. (1979) *The Language of Thought*, Cambridge, MA: Harvard University Press.

Foster, G.M. (1953) 'What is folk culture?', *American Anthropologist*, 55: 159–73.

Foster, S.C. (1984) 'The folk environment: Some methodological considerations', in D.F. Ward (ed.) *Personal Places: Perspectives on Informal Art Environments*, Bowling Green, OH: Bowling Green State University Popular Press.

Foy, M. (1976) *The Sugar Industry in Ireland*, Dublin: Cómhlucht Siúcre Éireann TEO.

Frantz, C. (1978) 'Ecology and social organization among Nigerian Fulbe (Fulani)', in W. Weissleder (ed.) *The Nomadic Alternative*, The Hague and Paris: Mouton Publishers.

Frye, H.N. (1957) *The Anatomy of Criticism*, Princeton: Princeton University Press.

Gallier, F. (2002) 'Analyse du Projet de Barrage de Talo et ses consequences prévisibles sur les systèmes de production ruraux du Djenneri', *Djenne Patrimoine Informations*, 13: 15–27.

Gao, K.-Q. and Shubin, N.H. (2003) 'Earliest known crown-group salamanders', *Nature*, 422 (6930): 424–6.

Garcia Chavez, J.R. (2004) 'Application of sustainable strategies for a new ecological community', in M.H. de Wit (ed.) *Built Environments and Environmental Buildings: Proceedings of the 21st PLEA International Conference, Eindhoven*, Eindhoven: Technische Universiteit.

Gardi, B., Maas, P. and Mommersteeg, G. (eds) (1995) *Djenné, il y a cent ans*, Eindhoven: Lecturis.

Gellner, E. (1973) 'Introduction to nomadism', in C. Nelson (ed.) *The Desert and The Sown* (Research Series 21, Institute of International Studies), Berkeley: University of California Press.

Giere, R.N. (ed.) (1992) *Cognitive Models of Science*, vol. 15, Minneapolis, MN: University of Minnesota.

Glaser, P. (2004) 'A digital gulf divides the world', *World Press Review*, February 2004: 25.

Glassie, H. (1968) *Pattern in the Material Folk Culture of the Eastern United States*, Philadelphia: University of Pennsylvania Press.

Bibliography

—— (1972) 'Eighteenth-century cultural process in Delaware Valley folk building', *Winterthur Portfolio*, 7: 29–57.
—— (1974) 'The variation of concepts within tradition: Barn building in Otsego County, New York', *Geoscience and Man*, 5: 177–235.
—— (1975) *Folk Housing in Middle Virginia: A Structural Analysis of Historical Artifacts*, Knoxville: University of Tennessee Press.
—— (1985) 'Artifact and culture, architecture and society', in S.J. Bronner (ed.) *American Material Culture and Folklife*, Ann Arbor: UMI Research Press.
—— (1993) *Turkish Traditional Art Today*, 2nd edn, Bloomington: Indiana University Press.
—— (2000) *Vernacular Architecture*, Bloomington: Indiana University Press.
Goldsmith, T.H. (1991) *The Biological Roots of Human Nature: Forging Links Between Evolution and Behavior*, New York: Oxford University Press.
Goldstein, S. (1992) 'Profile of American Jewry: Insights from the 1990 National Jewish Population Survey', in D. Singer (ed.) *American Jewish Yearbook 1992*, New York and Philadelphia: American Jewish Committee and the Jewish Publication Society.
Gould, S.J. (2002) *The Structure of Evolutionary Theory*, Cambridge, MA: Belknap Press of Harvard University.
Graf, D. (1986) 'Diagrams', *Perspecta*, 22: 43–57.
Graham, A. (1989) 'Railroad-tie architecture in Elko County, Nevada', in T. Carter and B.L. Herman (eds) *Perspectives in Vernacular Architecture*, vol. 3, Columbia: University of Missouri Press.
Greenfield, V. (1986) *Making Do or Making Art: A Study of American Recycling*, Ann Arbor: UMI Research Press.
Gregory, R.L. (1998) *The Oxford Companion to the Mind*, Oxford: Oxford University Press.
Griaule, M. (1949) 'L'Image du Monde au Sudan', *Journal de la Société des Africanists*, 19: 11.
—— (1975; 1st edn 1966) *Dien d'Eace: Entretiens avec Ogotemmêli*, Paris: Fayard.
Griaule, M. and Dieterlen, G. (1965) *Le Renard Pâle*, Paris: Institut d'Ethnologie.
Grillo, R.D. (1997) 'Discourses of development: The view from anthropology', in R.D. Grillo and R.L. Stirrat (eds) *Discourses of Development: Anthropological Perspectives*, Oxford: Berg.
Grunne, B. de (1995) 'An art historical approach to the terracotta figurines of the Inland Niger Delta', *African Arts*, 28 (4): 70–9.
Guidoni, E. (1975) *Primitive Architecture*, Milan: Electa.
Gulliver, P.H. (1975) 'Nomadic movements: causes and implications', in T. Monod (ed.) *Pastoralism in Tropical Africa*, London, Ibadan and Nairobi: Oxford University Press.
Gupta, R. (2000) 'Study on the optimization of the performance of vernacular housing in the Near East', unpublished dissertation, Oxford Brookes University.
Gutman, R. (1988) *Architectural Practice: A Critical View*, Princeton: Princeton Architectural Press.
Hakim, B.S. (1986) *Arabic-Islamic Cities: Building and Planning Principles*, London: KPI.
—— (1994) 'The "URF" and its role in diversifying the architecture of traditional Islamic cities', *Journal of Architectural and Planning Research*, 11 (2): 108–27.
—— (2001) 'Julian of Ascalon's treatise of construction and design rules from sixth-century Palestine', *Journal of the Society of Architectural Historians*, 60 (1): 4–25.
Hall, E.T. (1969) *The Hidden Dimension: Man's Use of Space in Public and Private*, London: Bodley Head.
Hannan, D.F. (1979) *Displacement and Development: Class, Kinship and Social Change in Irish Rural Communities*, Dublin: Economic and Social Research Institute.
Hannerz, U. (1989) 'Notes on the global ecumene', *Public Culture*, 1 (2): 66–75.
Hanson, J. (1998) *Decoding Homes and Houses*, Cambridge: Cambridge University Press.
Hasan, A. (1999) *Understanding Karachi*, Karachi: City Press.
Hasty, P., Campisi, J., Hooijmakers, J., Steeg, H. van and Vijg, J. (2003) 'Aging and genome maintenance: Lessons from the mouse', *Science*, 299 (5611): 1355–9.
Hatfield Dodds, S. (2000) 'Pathways and paradigms for sustaining human communities – some thoughts for community designers', in R. Lawrence (ed.) *Sustaining Human Settlement: A Challenge for the New Millennium*, Newcastle-upon-Tyne: Urban International Press.
He, S., Dong, W., Deng, Q., Weng, S. and Sum, W. (2003) 'Seeing more clearly: Recent advances in understanding retinal circuitry', *Science*, 302 (5644): 408–11.
Heath, K.W. (1988) 'Defining the nature of vernacular', *Material Culture*, 20: 1–8.

Bibliography

Heinrich, B. (2003) 'New buzz on the humble bee', *Science*, 302 (5644): 395–6.

Herdeg, K. (1983) *The Decorated Diagram: Harvard Architecture and the Failure of the Bauhaus Legacy*, Cambridge, MA: MIT Press.

Herman, B.L. (1985) 'Time and performance: Folk houses in Delaware', in S.J. Bronner (ed.) *American Material Culture and Folklife*, Ann Arbor: UMI Research Press.

Herrmann, W. (1962) *Laugier and the Eighteenth Century French Theory*, London: A. Zwemmmer.

—— (1966; 1st edn 1765) *Observations sur l'Architecture*, London: Gregg Press.

Hillier, B. (1996) *Space is the Machine: A Configurational Theory of Architecture*, Cambridge: Cambridge University Press.

Hillier, B. and Hanson, J. (1984) *The Social Logic of Space*, Cambridge: Cambridge University Press.

Hillier, B. and Leaman, A. (1976) 'Space syntax', *Environment and Planning Bulletin*, 3: 137–85.

Hirschfeld, Y. (1995) *Traditional Palestinian Dwelling in the Roman-Byzantine Period*, Jerusalem: Franciscan Press – Israel Exploration Society.

Hoare, A. (2002) 'Irish travellers: Landscape, society and the architecture of nomadic movement', unpublished MA dissertation, Oxford Brookes University.

Hobsbawn, E. and Ranger, T. (eds) (1983) *The Invention of Tradition*, Cambridge: Cambridge University Press.

Hoefel, F. and Elgar, S. (2003) 'Wave-induced sediment transport and sandbar migration', *Science*, 299 (5614): 1885–7.

Hostetler, J.A. (1963) *Amish Society*, Baltimore: Johns Hopkins University Press.

Hubka, T.C. (1984) *Big House, Little House, Back House, Barn: The Connected Farm Buildings of New England*, Hanover, NH: University Press of New England.

—— (1994) 'The Americanization of the barn', *Blueprints*, 12: 2–7.

—— (2003) *Resplendent Synagogue: Architecture and Worship in an Eighteenth-Century Polish Community*, Hanover, NH: University Press of New England.

Humphrey, C. and Sneath, D. (1999) *The End of Nomadism?*, Durham NC: Duke University Press.

Hymes, D. (1972) 'The contribution of folklore to sociolinguistic research', in A. Paredes and R. Bauman (eds) *Toward New Perspectives in Folklore*, Austin: University of Texas Press.

Ingold, T. (1986) *The Appropriation of Nature*, Manchester: Manchester University Press.

International Union for the Conservation of Nature (1980) *World Conservation Strategy*, New York: United Nations Environment Programme.

Irons, W. (1974) 'Nomadism as a political adaptation: The case of the Yomut Turkmen', *American Ethnologist* 1 (4): 635–58.

Jakobson, R. and Bogatyrev, P. (1980) 'Folklore as a special form of creation', *Folklore Forum*, 13: 1–21.

Jean-Baptiste, P. and Ducroux, R. (2003) 'Energy policy and climate change', *Energy Policy*, 74 (4): 155–66.

Jencks, C. and Silver, N. (1972) *Adhocism: The Case for Improvisation*, New York: Doubleday.

Johnson, D.L. (1969) *The Meaning of Nomadism* (Research Paper 118, Department of Geography), Chicago: University of Chicago.

Johnson-Laird, P.N. (1988) *The Computer and the Mind: An Introduction to Cognitive Science*, Cambridge, MS: Harvard University Press.

Jolly, M. (1992) 'Specters of inauthenticity', *Contemporary Pacific*, 4: 49–72.

Jones, M.O. (1987) *Exploring Folk Art: Twenty Years of Thought on Craft, Work, and Aesthetics*, Logan: Utah State University Press.

—— (1989) *Craftsman of the Cumberlands: Tradition and Creativity*, Lexington: University Press of Kentucky.

—— (1993) 'Why take a behavioral approach to folk objects?', in S. Lubar and W.D. Kingery (eds) *History from Things: Essays on Material Culture*, Washington, DC: Smithsonian Institution Press.

—— (1994) 'How do you get inside the art of outsiders?', in M.D. Hall and E.W. Metcalf Jr (eds) *The Artist Outsider: Creativity and the Boundaries of Culture*, Washington, DC: Smithsonian Institution Press.

—— (1995) 'Why make (folk) art?', *Western Folklore*, 54: 253–76.

—— (1997) 'How can we apply event analysis to material behavior, and why should we?', *Western Folklore*, 56: 199–214.

—— (2000) '"Tradition" in identity discourses and an individual's symbolic construction of self', *Western Folklore*, 59: 115–42.

Bibliography

—— (2001) 'The aesthetics of everyday life', in C. Russell (ed.) *In Self-Taught Art: The Culture and Aesthetics of American Vernacular Art*, Jackson, MS: University Press of Mississippi.
Jung, C.G. (1922) 'On the relation of analytical psychology to poetry', *The Spirit in Man, Art and Literature*, 15: 65–83.
Kahn, Ll. (2004) *Home Work: Handbuilt Shelter*, Bolinas: Shelter Publications.
Kandel, E.R. and Squire, L.R. (2001) 'Neuroscience: Breaking down scientific barriers to the study of brain and mind', in A.R. Damasio *et al.* (eds) *Unity of Knowledge: The Convergence of Natural and Human Science* (Annals, NYAS, 935) New York: New York Academy of Science.
Kant, I. (1952) *Critique of Judgement*; trans. J. Creed Meredith, Oxford: Clarendon Press.
Kausarul Islam, A.K.M. (2003) 'Patterns and changes of vernacular architecture in Bangladesh: An application of Amos Rapoport's theory of defining vernacular design', unpublished dissertation, Royal Institute of Technology, Stockholm.
Keech McIntosh, S. (ed.) (1995) *Excavations at Jenné-Jeno, Hambarketolo, and Kaniana: Inland Niger Delta, Mali: The 1981 Season*, Berkeley: University of California Press.
Kehoe, A.B. (2002) 'Theaters of power', in M. O'Donovan (ed.) *The Dynamics of Power* (Center for Archaeological Investigations, Occasional Paper 30), Carbondale: Southern Illinois University Press.
Kent, S. (1990) *Domestic Architecture and the Use of Space*, Cambridge: Cambridge University Press.
Khammash, A. (1986) *Notes on Village Architecture in Jordan*, Lafayette, LA: University Art Museum, University of Southern Louisiana.
Khan, A.H. (1996) *Orangi Pilot Project: Reminiscences and Reflections*, Karachi: Oxford University Press.
King, A.D. (1976) *Colonial Urban Development: Culture, Social Power and Environment*, Boston: Routledge & Kegan-Paul.
—— (1984) *The Bungalow: The Production of a Global Culture*, Boston: Routledge & Kegan Paul.
—— (1995) *The Bungalow: The Production of a Global Culture*, 2nd edn, New York and Oxford: Oxford University Press.
Kingston, K. (1996) *Creating Sacred Space with Feng Shui*, London: Judy Piatkus.
Kirshenblatt-Gimblett, B. (1983) 'The future of folklore studies in America: The urban frontier', *Folklore Forum*, 16: 175–234.
—— (1995) 'From the paperwork empire to the paperless office: Testing the limits of the "science of tradition"', in R. Bendix and R. Lévy Zumwalt (eds) *Folklore Interpreted: Essays in Honor of Alan Dundes*, New York: Garland.
Kopytoff, I. (1986) 'The cultural biography of things: Commodification as process', in A. Appadurai (ed.) *The Social Life of Things*, Cambridge: Cambridge University Press.
Kosslyn, S.M. (1994) *Image and Brain: The Resolution of the Imagery Debate*, Cambridge, MS and London: MIT Press.
Kraybill, D. (2001) *The Riddle of Amish Culture*, revised edn, Baltimore: Johns Hopkins University Press.
Kraybill, D. and Nolt, S.M. (1995) *Amish Enterprise: From Plows to Profits*, Baltimore: Johns Hopkins University Press.
Krinsky, C.H. (1996) *Contemporary Native American Architecture: Cultural Regeneration and Creativity*, New York: Oxford University Press.
Kristeller, P.O. (1983) '"Creativity" and "tradition"', *Journal of the History of Ideas*, 44: 105–14.
Lacey, M. (2003) 'With all the little wars, big peace is elusive', *New York Times*, 9 April 2003.
Lagopoulous, A.-P. (1975) 'Semiological urbanism: An analysis of the traditional western Sudanese settlement', in P. Oliver (ed.) *Shelter, Sign and Symbol*, London: Barrie & Jenkins.
Larson, M.S. (1993) *Behind the Postmodern Façade*, Berkeley: University of California Press.
Latter, R. and Oliver, P. (eds) (1997–2004) 'International Studies in Vernacular Architecture', unpublished course handbook, Oxford Brookes University.
—— (eds) (2002) 'Delving Valldigna: Cultural continuity of a Mediterranean valley', unpublished report, Oxford Brookes University.
—— (eds) (2003) 'Accommodating cultures: Supporting home building in development contexts', unpublished report, Oxford Brookes University.
Laughlin, C. and Brady, I. (eds) (1978) *Extinction and Survival in Human Populations*, New York: Columbia University Press.

Bibliography

Laugier, M.A.P. (1966; 1st edn 1753; 2nd edn 1755) *Essai sur l'Architecture*, Farnborough: Gregg Press.

Lave, J. and Wenger, E. (1991) *Situated Learning: Legitimate Peripheral Participation*, Cambridge: Cambridge University Press.

Lawrence, R. (1985) 'A more humane history of homes', in C.M. Werner, I. Altman and D. Oxley (eds) *Home Environments*, vol. 8, New York: Plenum Press.

—— (1987) *Housing, Dwelling and Homes: Design, Research and Practice*, Chichester: John Wiley & Sons.

—— (1990) 'Learning from colonial houses and lifestyles', in M. Turan (ed.) *Vernacular Architecture: Paradigms of Environmental Response*, Aldershot: Avebury.

—— (1995) 'Meeting the challenge: Barriers to integrate cross-sectoral urban policies', in M. Rolén (ed.) *Urban Policies for an Environmentally Sustainable World*, Stockholm: Swedish Council for Planning and Co-ordination of Research.

—— (1996) 'Urban environment, health and the economy: Cues for conceptual clarification and more effective policy implementation', in C. Price and A. Tsouros (eds) *Our Cities, Our Future: Policies and Action Plans for Health and Sustainable Development*, Copenhagen: World Health Organization Regional Office for Europe.

—— (1999) 'House, form and culture: What have we learnt in 30 years?'. Unpublished draft.

—— (ed.) (2000) *Sustaining Human Settlement: A Challenge for the New Millennium*, Newcastle-upon-Tyne: Urban International Press.

—— (2001) 'Human ecology', in M.K. Tolba (ed.) *Our Fragile World: Challenges and Opportunities for Sustainable Development*, vol. 1, Oxford: Eolss Publishers.

Le Corbusier (1987) *The Journey to the East*; edited and annotated by Ivan Zaknic; trans. Ivan Zaknic in collaboration with Nicole Pertuiset, Cambridge, MA: MIT Press.

Lewcock, R.B. (1975) 'The boat as symbol of the house', in P. Oliver (ed.) *Shelter, Sign and Symbol*, London: Barrie & Jenkins.

Lewis-Williams, J.D. (2002a) *The Mind in the Cave: Consciousness and the Origins of Art*, London: Thames and Hudson.

—— (2002b) *A Cosmos in Stone*, Walnut Creek, CA: Altamira Press.

Liao, C.J. and Cech, I.I. (1977) 'Effect of abrupt exposure to outdoor heat on humans accustomed to air-conditioning', *Urban Ecology*, 2: 355–70.

Ligers, Z. (1964) *Les Sorko (Bozo): Maîtres du Niger*, Paris: Librarie des Cinq Continents.

Lobell, M. (1983) 'Spatial Archetypes', *ReVISION* 6–2: 69–82.

Lockwood, W.G. (1986) 'East European gypsies in western Europe: The social and cultural adaptation of the Xoraxané', *Nomadic Peoples*, 21–2: 63–70.

Lockwood, Y.R. (1984) 'The joy of labor', *Western Folklore*, 43: 202–11.

Loevenstein, H.M., Berliner, P.R. and Keulen, H. van (1991) 'Runoff agroforestry in arid lands', *Forest Ecology and Management*, 45 (1–4): 59–70.

Logsdon, G. (1989) 'The barn raising', in J.A. Hostetler (ed.) *Amish Roots*, Baltimore: Johns Hopkins University Press.

Lomax, J.F. (1985) 'Some people call this art', in F.E. Abernethy (ed.) *Folk Art in Texas*, Dallas: Southern Methodist University Press.

Lopreato, J. and Crippen, T. (1999) *Crisis in Sociology: The Need for Darwin*, New Brunswick, NJ: Transaction Publishers.

Low, S. and Chambers, E. (eds) (1989) *Housing Culture and Design: A Comparative Perspective*, Pennsylvania: University of Pennsylvania Press.

Low, S.M. and Ryan, W.P. (1985) 'Noticing without looking: A methodology for the integration of architectural and local perceptions in Oley, Pennsylvania', *Journal of Architectural and Planning Research*, 2 (1): 3–22.

Lynch, K. (1981) *Good City Form*, Cambridge, MA: MIT Press.

Lyon, B.E. (2003) 'Egg recognition and counting reduce costs of avian conspecific brood parasitism', *Nature*, 422 (6931): 495–9.

Maas, P. and Mommersteeg, G. (1992) *Djenne: Chef-d'Oeuvre Architectural*, Bamako: Institut des Sciences Humaines and Amsterdam: Institut Royal des Tropiques.

McCoy, E. (1974) 'Grandma Prisbrey's bottle village', in Walker Art Center (ed.) *Naives and Visionaries*, New York: E.P. Dutton.

Bibliography

McCullough, M. (1996) *Abstracting Craft: The Practiced Digital Hand*, Cambridge, MA: MIT Press.
McIntosh, R. and Keech McIntosh, S. (1995) 'Background to the 1981 Research', in S. Keech McIntosh (ed.) *Excavations at Jenné-Jeno, Hambarketolo, and Kaniana (Inland Niger Delta, Mali): The 1981 Season*, Berkeley: University of California Press.
McMichael, A. (1993) *Planetary Overload: Global Environmental Change and the Health of the Human Species*, Cambridge: Cambridge University Press.
Macsai, J. (1985) 'Architecture as opposition', *Journal of Architecture Education*, 38 (4): 8–14.
Mangin, W. (1967) 'Latin American squatter settlements: A problem and a solution', *Latin American Research Review*, 2 (3): 65–98.
—— (1970) *Peasants in Cities: Readings in the Anthropology of Urbanization*, New York: Houghton Mifflin.
Magnuson, C. (1999) 'Aesthetics and grassroots art: A folklorist's perspective', in B. Brackman and C. Dwigans (eds) *Backyard Visionaries: Grassroots Art in the Midwest*, Lawrence: University Press of Kansas.
Mandler, G. (1975) *Mind and Emotion*, New York: Wiley.
Manley, R. (1989) *Signs and Wonders: Outsider Art Inside North Carolina*, Raleigh: North Carolina Museum of Art.
Marans, R. and Stokols, D. (eds) (1993) *Environmental Simulation: Research and Policy Issues*, New York: Plenum Press.
Marchand, T.H.J. (2001) *Minaret Building and Apprenticeship in Yemen*, London: Curzon.
Marcus, C. Cooper (1995) *House as a Mirror of Self*, Berkeley: Conari Press.
Marcuse, P. (1998) 'Sustainability is not enough', *Environment and Urbanization*, 10 (2): 103–11.
Marshall, H.W. (1982) *Missouri Artist Jesse Howard, with a Contemplation on Idiosyncratic Art*, Columbia: University of Missouri.
Marshall, Y. (2000) 'Transformations of Nuu-chah-nulth houses', in R.A. Joyce and S.D. Gillespie (eds) *Beyond Kinship: Social and Material Reproduction in House Societies*, Philadelphia: University of Pennsylvania Press.
Martin, C.E. (1983) 'Howard Acree's chimney: The dilemma of innovation', *Pioneer America: Journal of Historic American Material Culture*, 15: 35–50.
Marx, E. (1978) 'Ecology and politics of Middle Eastern pastoralists', in W. Weissleder (ed.) *The Nomadic Alternative*, The Hague and Paris: Mouton Publishers.
Marx, J. (2003) 'Building better mouse models for studying cancer', *Science*, 299 (5615): 1972–5.
Mathews, M.M. (1966) *Americanisms*, Chicago: University of Chicago Press.
Mayer, E., Kosmin, B. and Keysar, A. (2003) *American Jewish Identity Survey 2001*, New York: Center for Cultural Judaism.
Meir, I.A. (2000) 'Courtyard microclimate: A hot arid region case study', in K. Steemers and S. Yannas (eds) *Architecture-City-Environment: Proceedings of the 17th PLEA International Conference*, London: James & James.
—— (2002) 'Building technology in the Negev in the Byzantine Period (4–7 c. CE) and its adaptation to the desert environment' (in Hebrew), unpublished thesis, Ben-Gurion University.
Meir, I.A. and Gilead, I. (2002) 'Underground dwellings and their microclimate under arid conditions', in GRECO and ACAD (eds) *Design With the Environment: Proceedings of the 19th PLEA International Conference, Toulouse*, vol. 2, Toulouse: Ecole d'Architecture.
Meir, I.A. and Roaf, S.C. (2002) 'Thermal comfort, thermal mass: Housing in hot dry climates', in H. Levin (ed.) *Indoor Air 2002: Proceedings of the 9th International Conference Indoor Air Quality and Climate*, Santa Cruz, CA.
Meir, I.A., Gilead, I., Runsheng, T., Mackenzie Bennett, J. and Roaf, S.C. (2003) 'A parametric study of traditional housing types from the Middle East', in K.W. Tham, C. Sekhar and D. Cheong (eds) *Energy-Efficient Healthy Buildings: Proceedings of the 7th HB International Conference*, vol. 2, Singapore: NUS and ISIAQ.
Meir, I.A., Mackenzie Bennett, J. and Roaf, S.C. (2001) 'Learning from the past, shaping the future: Combining archaeology and simulation tools to teach building physics and appropriate solutions', in Pereira F.O.R et al. (eds) *Renewable Energy for a Sustainable Development of the Built Environment: Proceedings of the 18th PLEA International Conference Florianopolis, Brazil*, vol. 2, Florianapolis, Brazil: Org. Committee of PLEA 2001.
Meir, I.A., Pearlmutter, D. and Etzion, Y. (1995) 'On the microclimatic behavior of two semi-enclosed attached courtyards in a hot dry region', *Building and Environment*, 30 (4): 563–72.

Bibliography

Middleton (2001) 'Troubled waters', *Building Design*, October 2001: 18–25.
Mihailović, K. (1989) 'Tradition and industrialization', in S. Gustavsson (ed.) *Tradition and Modern Society*, Stockholm: Almqvist & Wiksell.
Mill, J.S. (1969; 1st edn 1873) *Autobiography*, Boston: Houghton Mifflin.
Miller, D. (1995) 'Consumption studies as the transformation of anthropology', in D. Miller (ed.) *Acknowledging Consumption: A Review of New Studies*, London and New York: Routledge.
Milspaw, Y.J. (1983) 'Reshaping tradition: Changes to Pennsylvania German folk houses', *Pioneer America*, 15: 67–84.
Ministry of Housing and Local Government Report (1969a) *Family Houses at West Ham: An Account of the Project with an Appraisal*, London: HMSO.
Ministry of Housing and Local Government Report (1969b) *The Family at Home: A Study of Houses in Sheffield*, London: HMSO.
Mithen, S. (1996) *The Prehistory of the Mind: The Cognitive Origins of Art and Science*, London: Thames & Hudson.
Monteil, C. (1932) *Une Cité Soudanaise: Djénné, Métropole du Delta Central du Niger*, Paris: Édition Anthropos.
Morell, V. (2002) 'Placentas may nourish complexity studies', *Science*, 298 (5595): 945.
Morgan, L.H. (1965; 1st edn 1881) *Houses and House Life of the American Aborigines*, Chicago: University of Chicago Press.
Morgenthaler, F. (1977) 'Reflex-modernization in tribal societies', in P. Oliver (ed.), *Shelter, Sign and Symbol*, New York: The Overlook Press.
Morris, A.E.J. (1972; 2nd edn 1979; 3rd edn 1994) *History of Urban Form: Before the Industrial Revolutions*, Harlow: Longman.
Munro, M. and Madigan, R. (1999) 'Negotiating space in the family home', in I. Cieraad (ed.) *At Home: An Anthropology of Domestic Space*, New York: Syracuse.
Nature (2003) 'Crick's immodest ambitions', 422 (6931): 455.
Nazaroff, W.W., Weschler, C.J. and Corsi, R.L. (2003) 'Indoor air chemistry and physics', *Atmospheric Environment*, 37 (39–40): 5451–3.
Negev, A. (1980) 'House and city planning in the ancient Negev and the Provincia Arabia', in G. Golany (ed.) *Housing in Arid Lands: Design and Planning*, London: The Architectural Press.
Nemeth, D.J. (1987) *The Architecture of Ideology: Neo-Confucian Imprinting on Cheju Island, Korea* (Geography 26), Berkeley: University of California Publication in Geography.
Netting, R. (1981) *Balancing on an Alp: Ecological Change and Continuity in a Swiss Mountain Community*, New York: Cambridge University Press.
Nicolaisen, W.F.H. (1979) '"Distorted function" in material aspects of culture', *Folklore Forum*, 12: 223–36.
Niles, S.A. (1997) *The Dickeyville Grotto: The Vision of Father Mathias Wernerus*, Jackson, MS: University Press of Mississippi.
Nitschke, G. (1964a) 'The Metabolists of Japan', *Architectural Design*, 34: 509–24.
—— (1964b) 'The work of Kikutake', *Architectural Design*, 34: 608–11.
—— (1966) 'Ma, the Japanese sense of place', *Architectural Design*, 36: 115–56.
Noble, A.G. (1984) *Wood, Brick, and Stone: The North American Settlement Landscape*, 2 vols, Amherst: University of Massachusetts Press.
Norton, J. (1997) 'Woodless constructions: Unstabilised earth brick vaulted dome roofing without formwork', *Building Issues*, 9 (2): 3–26.
Nuñez, L., Grosjean, M. and Cartajena, L. (2002) 'Human occupations and climate change in the Puna de Atacama, Chile', *Science*, 298 (5594): 821–4.
Odum, H.T. and Odum, E.C. (2004) 'The prosperous way down', *Energy*. Online. Available HTTP: http://www.sciencedirect.com/science?.
Ohrn, S. (1984) 'Faith into stone: Grottoes and monuments', in S. Ohrn (ed.) *Passing Time and Traditions: Contemporary Iowa Folk Artists*, Ames: Iowa State University Press.
Okely, J. (1983) *The Traveller-Gypsies*, Cambridge: Cambridge University Press.
Oliver, P. (ed.) (1969) *Shelter and Society*, London: Barrie & Rockliff.
—— (ed.) (1971) *Shelter in Africa*, London: Barrie & Jenkins.
—— (ed.) (1975) *Shelter, Sign and Symbol*, London: Barrie & Jenkins.
—— (1981) 'The cultural context of shelter provision', in I. Davis (ed.) *Disasters and the Small Dwelling*, Pergamon Press: Oxford.

Bibliography

—— (1979) 'The anthropology of shelter', in M. Keniger (ed.) *Market Profiles*, Brisbane: Architectural Education Conference.
—— (1984) 'Round the houses', in A. Papadakis (ed.) *British Architecture, 1984*, London: Architectural Design.
—— (1986) 'Vernacular know-how', *Material Culture*, 18: 113–26.
—— (1987) *Dwellings: The House Across the World*, Oxford: Phaidon.
—— (1989) 'Handed down architecture: Tradition and transmission', in J.-P. Bourdier and N. AlSayyad (eds) *Dwellings, Settlements and Tradition: Cross-Cultural Perspectives*, Lanham, MD: University Press of America.
—— (1990) 'Vernacular know-how', M. Turan (ed.) *Vernacular Architecture: Paradigms of Environmental Response*, Aldershot: Avebury.
—— 'Culture, dwellings and design', paper presented for an International Study Seminar on Design and New Strategies for the Contemporary Residence, March 1992.
—— (ed.) (1997a) *Encyclopedia of Vernacular Architecture of the World*, 3 vols, Cambridge: Cambridge University Press.
—— (1997b) 'Introduction', in P. Oliver (ed.) *Encyclopedia of Vernacular Architecture of the World*, Cambridge: Cambridge University Press.
—— (1997c) 'Tradition and transmission', in P. Oliver (ed.) *Encyclopedia of Vernacular Architecture of the World*, Cambridge: Cambridge University Press.
—— (1999) 'Vernacular architecture in the twenty-first century', Hepworth Lecture, London: Prince of Wales Institute.
—— (2003) *Dwellings: The Vernacular House World Wide*, London: Phaidon.
Oliver, P. and Doumanis, O. (1975) *Shelter in Greece*, Athens: Architecture Greece.
Oliver, P. and Vellinga, M. (forthcoming) *Atlas of Vernacular Architecture of the World*, London: Routledge.
Oliver, P., Davis, I. and Bentley, I. (1981) *Dunroamin: The Suburban Semi and its Enemies*, London: Barrie & Jenkins.
Ó Tuathaigh, G.G. (1998) 'Life on the land', in K.A. Kennedy (ed.) *From Famine to Feast: Economic and Social Change in Ireland 1847–1997*, Dublin: Institute of Public Administration.
Panofsky, E. (1948) 'Gothic architecture and scholasticism', Wimmer Lecture, Latrobe, PA: Archabbey Press.
Park, R., Burgess, E. and McKenzie, R. (1925) *The City*, Chicago: Chicago University Press.
Payne, G. (1968) 'Japan Rebuilds', *Arena*, 83 (921): 19–33.
—— (1969) 'Housing and urban growth', *Official Architecture and Planning*, 32 (4): 427–38.
—— (1977) *Urban Housing in the Third World*, Boston: Leonard Hill and London: Routledge & Kegan Paul.
Pearlmutter, D. (1993) 'Roof geometry as a determinant of thermal behavior: A comparative study of vaulted and flat roof surfaces in an arid zone', *Architectural Science Review*, 36 (2): 75–86.
—— (1998) 'Street canyon geometry and microclimate: Designing for urban comfort under arid conditions', in E. Maldonado and S. Yannas (eds) *Environmentally Friendly Cities: Proceedings of the 14th PLEA International Conference*, London: James & James.
Pearlmutter, D. and Meir, I.A. (1995) 'Assessing the climatic implications of lightweight housing in a peripheral arid region', *Building and Environment*, 30 (3): 441–51.
Pearson, B.J. and Doe, C.Q. (2003) 'Regulation of neuroblast competence in Drosophila', *Nature*, 425 (6958): 624–8.
Peeters, A. and Meir, I.A. (2002) 'More than a hole in the ground? Fusing object and context in subterranean architecture', *Open House International*, 25 (3): 47–57.
Pegram, J. (2002) *The Vastu Home*, London: Duncan Baird Publishers.
Peifer, M. (2002) 'Colon construction', *Nature*, 410 (6913): 274–6.
Peters, W. (1997) 'Nama', in P. Oliver (ed.) *Encyclopedia of Vernacular Architecture of the World*, Cambridge: Cambridge University Press.
Pinker, S. (1997) *How the Mind Works*, New York: W.W. Norton.
Posen, I. Sheldon and Ward, D.F. (1985) 'Watts Towers and the "Giglio" tradition', in A. Jabbour and J. Hardin (eds) *Folklife Annual 1985*, Washington, DC: Library of Congress.
Price, H.W. and Walters Jr, W.D. (1989) 'Barn raising at Metamora: A photographic essay', *Material Culture*, 21: 47–56.
Prussin, L. (1970) 'Sudanese architecture and the Manding', *African Arts*, 3 (4): 13–19, 64–7.

Bibliography

—— (1986) *Hatumere: Islamic Design in West Africa*, Berkeley: University of California Press.
—— (1995) *African Nomadic Architecture: Space, Place and Gender*, Washington, DC: Smithsonian Institution Press.
—— (1996) 'When nomads settle', in M.J. Arnoldi, C.M. Geary and K.L. Hardin (eds) *African Material Culture*, Bloomington and Indianapolis: Indiana University Press.
Putnam, R.D. (2000) *Bowling Alone: The Collapse and Revival of American Community*, New York: Touchstone.
Quatremere de Quincy, A.C. (1999; 1st edn 1789) *The True, the Fictive and the Real: The Historical Dictionary of Architecture of Quatremere de Quincy*, trans. S. Younes, London: Papadakis.
Ragette, F. (1980) *Architecture in Lebanon: The Lebanese House During the Eighteenth and Nineteenth Centuries*, Delmar and New York: Caravan Books.
Ramirez, M. (2005) 'Architecture class to create shelters for tsunami victims', *Office of International Communications*, Houston: University of Houston.
Rao, A. (1987) 'The concept of peripatetics: An introduction', in A. Rao (ed.) *The Other Nomads*, Köln: Böhlau Verlag.
Rapoport, A. (1969) *House, Form and Culture*, Englewood Cliffs, NJ: Prentice-Hall.
—— (1977) *Human Aspects of Urban Form*, Oxford: Pergamon Press.
—— (1979) 'An approach to designing Third World environments', *Third World Planning Review*, 1 (1): 23–40.
—— (1980) 'Cross-cultural aspects of environmental design', in I. Altman, A. Rapoport and J.F. Wohlwill (eds) *Human Behaviour and Environment*, vol. 4, New York: Plenum Press.
—— (1982a) 'An approach to vernacular design', in J.M. Fitch (ed.) *Shelter: Models of Native Ingenuity*, Katonah, NY: Katonah Gallery.
—— (1982b) 'Urban design and human systems: On ways of relating buildings to urban fabric', in P. Laconte, J. Gibson and A. Rapoport (eds) *Human and Energy Factors in Urban Planning: A Systems Approach* (NATO Advanced Institute Series, series D: Behavioral and Social Sciences 12), The Hague: Nijhoff.
—— (1983a) 'Development, culture-change and supportive design', *Habitat International*, 7 (5–6): 249–68.
—— (1983b) 'The effect of environment on behavior', in J.B. Calhoun (ed.) *Environment and Population: Perspectives on Adaptation, Environment and Population*, New York: Praeger.
—— (1984) 'Culture and the urban order', in J. Agnew, J. Mercer and D.E. Sopher (eds) *The City in Cultural Context*, London: Allen & Unwin.
—— (1987) 'Learning about settlements and energy from historical precedents', *Ekistics*, 54 (325–7): 262–8.
—— (1988a) 'Spontaneous settlements as vernacular design', in C.V. Patton (ed.) *Spontaneous Shelter: International Perspectives and Prospects*, Philadelphia: Temple University Press.
—— (1988b) 'Levels of meaning in the built environment', in F. Poyatos (ed.) *Cross-Cultural Perspectives in Non-Verbal Communication*, Toronto: H.J. Hogrefe.
—— (1989) 'On the attributes of tradition', in J.-P. Bourdier and N. AlSayyad (eds) *Dwellings, Settlements and Tradition: Cross-Cultural Perspectives*, Lanham, MD: University Press of America.
—— (1990a) *History and Precedent in Environmental Design*, New York: Plenum.
—— (1990b) *The Meaning of the Built Environment*, Tucson: University of Arizona Press.
—— (1990c) 'Defining vernacular design', in M. Turan (ed.) *Vernacular Architecture: Paradigms of Environmental Response*, Aldershot: Avebury.
—— (1990d) 'Levels of meaning and types of environments', in Y. Yoshitake et al. (eds) *Current Issues in Environment-Behavior Research: Proceedings of the Third Japan–US seminar, Kyoto 1990*, Tokyo: University of Tokyo.
—— (1990e) 'Systems of activities and systems of settings', in S. Kent (ed.) *Domestic Architecture and the Use of Space: An Interdisciplinary Cross-Cultural Study*, Cambridge: Cambridge University Press.
—— (1990f) 'Science and the failure of architecture: An intellectual history', in I. Altman and K. Christensen (eds) *Environment and Behavior Studies: Emergence of Intellectual Traditions*, vol. 11, New York, Plenum.
—— (1992) 'On regions and regionalism', in N.C. Markovich, W.F.E. Preiser and F.G. Sturm (eds) *Pueblo Style and Regional Architecture*, New York: Van Nostrand Reinhold.

Bibliography

—— (1993a) 'On cultural landscapes', *Traditional Dwellings and Settlements Review*, 3 (2): 33–47.
—— (1993b) 'On the nature of capitals and their physical expression', in J. Taylor, J.G. Lengellé and C. Andrew (eds) *Capital Cities: International Perspectives/Les Capitals: Perspectives Internationales*, Ottawa: Carleton University Press.
—— (1993c) *Cross-Cultural Studies and Urban Form*, College Park, MD: University of Maryland, Urban Studies and Planning Program.
—— (1994) 'Sustainability, meaning and traditional environments', IASTE Working Papers Series, vol. 75, Berkeley: Center of Environmental Design Research, University of California.
—— (1995a) 'Settlements and energy: Historical precedents', in A. Rapoport, *Thirty-Three Papers in Environmental-Behavior Research*, Newcastle: Urban International Press.
—— (1995b) 'Environmental quality and environmental quality profiles', in A. Rapoport *Thirty-Three Papers in Environmental-Behavior Research*, Newcastle: Urban International Press.
—— (1995c) 'Culture and built form: A reconsideration', in A. Rapoport, *Thirty-Three Papers in Environmental-Behavior Research*, Newcastle: Urban International Press: 399–436.
—— (1995d) 'Flexibility, open-endedness and design', in A. Rapoport, *Thirty-Three Papers in Environmental-Behavior Research*, Newcastle: Urban International Press: 529–62.
—— (1995e) 'Culture and environment', in A. Rapoport, *Thirty-Three Papers in Environmental-Behavior Research*, Newcastle: Urban International Press: 269–82.
—— (1997) 'Theory in environment-behavior studies: Transcending times, settings and groups', in S. Wapner *et al.* (eds) *Handbook of Japan-US Environment-Behavior Research*, New York: Plenum.
—— (1998) 'Using "culture" in housing design', *Housing and Society*, 25 (1–2): 1–20.
—— (1999a) 'A framework for studying vernacular design', *Journal of Architecture and Planning Research*, 16 (1): 52–64.
—— (1999b) 'Archaeological inference and environment-behavior studies', in F. Braemer *et al.* (eds) *Habitat et Societiété*, Antibes: Éditions APOCA.
—— (1999–2000) 'On the perception of urban landscapes', *Urban Design Studies*, 5–6: 129–48.
—— (2000a) 'Theory, culture and housing', *Housing, Theory and Society*, 17 (4): 145–65.
—— (2000b) 'Science, explanatory theory and environment-behavior studies', in S. Wapner *et al.* (eds.) *Theoretical Perspectives in Environment-Behavior Research*, New York: Kluwer Academic/Plenum Publishers.
—— (2001) 'Architectural anthropology or environment-behavior studies', in M-J. Amerlinck (ed.) *Architectural Anthropology*, Westport, CT: Bergin and Garvey.
—— (2002a) 'On the size of cultural groups', *Open House International*, 27 (3): 7–11.
—— (2002b) 'Environment-behavior research in an Asian-Pacific context', *Journal of Asian Urban Studies*, 3 (3): 17–20.
—— (2002c) 'The role of neighborhoods in the success of cities', *Ekistics*, 69 (412–14): 145–51.
—— (2004a) *Culture, Architecture and Design*, Chicago: Locke Science Publishing.
—— (2004b) 'How can we use "culture" in analysis and design', *Ekistics*.
Rapoport, A. and Watson, N. (1972) 'Cultural variability in physical standards', in R. Gutman (ed.) *People and Buildings*, New York: Basic Books.
Rapoport, M. (2003) 'Nomads within their homes: The forces that shape intramural migration in arid climates', in W.G. Bustamante and E.B. Collados (eds) *Rethinking Development: Proceedings of the PLEA 2003 International Conference Santiago: Escuela de Construction Civil, Pontifica Universidad Catolica de Chile*.
Redfield, R. (1947) 'The folk society', *American Journal of Sociology*, 52: 293–308.
—— (1960) *The Little Community and Peasant Society and Culture*, Chicago: University of Chicago Press.
Restak, R.M. (1985) *The Brain*, Toronto and New York: Bantam Books.
RIBA (2003) *Criteria for Validation*, London: RIBA Publications.
Rich, T. (2002) 'Sukkot', *Judais*, 101. Online. Available HTTP: http://www.jewfaq.org/holiday5.htm.
Riesman, D. (1961) *The Lonely Crowd*, New Haven: Yale University Press.
Rice, M. (2003) *Village Buildings of Britain*, London: Time Warner Books.
Roaf, S. (1989) 'The windcatchers of Yazd', unpublished thesis, Oxford Polytechnic.
—— (1990) 'The traditional technology trap: Stereotypes of the Middle Eastern traditional building types and technologies', *Trialog*, 25: 26–33.

Roaf, S. with Crichton, D. and Nicol, F. (2004a) *Adapting Buildings and Cities to Climate Change*, Oxford: Architectural Press.

Roaf, S., Fuentes, M. and Gupta, R. (2003) *Ecohouse 2: A Design Guide*, Oxford: Architectural Press.

Roaf, S. with Horsley, A. and Gupta, R. (2004b) *Closing the Loop: Benchmarks for Sustainable Buildings*, London: RIBA Publications.

Rosen, S. (1979) *In Celebration of Ourselves*, San Francisco: California Living in association with the San Francisco Museum of Modern Art.

Rowntree, L.B. and Conkey, M.W. (1980) 'Symbolism and the cultural landscape', *Annals of the Association of American Geographers*, 70 (4): 459–74.

Rudofsky, B. (1964) *Architecture without Architects*, New York: Doubleday & Co.

—— (1977) *The Prodigious Builders*, New York: Harcourt Brace Jovanovich.

Runsheng, T., Meir, I.A. and Etzion, Y. (2003a) 'Thermal behavior of buildings with curved roofs as compared with flat roofs', *Solar Energy*, 74 (4): 273–86.

—— (2003b) 'An analysis of absorbed radiation by domed and vaulted roofs as compared with flat roofs', *Energy and Buildings*, 35 (6): 539–48.

Rykwert, J. (1988) 'On the oral transmission of architectural theory', in J. Guillaume (ed.) *Les Traites d'Architecture de la Renaissance*, Paris: Picard.

Rzadkiewicz, J. and J. Young (2003) 'The bottle houses of Edouard Arsenault', *Valence*. Online. Available HTTP: http://www3.sympatico.ca/valence/bottlehouses.

Sahlins, M. (1999) 'What is anthropological enlightenment? Some lessons of the twentieth century', *Annual Review of Anthropology*, 28: i-xxiii.

Salama, R. (1995) 'User transformation of government housing projects: Case study, Egypt', unpublished dissertation, McGill University.

Salo, M.T. (1986) 'Peripatetic adaptation in historical perspective', *Nomadic Peoples*, 21–2: 7–36.

Salzman, P. (1967) 'Political organization among nomadic peoples', *Proceedings of the American Philosophical Society* 111 (2): 115–31.

—— (1972) 'Multi-resource nomadism in Iranian Baluchistan', in W. Irons and N. Dyson-Hudson (eds) *Perspectives on Nomadism*, Leiden: E.J. Brill.

—— (ed.) (1980) *When Nomads Settle*, New York: Praeger Publishers.

—— (1995) 'Studying nomads: An autobiographical reflection', *Nomadic Peoples* 36–7: 157–66.

Santamouris, M. *et al.* (2001) 'On the impact of urban climate on the energy consumption of buildings', *Solar Energy*, 70 (3): 201–16.

Santino, J. (1986) 'The folk "assemblage" of autumn: Tradition and creativity in Halloween folk art', in J.M. Vlach and S.J. Bronner (eds) *Folk Art and Art Worlds*, Ann Arbor: UMI Research Press.

Sastrosasmita, S. and Nurul Amin, A.T.M. (1990) 'Housing needs of informal sector workers: The case of Yogyakarta, Indonesia', *Habitat International*, 14 (4): 75–88.

Sauerhaft, B., Berliner, P.R. and Thurow, T.L. (1998) 'The fuelwood crisis in arid zones: Runoff agriculture for renewable energy production', in H.J. Bruins and H. Lithwick (eds) *The Arid Frontier. Interactive Management of Environment and Development*, Dordrecht and London: Kluwer.

Schildkrout, E. (1978) 'Age and gender in Hausa Society: Socio-economic roles of children in urban Kano', in J.S. La Fontaine (ed.) *Sex and Age as Principles of Social Differentiation*, London: Academic Press.

Schopenhauer, A. (1966; 1st edn 1818) *The World as Will and Representation*; trans. E.F.J. Payne, New York: Dover Publications.

Science (2002a) 298 (5594).

—— (2002b) 298 (5595).

—— (2003a) 299 (5613).

—— (2003b) 302 (5644).

Sciorra, J. (1989) 'Yard shrines and sidewalk altars of New York's Italian Americans', in T. Carter and B.L. Herman (eds) *Perspectives in Vernacular Architecture*, vol. 3, Columbia: University of Missouri Press.

Segal, H. (1983) 'Kleinian analysis', in J. Miller (ed.) *States of Mind: Conversations with Psychological Investigators*, London: BBC.

Segerstråle, U. (2000) *Defenders of the Truth: The Sociobiology Debate*, New York: Oxford University Press.

Bibliography

Seltzer, D.J. (2000) 'Bottle houses', *Roadside Architecture*. Online. Available HTTP: http://www.agilitynut.com/bh.html.

Semper, G. (1851) *Die vier Elemente der Baukunst: Ein Beitrag zur Vergleichenden Baukunde*, Braunschweig.

Sewell, E.P. and Linck Jr, C.E. (1985) 'The rural mailbox', in F.E. Abernethy (ed.) *Folk Art in Texas*, Dallas: Southern Methodist University Press.

Shapiro, M. (1997) *Impressionism: Reflections and Perceptions*, New York: George Braziller.

Sheehy, C.J. (1998) *The Flamingo in the Garden: American Yard Art and the Vernacular Landscape*, New York: Garland.

Shils, E. (1971) *Tradition*, Chicago: University of Chicago Press.

Sider, G. (1986) *Culture and Class in Anthropology and History: A Newfoundland Illustration*, Cambridge: Cambridge University Press.

Silas, J. (1984) 'The Kampung Improvement Programme in Indonesia: A comparative case study of Jakarta and Surabaya', in G. Payne (ed.) *Low-Income Housing in the Developing World: The Role of Sites and Services and Settlement upgrading*, Chichester and New York: John Wiley.

Sillitoe, P., Bicker, A. and Pottier, J. (2002) *Participating in Development: Approaches to Indigenous Knowledge*, London: Routledge.

Silva, K.D. (2001) 'Advances in environmental cognition research 1980–2000', unpublished thesis, University of Wisconsin-Milwaukee.

Skinner, T. (2003) *Log and Timber Frame Homes*, Atglen, PA: Schiffer.

Smith, E.L. (1960) *The Amish Today: An Analysis of Their Beliefs, Behavior and Contemporary Problems*, Allentown: Pennsylvania German Society.

Smith, E.L. and Stoltzfus, G.M. (1959) 'The community barn-raising', *Historical Review of Berks County*, 24: 37–43.

Smith, K.R. (2003) 'The global burden of disease from unhealthy buildings: Preliminary results from comparative risk assessment', in K.W. Tham, C. Sekhar and D. Cheong (eds) *Energy-Efficient Healthy Buildings: Proceedings of the 7th HB International Conference*, vol. 1, Singapore: NUS and ISIAQ.

Sokolowski, M.B. (2002) 'Social eating for stress', *Nature*, 419 (6910): 893–4.

Sontag, S. (1982) 'Writing itself: On Roland Barthes', in S. Sontag (ed.) *Barthes: Selected Writings*, London: Fontana/Collins.

Sordinas, A. (1976) 'Traditional building materials in rural Corfu, Greece: A technological and socio-cultural analysis 1800–1950', Paper presented at 4th International Congress of Agricultural Museums, Reading, April.

Spalding, S. (1992) 'The myth of the classic slum: Contradictory perceptions of Boyle Heights Flats, 1900–1991', *Journal of Architectural Education*, 45 (2): 107–19.

Sphere Project (2004) *Humanitarian Charter and Minimum Standards in Disaster Response*, Oxford: Oxfam Publishing.

Spooner, B. (1972) 'The status of nomadism as a cultural phenomenon in the Middle East', in W. Irons and N. Dyson-Hudson (eds) *Perspectives on Nomadism*, Leiden: E.J. Brill.

Stanley, T. (1984) 'Two South Carolina folk environments', in D.F. Ward (ed.) *Personal Places: Perspectives on Informal Art Environments*, Bowling Green, OH: Bowling Green State University Popular Press.

Steemers, K. (2003) 'Energy and the city: Density, buildings and transport', *Energy and Buildings*, 35 (1): 3–14.

Steen, B., Steen, A. and Komatsu, E. (2003) *Built by Hand: Vernacular Buildings Around the World*, Salt Lake City: Gibbs Smith.

Stenning, D.J. (1994) *Savannah Nomads*, Munster and Hamburg: LIT Verlag and the International African Institute.

Stern, P. (2002) 'Neuroscience: A vibrant connection', *Science*, 298 (5594): 769.

Stevens, A. (1982) *Archetype: A Natural History of the Self*, London: Routledge & Kegan Paul.

Stevenson, G. (1991) *Common Property Economics: A General Theory and Land Use Applications*, Cambridge: Cambridge University Press.

Stone, L., Zanzi, J. and Iversen, E. (1999) 'In imitation of nature: Father P.M. Dobberstein's Grottoes in Iowa and Wisconsin', in B. Brackman and C. Dwigans (eds) *Backyard Visionaries: Grassroots Art in the Midwest*, Lawrence: University Press of Kansas.

Sunstein, C.R. (2003) 'Sober lemmings', *The New Republic*, 228 (4606): 34–7.

Syntichaki, P., Xu, K., Driscoll, M. and Tavernakaris, N. (2002) 'Specific aspartyl and calpain proteases are required for neurodegeneration in C. elegans', *Nature*, 419 (6910): 939–44.

Szabo, A. and Barfield, T.J. (1991) *Afghanistan: An Atlas of Indigenous Domestic Architecture*, Austin: University of Texas Press.

Tapper, R.L. (1979) 'The organization of nomadic communities of the Middle East', in L'Equipe écologie et anthropologie des sociétés pastorales (ed.) *Pastoral Production and Society*, Cambridge: Cambridge University Press.

Taylor, A. (1994) 'The wisdom of many and the wit of one', in W. Mieder and A. Dundes (eds) *The Wisdom of Many: Essays on the Proverb*, Madison: University of Wisconsin Press.

Theis, S. (1998) 'The drive to create: Three handmade personal spaces in Houston', *Folk Art Messenger*, 11 (2). Online. Available HTTP: http://www.folkart.org/mag/drive/drivetocreate.html.

Time (1976) 'Moss the Tentmaker', 26 July 1976: 60.

Tipple, G. (2000) *Extending Themselves: User-Initiated Transformations of Government-Built Housing in Developing Countries*, Liverpool: Liverpool University Press.

Tishler, W.H. (1982) 'Stovewood construction in the Upper Midwest and Canada: A regional vernacular architectural tradition', in C. Wells (ed.) *Perspectives in Vernacular Architecture*, Annapolis, Maryland: Vernacular Architecture Forum.

Tortora, V.R. (1980) *The Amish Folk of Pennsylvania Dutch Country: Their Life, Manners, Customs and Costumes*, Manheim, PA: Photo Arts Press.

Tuan, Y.F. (1977) *Space and Place: The Perspective of Experience*, London: Edward Arnold.

—— (1989) 'Traditional: What does it mean?', in J.-P. Bourdier and N. AlSayyad (eds) *Dwellings, Settlements, and Tradition: Cross-Cultural Perspectives*, Lanham, MD: University Press of America.

Tuleja, T. (ed.) (1997) *Usable Pasts: Traditions and Group Expressions in North America*, Logan: Utah State University Press.

Turan, M. (ed.) (1990) *Vernacular Architecture: Paradigms of Environmental Response*, Aldershot: Avebury.

Turner, J.F.C. 'Uncontrolled urban settlement: Problems and policies', paper presented at United Nations Seminar on Development Policies and Planning in Relation to Urbanization, Pittsburgh, PA, 1965.

—— (1967) 'Barriers and channels for housing development in modernizing countries', *American Institute of Planners*, 33 (3).

—— (1976) *Housing by People: Towards Autonomy in Building Environments*, London: Mario Boyars.

Turner, J.F.C and Fichter, R. (1972) *Freedom to Build: Dweller Control of the Housing Process*, New York: Macmillan.

Tyrwhitt, J. (1947) *Patrick Geddes in India*, London: Lund Humphries.

United Nations (1993) *Agenda 21: Programme of Action for Sustainable Development*, New York: United Nations.

—— (2001) *Johannesburg Summit 2002: World Summit on Sustainable Development*, New York: United Nations.

United Nations Commission on Human Settlements (2001) *The State of the World's Cities*. (document HS/619/01[E]), Nairobi: UNCHS (Habitat).

United Nations Disaster Relief Organisation (1982) *Shelter after Disaster: Guidelines for Assistance*, New York: United Nations.

United Synagogue of Conservative Judaism (2003) 'Sukkot'. Online. Available HTTP: http://www.uscj.org/scripts/uscj/paper/ARticle.asp?ArticleID=286.

Upton, D. (1979) 'Toward a performance theory of vernacular architecture: Early Tidewater Virginia as a case study', *Folklore Forum*, 12: 173–98.

—— (1985) 'The preconditions for a performance theory of architecture', in S.J. Bronner (ed.) *American Material Culture and Folklife*, Ann Arbor: UMI Research Press.

—— (1990) 'Outside the academy: A century of vernacular architecture studies, 1890–1990', in E.B. MacDougall (ed.) *The Architectural Historian in America*, Washington: National Gallery of Art.

—— (1993) 'The tradition of change', *Traditional Dwellings and Settlements Review*, 5 (1): 9–15.

Bibliography

Upton, D. and Vlach, J.M. (eds) (1986) *Common Places: Readings in American Vernacular Architecture*, Athens and London: University of Georgia Press.
Vellinga, M. (2004a) *Constituting Unity and Difference: Vernacular Architecture in a Minangkabau Village*, Leiden: KITLV Press.
Vellinga, M. (2004b) 'The attraction of the house: Architecture, status and ethnicity in West-Sumatra', in P.J.M. Nas, G. Persoon and R. Jaffe (eds) *Framing Indonesian Realities: Essays in Symbolic Anthropology in Honour of Reimar Schefold*, Leiden: KITLV Press.
Vernez Moudon, A. (1986) *Built for Change*, Cambridge, MA: MIT Press.
Vernez Moudon, A. *et al.* (1980) 'The development of San Francisco zoning legislation for urban environs', in R.R. Stough and A. Wandersman (eds) *Optimizing Environments* (EDRA 11) Washington, DC: EDRA.
Vlach, J.M. (1985) 'The concept of community and folklife study', in S.J. Bronner (ed.) *American Material Culture and Folklife*, Ann Arbor: UMI Research Press.
—— (1986) '"Properly speaking": The need for plain talk about folk art', in J.M. Vlach and S.J. Bronner (eds) *Folk Art and Art Worlds*, Ann Arbor: UMI Research Press.
—— (2003) *Barns*, New York: W.W. Norton.
Wackernagel, M. and Rees, W. (1996) *Our Ecological Footprint: Reducing Human Impact on Earth*, Gabriola Island Canada: New Society Publishers.
Walbert, D. (2002) *Garden Spot: Lancaster County, the Old Order Amish, and the Selling of Rural America*, New York: Oxford University Press.
Walker Art Center (1974) *Naives and Visionaries*, New York: E.P. Dutton.
Wallman, S. (1984) *Eight London Households*, London and New York: Tavistock Publications.
Wampler, J. (1976) *All Their Own: People and the Places They Build*, Oxford: Oxford University Press.
Ward, D.F. (1984) *Personal Places: Perspectives on Informal Art Environments*, Bowling Green, OH: Bowling Green State University Popular Press.
Warner, H.R. (2003) 'Use of model organisms in the search for human aging genes', *Science*, 299 (5609): 971.
Warnock, M. (1994) *Imagination and Time*, Oxford: Blackwell.
Waterson, R. (1990) *The Living House: An Anthropology of Architecture in South-East Asia*, Singapore: Oxford University Press.
Weaver-Zercher, D. (2001) *The Amish in the American Imagination*, Baltimore: Johns Hopkins University Press.
Weissleder, W. (1978) 'Introduction' in W. Weissleder (ed.) *The Nomadic Alternative*, The Hague/Paris: Mouton Publishers.
Welsch, R.L. (1970) 'Sandhill baled-hay construction', *Keystone Folklore*, 15: 16–34.
—— (1976) 'Railroad-tie construction on the pioneer Plains', *Western Folklore*, 35: 149–56.
Werner, C.M., Altman, I., Oxley, D. (eds) (1985) *Home Environments*, vol. 8, New York: Plenum Press.
Weslager, C.A. (1969) *The Log Cabin in America: From Pioneer Days to the Present*, New Brunswick: Rutgers University Press.
World Health Organization (1999) 'Air Pollution' (Fact Sheet 187), revised Sept. 2000. Online. Available HTTP: http://www.who.int/inf-fs/en/fact187.html.
Williams, M.A. and Morrisey, L. (2000) 'Constructions of tradition: Vernacular architecture, country music, and auto-ethnography', in S. McMurry and A. Adams (eds) *People, Power, Places: Perspectives in Vernacular Architecture*, vol. 8, Knoxville: University of Tennessee Press.
Williams, N., Kellogg, E.H. and Lavigne, P.M. (1987) *Vermont Townscape*, New Brunswick, NJ: Center for Urban Policy Research, Rutgers University.
Williamson, T., Radford, A. and Bennetts, H. (2002) *Understanding Sustainable Architecture*, London: Routledge.
Wilson, E.O. (1998) *Consilience: The Unity of Knowledge*, New York: Knopf.
Wobst, H.M. (1977) 'Stylistic behavior and information exchange', in C.E. Cleland (ed.) *For the Director: Research Essays in Honor of James B. Griffin*, (Museum of Anthropology, Anthropological Papers 61), Ann Harbor: Museum of Anthropology, University of Michigan.
Wolff, E.R. (1982) *Europe and the People Without History*, Berkeley: University of California Press.
Woods, M.N. (1999) *From Craft to Profession: The Practice of Architecture in Nineteenth Century America*, Berkeley: University of California Press.

World Commission on Environment and Development (1987) *Our Common Future: The Bruntland Report*, Oxford: Oxford University Press.

Yethiraj, A. and Blaaderen, A. van (2003) 'A colloidal model system with an interaction tunable transition from hard sphere to soft and dipolar', *Nature*, 421 (6922): 513–16.

Young, M. and Wilmott, P. (1957) *Family and Kinship in East London*, London: Routledge & Kegan Paul.

Zug, C.G., III (1994) 'Folk art and outsider art: A folklorist's perspective', in M.D. Hall and E. Metcalf Jr (eds) *Artist Outsider: Creativity and the Boundaries of Culture,* Washington, DC: Smithsonian Institution Press.

Index

Aalto, Alvar 103
Aaron 30
aboriginal, Australian 113, 195
Abraham 30, 52
Aceh 13, 148
Adam 101–2
adobe *see* brick, sun-dried
Aegean Islands 206
Afghanistan 15, 86, 154, 160, 206
Africa (African) 71, 84, 107, 185, 195, 205–6, 211, 240, 254, 266; North 49, 99, 205; South 232, 240, 253; Southern 186; sub-Saharan 106, 173, 226; West 1, 48–9, 89, 91, 211
Afshar, Farokh 106
Aga Khan: Award for Architecture 105; Foundation 171
Agenda 21 125
Akkadian empire 197
Alexandria 107
Alps (Alpine) 110–11, 113, 116–26
AlSayyad, Nezar 100
America (American) 7, 29–30, 32, 34, 42, 146, 156, 206, 234; Latin 84; North 86, 91, 128, 174, 204, 206; South 99, 206
Amish 7, 29, 33–7, 44
Amman 169
Anasazi 197
Anatolia 204, 206; Anatolian Plateau 163
Andaman Islands 13
Ando, Tadao 103
Angola 106
Ankara 163
anthropology, field of 7, 14, 18, 99, 128, 130–2, 137, 243, 247–8, 250, 257, 268; architectural anthropology 100
anthropomorphic metaphor 16, 210–11, 264
Antoniou, Jim 171
Aosta 118

apprenticeship-style learning 8, 46–7, 50–5, 58–62, 233, 236
Arabia (Arab, Arabic, Arabian) 170, 174, 208, 211; Saudi Arabia 105, 208
archaeology, field of 7, 191, 195
Archigram Group 103, 158
Architectural Association 162–3, 200, 205
Architectural Design 156
architecture, field of 1, 14, 18, 97–9, 101, 108, 128, 130–1, 137, 243, 247–8, 268
Arctic 206
Arts and Crafts Movement 3
Ascalon, Julian of 193
Ashmore, Joe 154
Asia (Asian) 84, 146, 174, 206, 226, 266–7; Central 206
Asian Development Bank 150
Association for the Development of an African Urbanism and Architecture (ADAUA) 107
Australia (Australian) 146, 174
Austria 118
authenticity 9–10, 29, 49, 55, 84–91, 94

Bachelard, Gaston 199, 211, 214
Baluchi 68
Bangladesh 240
Bani, River 48, 52–3, 61
Bariz 104–5
Barker Report 255
barn 84, 90, 118, 120–1; barn-raisings 7, 33–7, 44
Barragan, Luis 103
barrios see informal settlements
Barthes, Roland 92
Basseri 68–9
basti see informal settlements
Bath 252
Bauhaus 237
Bawa, Geoffrey 103

Index

Beaux Arts 236–7
Beer Can House 39–42
Beijing 252
Bellever 258
Bengal (Bengali) 89, 163
Benghazi 68
Benjamin, Walter 201
Bicol 149
bidon-villes see informal settlements
biology, field of 181, 184, 196
Bonehill 258
Botswana 250
Botta, Mario 212
Bougainville 172
Bourdier, Jean-Paul 100
Bourdieu, Pierre 69
Bozo 48, 53
Brahms, Johannes 105
Brazil 267
brick: fired 73, 78, 86, 124, 174, 263, 267; sun-dried 16, 48–9, 51–2, 54–8, 61, 104, 106–7, 167, 204, 211, 217, 219, 223–2, 263
bricolage 38–43
Buckinghamshire 254
Buffalo, New York 194
Building and Social Housing Foundation (BSFH) 171
bungalow 82, 89, 163
Burgess, Ernest 113
Burkina Faso 54, 105, 107
Byzantine 226

Cain, Alan 106
Cairo 105, 107
Calatrava, Santiago 103
Calcavodo del Param 205
California 204
Cambridge 152
Cameroon 67
Canada 33, 91
Caribbean 194
Carola, Fabrision 105
Çatal Hüyük 204
cave, concept of 15, 202–12
cement 49, 127
Centre for Alternative Technology, Machynlleth, Wales 256
Centre for Environmental Design Research, University of California, Berkeley 100
Centre Pompidou 106
Cervantes, Miguel de 104

Cézanne, Paul 201
Chadirji, Rifat 103
Chagford 257
Chandigarh 162
Cheju Island 193
China (Chinese) 156, 158, 174, 193, 206, 217, 252
Chomsky, Noam 199
ClaroMundi 253–4
Classicism 105
climate change 1–2, 6, 14, 16, 18, 93, 196–7, 216, 219, 229, 267
concrete 49, 62, 90–1, 169, 226, 234, 267
conservation 3, 8, 18, 47–50, 60–2, 82, 91, 108, 111, 123, 125, 129, 247, 254–5, 259, 260, 265, 268
Correa, Charles 103
corrugated iron 39, 49, 62, 223, 226, 229
Corsellis, Tom 152
Cotswolds (Cotswold) 82, 86, 90
courtyard 15–16, 68, 202, 208–11, 215–17, 223, 226
CRATerre (Centre for Research in Earthen Architecture) 11, 106–7
Creta (Cretan) 206
cultural change 81, 85–6, 92, 184, 189–90, 192, 196, 245
cultural geography, field of 7, 18, 99, 130, 268

Dartmoor 257–8
David 30
Declaration on Environment and Development 125
Delos 210
Department of Architecture, Oxford Brookes University 255
Desert Architecture and Urban Planning Unit, Ben-Gurion University 220
desertification 106, 226, 229
Dethier, Jean 106
Development Planning Unit, University College London 167
Development Workshop 11, 106–7
Devon 257
Djenne 8, 47–55, 59–62; Djenne-Djeno 48–9
Doat, Patrice 106
Dogon 49, 211
dome (domical) 107, 204, 206, 218, 223–4
Doshi, Balkrishna 103

289

Index

earth 206, 218–19, 223, 226–7, 263; compacted 106; rammed 219, 263
Eco, Umberto 202
economics, field of 14
Ecuador 91
education, importance of 6, 14, 16–19, 128–9, 231–61, 267–8
Egypt (Egyptian) 29, 104–5, 164, 166, 208, 232
Eisenman, Peter 103
Eldern, Sedad 103
El Hekr 164–5
Elkhart County, Indiana 33
El-Wakil, Abdelwahed 105
Energy Efficient Buildings Programme, Oxford Brookes University 220
England 34
England, Richard 103
Environment-Behaviour Studies (EBS) 128, 130–2, 137, 179–83, 185–6, 192, 195–6
Eskimo *see* Inuit
Etruscan (Etruscans) 206, 209–10
Europe (European) 29, 34, 89, 117–18, 122–3, 128, 146, 156, 162, 170, 174, 195, 203–6, 218, 248, 268
European Union 259
Eve 101–2

Fathy, Hassan 11, 104–9
favellas see informal settlements
feng shui 174–5, 193
Fernworthy 258
Filarete di Averlino, Antonio 101–2
Finland 238
Florence 101
folklore, field of 99
Foster, Norman 103
France (French) 48–9, 53, 58, 67, 101, 106, 117, 166, 205, 236
Fundación de Jaume II el Just 259

Gambian 68
Gandhi, Indira 160
Gaudi, Antonio 103
Gbaya 67
gecekondu see informal settlements
Geddes, Patrick 159, 161
Gehry, Frank 103
gender 132–4, 136, 140, 248, 263
generative concepts 15–16, 200–3, 208–13
Geneva 107
Georgian 252

Germany 247, 259
Gibraltar 185
Giglio 43
Glassie, Henry 25
globalization, process of 2, 5–6, 10–11, 18, 83–5, 92, 108, 155, 174–5, 234, 240, 246, 251, 253, 260, 265–6
Goethe, Johann Wolfgang von 212
Goff, Bruce 103
Goffman, Irving 124
Gothic 98
governance 172–4
Graf, Douglas 200
Graubünden 118
Great Plains 43
Greece (Greek) 99, 113, 206, 210–11
Greenland 197
Grimspound 257
Gropius, Walter 102

Haddenham 254
Hadid, Zaha 103
Haeckel, Ernst 113
Hall, Edward 124
Hanukkah 31
Hatfield Dodds, Steve 112
hearth 15, 204–5, 210, 213
Hellenic (Hellenistic) 206, 210, 226
Hepworth Lecture 1, 18
Herculaneum 210
Hillier, Bill 162–3, 200
Hindu 211
Hiroshima 162
history, field of 7, 130, 268
Hite, Mary 40, 42
Hohokam 197
Holmes County, Ohio 33
Holy Quran 47, 52
Honda, Tomotsune 205
Hong Kong 174, 218, 229
Houben, Hugo 106
Houndtor, Dartmoor 258
housing: field of 1–2, 12, 14, 20, 125, 128–30, 131–3, 135, 142, 144, 162, 176; problem and provision of 1–3, 6, 12, 44, 98–9, 112, 129–30, 132, 160, 162, 173, 216, 229, 246, 252, 254–6, 260, 262–7
Housing (Traveller Accommodation) Act 1998 73
Houston 7, 29, 37–43
human ecology, field of 11, 110, 113–15, 121–2, 126, 268

Index

Hunza 206
Hutterites 34
hybridisation 87–8, 91–2

igloo 206
Ile d'Abeau 106–7
India (Indian) 13, 15, 99, 156, 159, 162, 174, 207, 211, 232
Indiana 33
Indian Ocean 148
indigenisation, process of 86–8
indigenous knowledge 10, 46–7, 59, 111, 127
Indonesia (Indonesian) 90–1, 99, 148–9, 156, 171, 250, 267
industrialization, process of 17, 35–6, 81, 126, 163, 234, 255, 260
Industrial Revolution 82
informal settlements 13–4, 36, 88–91, 94, 99, 155, 160–1, 163–4, 166, 169, 171, 173, 175, 181, 183, 189–91, 195–6, 232, 234, 240, 266
Inland Niger Delta 48, 61
integrated and multi-disciplinary approach 12, 14, 20, 100, 115, 129–30, 132, 138, 142, 155, 167, 176, 268
International Foundation for Architectural Syntheses (FISA) 107
internationalism 104
International Studies in Vernacular Architecture (ISVA) Programme, Oxford Brookes University 247–8, 253–4, 258
International Union for the Conservation of Nature (IUCN) 111
Internet 146
Inuit 85, 113
Iran 67, 106, 160, 217
Iraq 208
Ireland, Republic of (Irish) 9, 66, 72–4, 78–80
Isaac 30
Ise 159
Islam (Islamic) 47, 156, 193–4, 218
Ismailia 164–9; Ismailia Land and Housing Development Agency 168
Israel 29, 164, 191, 220, 225
Italy 206, 238
Izumo 159

Jacob 30
Japan (Japanese) 89, 91, 154, 156–9, 161, 175, 205, 250
Java 90
Jeddah 105

Jencks, Charles 103
Jericho 218
Jewish (Jews) 30–3, 44
Johannesburg 2
Jordan 169
Joseph 30
Jung, Carl 200, 202, 212–13

Kaba 52
Kaedi 105
kampung see informal settlements
Kampung Improvement Programme 171
Kant, Immanuel 202
Karachi 171
Katsura Palace, Kyoto 157–8
King, Anthony 162–3
Kizil 206
Kobe 154
Korea 193
Krier, Leon 103, 105
Krier, Rob 103
Kyoto 156–7, 160

Lagos 232
Lake Léman 117
Lakota 90
Lancaster County, Pennsylvania 33
Laugier, Marc Antoine 101–2
Lawrence, Roderick J. 130
Leaman, Adrian 200
Lebanon 208
Le Corbusier 102, 161–2, 239, 242
Legoretta, Ricardo 103
Lesotho 173
Leviticus 29
Libya 205
Liebeskind, Daniel 103
Lima 232
limestone 82
log construction 84, 91, 120–1, 206
London 68, 90, 247, 252
longhouse 1, 258
Loos, Adolf 98
Los Angeles 43
Lyon 107

Mali (Malian) 8, 15, 47–50, 60, 62, 91, 105–7
Malta 206, 211
Mamluk 105
Mangin, William 162
Maori 86

291

Index

Marcuse, Peter 112
Marrakesh 167
Martin Centre, Cambridge University 154
Mary 77
Maseru 173
Maslack 154
Mauritania 68, 105–7
Maya 197, 206
Mayotte 107
Mbororo 67
Mediterranean 16, 102, 205–6, 209, 216–17, 220–1, 226
Merrivale 258
Merton, Robert 124
Mesopotamia 208
Metabolists 158
Mexico 91
Middle East (Middle Eastern) 16, 30, 65, 166, 170, 216–18, 220–1
Mies van der Rohe, Ludwig 102
migration 2, 6, 66, 68–71, 74, 80, 163, 169, 187, 191, 196, 250
Mill, John Stuart 201
Milkovisch: Guy 42; John M. 39–42; Ronny 42
Minangkabau 90
Mirhas 107
mobility, concept of 63–8, 71, 73–5, 77, 79–80
model system 184–5; vernacular design as 15, 185, 189–90
Modernism 10–1, 98, 102–4
modernity, concept of 5–6, 9, 60–2, 84, 86, 88, 91, 94, 170
modernization, process of 4–5, 10, 34–6, 83, 85, 92, 163, 187, 189, 234, 254
Mongolia (Mongolian) 90–1
Moore, Charles 103
Morocco (Moroccan) 49, 167; Moroccan-style 50–1, 55, 60
Moses 30
Mumbai 232
Muneo, Rafael 103
Museum of Modern Art 99
Mycenaen 15, 206

Nabatean 216
Nara 156
Native Americans 90–1
natural disasters 6, 18, 44, 121, 126, 145–54, 219, 258–9
Negev: desert 220; Highlands 225
Neo-Classicism 98

Nepal 232
Netting, Robert 21
New Delhi 160–1, 163
New Guinea 15, 206; Papua New Guinea 172–3
New Gourna 11, 107
New York 33, 213
Nicobar Islands 13
Nigeria 250
Nijo Castle, Kyoto 157
Nitschke, Gunter 156–8, 175
nomadism, concept of 8–9, 63–71, 80
Nootka (Nuu-chah-nulth) 86
Norton, John 106
Nottingham 160–1
Nouvel, Jean 103
Nubian 104; vaulting 106

Ohio 33
Old Order Mennonites 34
Oliver, Paul 1–2, 11, 18, 24–5, 44, 99–100, 108–9, 162, 175
Oman 68
Ontario 33
Orange Show Foundation 42
Orangi Pilot Project 171
Ottoman Land Act 163, 166
Ouagadougou 105

Pacific 91, 173, 210
Pakistan 171, 206
Palestine 169
Palladian 209
palm 54–6
Panofsky, Erwin 203
Paris 68, 106
Park, Robert 113
participatory approach 125–6, 153, 169, 171–2, 176
Pennsylvania 33, 194; barn 34, 86
performance 215–16, 218–29, 254
Persian 68
Peru 163, 173
Pessac 242
Philippines 149, 250
Piano, Renzo 103
place, concept of 65, 70–1, 80, 156–9
planning, field of 2, 14, 243, 268
plaster 204, 234
Pompeii 210
population growth 1–2, 12, 93, 115, 229, 245, 263–4, 266

Index

Portogheze, Paolo 103
Port Said 164
post-disaster shelter 13, 145–54, 246
post-modernism 10, 84, 98
Poundsgate 257
Price, Bill 146–7
Prince of Wales 103; Institute 1; Prince's Trust 171
Prix, Wolf 103
prosphika see informal settlements
pueblo 1, 204
Pukhtuns 67

Rajasthani 240
Rapoport, Amos 26, 132, 162–3, 175
Ravereau, André 105
reconstruction 149–50, 152–4
recycled houses 7, 29, 37–44
Red Cross 146, 246
Redfield, Robert 29, 43
regeneration 18, 99
renaissance 98, 102, 218
renovation 123, 126
resources, depletion of 6, 14, 93, 219, 229, 256, 263–4, 267
Reza Shah 68
Rhône, River 117–18, 121
Rice, Matthew 81–3, 85–7
Riyadh 106
Rodia, Simon 43
Rogers, Richard 103
Roman (Romans) 98, 120, 209–10, 225–6
roof: flat 54, 223, 225–6; pitched 223
Rossi, Aldo 103
Rosso 107
Rotary Club 146
Royal Institute of British Architects (RIBA) 247–8
Rudofsky, Bernard 99, 106
Russia 15, 203

Sacramento Valley 204
Sadat, Anwar 168
Sahel 49, 60
Sahlins, Marshall 85–7
St Lawrence Island 85
Samuel, Joshua 42
San 186, 195
San Francisco 194
Sarajevo 193
Sardinia 206

Scharoun, Hans 103
Scholasticism 203
Schopenhauer, Arthur 201
Scorhill 258
sedentarization, process of 71–4, 79, 256
Semper, Gottfried 204, 210
Senegal 107
Seville 107
Sforza, Lord 101; Sforzinda 101
Shinto 157
Shovel Down 258
Sinai 164
Singapore 91
Sisa, Alvaro 103
slate 121, 256
sociology, field 128, 130–2, 137
Soleri, Paolo 103
Somalia (Somali) 68, 106
Somerset 145
South Carolina 42
South China Sea 218
space, concept of 155–8, 160, 162, 174–5
space syntax theory 132, 137, 163, 201
Spain 205, 259
spatial: configuration 138, 142, 144; configuration diagram 137–8; mapping 136–7
Sphere Project 151
squatter settlements *see* informal settlements
Sri Lanka (Sri Lankan) 13, 149–51
Stiniel 257
stone 105, 120–1, 126, 218, 223–6; corbelled stone 206, 218
style-Soudanaise 8, 49, 60–1
Sudanese 68
Suez 164, 166
Sukkah (*Sukkot*) 7, 29–33, 35–6, 44
Sullivan, Louis 98
Sumatra 148–9; West Sumatra 90–1
Surabaya 171
sustainability 10, 47, 108, 110–19, 129, 183, 190, 196, 216, 235, 245, 249, 254; concept of 3, 11, 93, 112, 263, 265; sustainable development 125, 189, 228, 253
Switzerland (Swiss) 11, 110–13, 116–18, 121–6

Taj Mahal 211
Talmud 30–1
Talo Dam project 61
technological change 2, 3, 6, 14, 72, 238, 240, 245, 265

293

Index

tent 72, 75–8, 150–1, 211
Terry, Quinlan 103, 105
Texas 37
Thailand 13, 250
thatch 207
Ticino 118
tile 226, 234
timber 90, 106, 120–1, 126, 207, 218, 256; frame 89
Timbuktu 49
time diary 135–8, 140
Tokyo 175
Topkapi 211
Törbel 121
Toucoulour 68
tourism 2, 29, 49, 90, 123–4, 149, 246, 259
tradition: authority and control in 7–8, 23, 25, 29, 44, 60; concept of 5–8, 16, 19, 23–9, 44, 60–1, 83, 86, 89–91, 108, 181, 203; end of 5; revitalisation of 36, 265; sustainability of 8; transmission of 7–8, 11, 23–5, 27–9, 44, 47, 93, 98, 108, 111, 116, 121–3, 134, 242–3, 253, 260, 265
traditionalism, concept of 10, 99, 103–4
Travellers 8–9, 66–80
tsunami 13, 146–52
Tuan, Yi-Fu 26
Tuareg 68
Tukolor-style 49–50
Tunisia 205
Turkey 99, 163–4, 166
Turner, John F.C. 89, 99, 162–3, 175, 232
Tyrol 118

UK (Britain, British, England, English) 9, 12, 66, 72–3, 79, 81–2, 87–91, 129, 135, 145–6, 156–8, 161–4, 193, 217–20, 230, 245, 248–50, 253, 255–7
UNESCO 60; World Heritage Site 8, 49
United Nations 146, 150, 246; United Nations Office of Conflict and Humanitarian Affairs (UNOCHA) 147; United Nations Organization for the Conservation of Nature 106
United States 30–3, 44, 89, 129, 193–5, 250
University College London 163
University of Grenoble 106
University of Houston 146
Upland South 33
Upton, Dell 84–5
urban design, field of 20, 156, 162

Urban Design International 156
urban development, field of 160, 167, 176
urbanization, process of 6, 36, 81, 92, 163, 170, 187, 229, 245, 253, 255
urban planning, field of 155

Valais, Canton of (Wallis) 11, 117–20
Valencia 259
Valldigna 259
vastu 174–5
vault 202, 207, 217–18, 223–4; vaulting 104, 107
Vauthrin, Jak 105, 107
Venetian 209
Venezuela 247
Venturi, Robert 103
Vermont 194
vernacular: as model system 15, 180, 184–5, 189–90, 192, 194–7; concept of 9–10, 14, 19–20, 23–5, 28, 83–4, 86–9, 93–4, 100, 180–1; reification of 88
vernacular architecture studies, field of 2–4, 10, 12, 14–15, 19, 82–4, 87–8, 91, 93–4, 99–100, 115, 129–30, 179–81, 249, 253, 257, 262–4
vernacularization, process of 87–8, 90–3
Victorian 211
Vietnam 106
Vikings 197
Virginia 193
Visp, River 121
Vitruvian 102

Walterboro 42
Watts Towers 43
West Indies 210
Widecombe 258
windcatchers 16, 217–19, 229
World Bank 169, 252
World Commission on Environment and Development (WCED) 112
World Summit on Sustainable Development 2
Wright, Frank Lloyd 213

Xoraxané 69

Yangtze River Dam 252
Yazd 217–18, 229
Yomut Turkmen 67
Yupik, Siberian 85–7
yurt 86, 90